Constance Reid

NEYMAN
—from life

Springer-Verlag
New York   Heidelberg   Berlin

Library of Congress Cataloging in Publication Data
Reid, Constance.
  Neyman—from life.
  Includes index.
  1. Neyman, Jerzy, 1894–1981
2. Mathematicians—United States—Biography.
I. Title.
QA29.N43R44 1982 519.2'092'4 [B] 82-10539

With 51 Photographs

Typeset by Publication Services, Urbana, IL.
Printed and bound by Halliday Lithograph, Plympton, MA.
Printed in the United States of America.

9 8 7 6 5 4 3 2 1

ISBN 0-387-90747-5 Springer-Verlag   New York   Heidelberg   Berlin
ISBN 3-540-90747-5 Springer-Verlag   Berlin   Heidelberg   New York

Preface

This book is the result of a suggestion made to me by Erich Lehmann of the Department of Statistics of the University of California at Berkeley. Throughout the writing of it I have enjoyed his assistance and companionship and that of his wife, Juliet Shaffer.

In the beginning, writing about a man like Neyman, who was alive and still very active, I planned to limit myself to his recollections and records and to my personal observations of him during his eighty-fifth year. It soon became apparent, however, that I would need to supplement these with the recollections of his colleagues and former students. With his consent I consulted them.

I am most grateful for the time and effort which so many people expended in talking and writing to me. I would like to mention especially the assistance I received from Lucien Le Cam and Elizabeth Scott and the very friendly cooperation of Egon S. Pearson. Without the latter's careful preservation of Neyman's early letters I would have found it impossible to write in any detail of the years from 1926 to 1938.

A number of libraries and archives also provided valuable material. These included the Bancroft Library and the various administrative archives of the University of California, the D.M.S. Watson Library at University College, the National Archives, and the Rockefeller Foundation Archives. The careful work of Kay Kewley, who was responsible for the organization of the Neyman papers, made my own work much easier.

Letters and papers of R.A. Fisher are quoted with permission of the University of Adelaide. Acknowledgment of other specific sources has been made where appropriate in the text.

Polish names and words which have been anglicized are given in that form; Russian names and words, in the form generally used by Neyman.

The manuscript was patiently read in almost every version by Erich Lehmann and Juliet Shaffer and in semi-final draft by David Blackwell, Joseph L. Hodges, Jr., Lucien Le Cam, Julia and Raphael Robinson, and Elizabeth L. Scott.

After all this help, any errors that remain are most definitely my own.

San Francisco, California   Constance Reid
August 23, 1982

Jerzy Neyman died on August 5, 1981, at the age of eighty-seven, following a series of mild heart attacks. At that time I had already completed the first draft of this book. I had promised him that whatever I wrote it would not be an "obituary"—he hated the thought—and I decided to end the book, as I had written it, with the celebration of his eighty-fifth birthday in the spring of 1979.

There are also others I have mentioned who have died in the interim, and I would like to list their names here:

Michel Loève (1907–1979)
Olga Solodovnikova Neyman (1899–1979)
Raymond T. Birge (1887–1980)
Egon S. Pearson (1895–1980)
Jack C. Kiefer (1924–1981)
J.O. Irwin (1898–1982)
Julius Blum (1922–1982)
William Dennes (1898–1982)
Ronald W. Shephard (1913–1982)

As Neyman would have written (although he would have written in Latin):

May the earth rest lightly on them.

Neyman—from life

Each morning before breakfast every single one of us approaches an urn filled with white and black balls. We draw a ball. If it is white, we survive the day. If it is black, we die. The proportion of black balls in the urn is not the same for each day, but grows as we become older.... Still there are always some white balls present, and some of us continue to draw them day after day for many years.

> J. Neyman and E.L. Scott—
> "The distribution of galaxies"

1978 It is spring in Berkeley. For Jerzy Neyman it is the eighty-fifth spring. As I enter his office on the fourth floor of Evans Hall at the University of California, he is studying a paper on cosmology in the *The Heritage of Copernicus,* the volume which he organized and edited for the National Academy of Sciences to commemorate the 500th anniversary of the birth of the great Polish astronomer. (He plans to nominate the authors for the Medal of Science, which he himself received for laying the foundations of modern statistics.) Because of failing eyesight, he has to read line by line with a long narrow magnifying glass, the venetian blinds adjusted to shut out the glare of daylight.

He gets up when I enter his office and comes around his desk to greet me. Stooped a little with age, he is no longer the dynamic, substantial figure of the nineteen fifties and sixties that I have seen in photographs from that period. But there is still a charming courtliness in his manner. His handshake is warm and strong, his color healthy, and his face, with the neat moustache, that of a man who receives great pleasure from what he is doing.

I explain that some of his colleagues would like me to write in a nontechnical way about him, his scientific work, and his efforts on behalf of statistics. He is wary. He has stubbornly refused to read a biographical sketch written at the time of his eightieth birthday, referring to it always as his "obituary." But still he says to me as he said to the earlier writer, the English statistician John Hammersley, "It's a free country [He obviously likes the expression.]—and if people want to write about me, I can't stop them."

1

Abruptly, he changes the subject by picking up *The Heritage of Copernicus* from his desk.

"I worked very hard on this book. I learned Latin in the gymnasium in Kharkov, where I went to school, but I had to learn Medieval Latin for this so that I could read Copernicus's papers. I found him a very attractive person. With a sense of humor."

He points to the sketch of the astronomer on the book's jacket and explains that he specifically requested the artist to show Copernicus "a little bit smiling."

Sketches of half a dozen more recent scientists appear around the sketch of Copernicus, all of them responsible to some degree, according to Neyman, for "quasi-Copernican revolutions" in their fields. By a quasi-Copernican revolution he means one which occurs when somebody breaks through "the routine of thought" and brings to a science a new idea which deepens understanding.

Would he classify the revolution in mathematical statistics brought about by the Neyman-Pearson collaboration as a quasi-Copernican revolution?

"To a degree," he concedes, a little embarrassed by the question.

He shuffles among the papers on his desk and produces a recent letter from Egon Pearson.

Pearson is also a survivor, eighty-three years old the previous November, but he has been retired for a number of years. He lives now in a home for the elderly in the south of England, surrounded by his papers and those of his father, Karl Pearson, who was also the father of mathematical statistics. Egon Pearson is absorbed in putting on paper for posterity his own knowledge and experience of the history of mathematical statistics in an effort to circumvent what he expects to be the ill-founded speculations of historians of science. He has written Neyman a long letter, commenting in detail—with some difference of opinion—upon a paper in which Neyman has made some remarks about the motivation of their joint work. Neyman's reply, which he also produces for me, is very brief. When the disparity in length between his letter and that of Pearson is commented upon, he says apologetically, "But I am very busy." Ever since he became officially emeritus in 1961 (when he was sixty-seven), he has been regularly "recalled to active duty" by the university. He is the director of the Statistical Laboratory, which he founded in 1939. He teaches a class almost every quarter and conducts a weekly seminar. He has a still very active interest in the applications of statistics—from carcinogenesis to weather modification. He is also currently involved in two "big fights" with the scientific establishment.

Just before five o'clock his secretary comes in with the day's letters for him to sign. She is followed by his long-time collaborator and constant companion, Professor Elizabeth "Betty" Scott—a large, pleasant woman in her middle sixties, attractively dressed in a pastel spring suit. The most recent Neyman-Scott work has been on weather modification, and she reminds Neyman that they are to go the next day to Sacramento to testify before the State Assembly's Committee on Water Resources.

He has invited me to have dinner with him, and he urges, "Come with us, Betty."

But she declines.

"Then come and have a little drink at least."

But she shakes her head again.

"Too bad!" he says, emphasizing and separating the words.

He goes to the closet for his overcoat and beret. As he pulls on his coat, he points to a small framed motto which sits on top of his file cabinet next to a large portrait of Copernicus, also "a little bit smiling."

Life is complicated but not uninteresting.

"I take credit for that sentiment," he says with satisfaction.

Professor Scott leaves to go back to her office, and Neyman and I start very slowly toward the Faculty Club. An informal game of soccer is in progress on the lawn in front of Evans Hall, and Neyman remarks that he used to play soccer.

"I was fast," he says. "Very fast."

1894
—
1906

If Neyman has a favorite book, it is *Penguin Island* by Anatole France. He read it first when he was a student at the University of Kharkov in the Ukraine, but the copy he owns and handles with affection is one which he bought on the Left Bank when he was a Rockefeller Fellow in Paris in 1926–27.

One year he wanted to give "Penguins" to friends and colleagues for Christmas, and he discovered that in paperback English editions the preface—an ironic treatment of history and historians, integral to the book—is omitted. He proceeded to make his own English translation and insert a copy in each gift. Since then, he has had an opportunity to compare his translation with that of an English author, and he is satisfied that his version stands up. "It is not unreasonable" is the way he puts it.

During 1978–79, when I am talking to him every Saturday morning for a number of months about his recollections of his life, he refers frequently to the difficulties of writing about the past as France describes them in the preface to *Penguin Island.*

"It is a very hard job to write history," the fictional historian of the Penguins complains. "One never knows exactly how things developed; and the embarrassment of the historian grows with the abundance of documents. If a fact is known through a single testimony, it is accepted without much hesitation. The perplexities begin when the events are reported by two or more witnesses—because their testimonies are always contradictory and nonreconcilable."

These lines are peculiarly apposite to an account of Neyman's life, which falls naturally into three periods, each of which illustrates in a different way

3

France's comments on the difficulties of writing history. The first period extends from 1894, when he was born in Russia, to 1921, when he left Russia for Poland in the exchange of nationals agreed upon in the Treaty of Riga. This is the Russian Period. The second extends from 1921 to 1938—the Polish-English Period, which is distinguished by his collaboration with Egon Pearson. The third begins in 1938, when he came to Berkeley—the American Period.

Access to the Russian Period can be gained, for the most part, only through Neyman's memory. He has a rather remarkable memory, constantly amazing his secretary with its exactness; but his recall of long ago events has not sharpened with age as is so often the case with older people. As he points out, he remembers some things from this early period very well. He will sing an old drinking song from university days in Kharkov, the tune German but the words Russian, and then recount how one time at the Faculty Club he heard that same tune and, to his amazement, saw everyone in the room stand up. "It is a patriotic song of the University of California!" he says delightedly. (It is, in fact, "Our Sturdy Golden Bear.") But he does not remember how the Bolshevik Revolution came to Kharkov, or how he heard about the execution of the czar and his family, or from whom he learned the calculus of variations with which he proved the famous Neyman-Pearson Lemma, the keystone of all subsequent work in the testing of statistical hypotheses. The fact is that Neyman at eighty-four is interested, not in the past, but in the present and the future. Thus the account of his Russian period will necessarily exhibit some of the vagaries of memory. Trivial events can be vividly described, some significant events have been lost forever. With the exception of two documents—one the handwritten manuscript of a prize-winning undergraduate paper, never published, and the other the published summary of the same work by a professor at Kharkov—he has in his possession no personal records from this Russian period. There is essentially only one memory other than his own that can be consulted—that of Olga Solodovnikova Neyman, whom he married in Kharkov in 1920.

The Polish-English period is better documented than the Russian, but it has its peculiar difficulties, too. Most important and interesting, there is an almost complete set of the letters which Neyman wrote to Egon Pearson during the years of their collaboration, when he was in Poland and Pearson in England. Unfortunately these constitute a dialogue in which only one voice is clearly heard. For Neyman has not saved a single one of Pearson's responding letters.

The American period presents entirely different problems. When Neyman came to Berkeley, he began to keep files. Essentially he began to keep everything. His files spread from his office to his secretary's office to yet another office where a former secretary is organizing his early papers for their eventual deposit in the archives of the Bancroft Library.

Anatole France has something appropriate to say for each of the three periods of Neyman's life. For the American: "The embarrassment of the historian grows with the abundance of documents." For the Polish-English:

"One never knows exactly how things developed." For the Russian: "If a fact is known through a single testimony, it is accepted without much hesitation. The perplexities begin when the events are reported by two or more witnesses—because their testimonies are always contradictory and nonreconcilable." To paraphrase France, it is a very hard job to write anyone's life—perhaps especially when that person is at hand to be questioned. Neyman recognizes my difficulties and is amused by them. When he is asked and cannot recall something about his childhood or his youth, he lifts his shoulders slightly and says apologetically, "It was a long time ago." And indeed it was. Sometimes his connections between personal events and known "historical" events are in error, and he stands corrected. We make an agreement that, since he is a truthful person, any incident which he has recalled in the past, in earlier letters or documents, will be accepted as fact, even if he doesn't remember it now.

With France's various caveats in mind, we begin, in the spring of 1978, to piece together Neyman's recollections of his early life in Russia.

This life began on April 16, 1894, in Bendery, "the gate of Bessarabia." Later that same year Nicholas II, who was to be the last czar of the Russians, succeeded his father, Czar Alexander III. Bendery was Russian at the time of Neyman's birth. Later, in the period between the two world wars, it was Romanian. It is now again Russian. When he was born, Poland as a nation had not existed for almost a century—what was nostalgically referred to by Poles as "Poland Proper" had been divided among Germany, Austria, and Russia. Neyman, however, considers that he was born a Pole, although born in Russia, and that his mother tongue was Polish, although he spoke Russian almost as early as he spoke Polish.

He has no personal knowledge of the migration which brought the Polish Neymans under the rule of the Russian czar. What he does know is that in 1863, at the time of the Second Polish Uprising in Russia, his grandfather, Hermogenes Neyman, was the owner of a good-sized estate near Kaniuv, a village south of Kiev, and possessed the Russian title *Dvoryanin*, a category of nobleman privileged to own land and serfs.

He describes his grandfather as having had thirteen children—two daughters and eleven sons, of whom his father, Czesław Neyman, was the youngest; however, two genealogical charts in his possession give figures which differ from his and also from one another's. In these the number of children is given as twelve and fourteen respectively, the girls and boys are more or less evenly divided, although not in the same proportion in each case. Names are also at odds; but on each chart Tytus is an elder son and Czesław, a younger.

For his participation in the Uprising of 1863, Hermogenes Neyman was burned alive in his house, his lands confiscated, and all his sons except Czesław sentenced to exile in Siberia. Neyman has no idea of the type of punishment meted out to the members of his family who were sent there. A wide range was possible. Some exiles suffered a slow death in the salt mines, others lived like officers at a distant post. Young Czesław was permitted to remain with his older sisters on condition that he would be reared in a community where there were not many Polish families. Neyman thinks that his father was twelve or thirteen years old at the time, and this figure is substantiated by dates given to me by Neyman's older brother's granddaughter.

Neyman does not know any details of his father's life between 1863 and his eighteenth year, when he became a student of law at the university in Kiev. By that time restrictions upon Poles' associating with one another had been relaxed. Czesław took a room in the home of a Polish widow, surnamed Lutosławska, and fell in love with her daughter Kazimiera, who attended a girls' gymnasium in Kiev.

Until Neyman, at my urging, questioned a first cousin, Roman Lutosławski, who still lives in Warsaw, he knew very little about his maternal background and was able to come up with his maternal grandfather's first name only by recalling the patronymic with which his mother's Russian friends addressed her: Kazimiera Kazimierzovna. According to Neyman's cousin Roman, the Lutosławskis were landowning gentry in Piotrkow in "Poland Proper." It appears that the maternal grandfather, Kazimierz Lutosławski, also participated in the Polish uprising, for about 1863 his lands were confiscated and he and his family were sent away from Piotrkow. He died in Tashkent, at the time a dismal outpost of empire recently taken over by the Russians from the Chinese. (Neyman finds it curious that these facts of his maternal grandfather's activities and his death in Tashkent were never mentioned by his mother or his grandmother in his presence. He thinks that perhaps they continued to be afraid to speak of any participation in the Polish uprising.) After what Neyman's cousin Roman refers to as the "amnesty," the other members of the Lutosławski family were permitted to leave Tashkent; but they were forbidden to return to the area where they had owned land. They settled for the most part in the Ukraine, in Kiev and Kharkov.

Czesław Neyman married Kazimiera Lutosławska before she received her diploma from the girls' gymnasium. Neyman thinks she may have been pregnant at the time. He remembers her recounting how at the commencement exercises she was teased until she wept about now having a different name from the one on her diploma. The couple had four children. The eldest, Karol, was born in late 1877; and the youngest, Jerzy, sixteen years later. In between, there were two daughters who died before Jerzy was born, presumably in one of the many epidemics which periodically swept the Ukraine. Neyman does not know his sisters' names, nor does he know where his parents lived between their marriage and his birth.

For a lawyer, Czesław Neyman seems to have been unusually peripatetic. From Jerzy's birthplace in Bendery, which lies several hundred miles south of Kiev, he immediately moved his family eastward, first to Kherson, where he had relatives, and then to Melitopol, which is in the Crimea. Neyman's earliest memory dates from Melitopol, where his nurse took him into a Russian Orthodox church because there was no Roman Catholic church in the town. By 1903 his brother, Karol, who like his father studied law at the University of Kiev, had married a girl from a German family, Marta Loesz, and was the father of a son—also Jerzy.

(Jerzy is the Polish form of the English George. It is not easy for Americans to pronounce correctly, being unlike "Jersey," and many of Neyman's long-time colleagues, following the lead of Professor Scott, have settled for "Jerry.")

Neyman, or "Jurek," as he was then called, grew up essentially as an only child in a household consisting of father, mother, and mother's mother. Living was comfortable. There were servants for his mother and for him a Ukrainian nurse, from whom he learned Ukrainian in addition to Polish and Russian.

By the time Jurek was eight, Czesław Neyman had moved the family yet again and was established still farther south in Simferopol, the apex of an equilateral triangle on the Crimean peninsula, the other two points of which are the coastal cities of Sevastopol and Yalta. Neyman thinks that in Simferopol his father had some sort of government position, "not exactly a judge but something like that." There were half a dozen or so Polish families of approximately the same social class as the Neymans, and these banded together to give their children a Polish education. The informal "school" was conducted in the homes of the various families with a certain amount of secrecy, but it is now Neyman's opinion that the government must have been aware of its existence.

In addition to the usual elementary school subjects, a proper education for Polish children included instruction in French and German. This required native-speaking governesses, French one year, German the next, who moved from household to household. Although he was equally exposed to both languages in his youth, Neyman thinks that he developed a lifelong preference for French as a result of his parents' attitude "that from Paris emanated all things attractive."

By the time he entered the local gymnasium at the age of ten, he could speak five languages reasonably well and, except for Russian history and Russian geography, knew much more than the other children in his class. "Polish School" continued in the late afternoons, and his best friend continued to be the Polish boy nearest himself in age. This was Kazimierz Rakowski, or "Kazik." The senior Rakowski owned a substantial farm just outside of town, and a favorite activity of the boys was hunting, the sport of Polish gentlemen. What did they hunt? "Birds mostly." Kazik had a double-barreled shotgun much superior to Jurek's "Monte Cristo," which had to be reloaded after each shot. A few years later, after Jurek and his family had left Simferopol, Kazik committed suicide. Neyman does not know the reason, or whether Kazik used his gun to kill himself. "Perhaps I was never told." But in a note left to his parents Kazik asked that they send his gun to his friend Jurek.

Other memories of childhood which Neyman can dredge up are few.

There was, of course, Christmas. He remembers real candles and mandarin oranges on the tree—a traditional dessert, semi-liquid and concocted from poppy seeds—the grownups having their game of *vint* (similar to bridge) as the children drowsed off. On New Year's Day everyone visited "and left their cards so that you would not forget that they had come." And Easter? "Somehow I was not impressed by Easter."

His father went each morning to his office, came home for the midday meal, kept files, had a typewriter, was interested in archeology, published some papers on that subject. He remembers being taken on archeological expeditions led by his father, being shown dirt-encrusted objects and being informed that they were made of gold, hearing his father say: "I am an archeologist who practices law for a living."

His mother—a short, energetic woman—comes through less vividly. She was a devout Roman Catholic, but her devoutness was tempered by her devotion to the health of her child. "On the day before Easter she had nothing except tea, but I was always fed."

Asked for more information about her, he is silent a long time.

"I felt, very strongly, her affection and care," he says slowly. "In the early days she tried hard to see that I learned languages, that I learned one year the contents of studies in the school for the next year. And this established a kind of habit. I got accustomed to it and tried to continue it, and she did whatever she could to help me."*

In the evenings in Simferopol the family sat around the dining room table. Sometimes Czesław Neyman read aloud something which he had written and typed on the typewriter. Neyman does not remember whether these things were law or archeology, he remembers only the typewriter. He himself would sit on his knees at the table, reading. For a long time his favorite book was *Swiss Family Robinson,* which his mother had given him as a present. He would read the book from beginning to end and then immediately turn back to the first page and start to read the story again. On several occasions his mother scolded him. Such rereading of the same book was a waste of time. Finally, one evening, she snatched the book from him, tore it in two, and tossed it into the fire.

(This long remembered act on his mother's part did not "break the habit" of reading a favorite book over and over. Still today, when Neyman sets out on a long airplane flight, he takes with him a book that has been a favorite for many years—*Penguin Island, Huckleberry Finn* or *Tom Sawyer,* a novel by Jack London or Joseph Conrad, or a volume of the Forsyte Saga, most often *The White Monkey.*)

During Neyman's childhood in Simferopol, Russia went to war against Japan. He is sure that as a ten-year-old he was aware of the declaration of war,

*Anyone who has argued with Neyman over the exact expression of a thought in English recognizes that he is very well informed on the correct use of that language. He is, however, cavalier in his treatment of it in ordinary discourse. Early on, I ask him how he feels about his conversations with me and his very early letters in English being quoted in print. Should errors in grammar and spelling be corrected? No, he says. Everything should be recorded as spoken or written. "Except in places where the meaning is not clear. Then you should put something like 'here Neyman apparently means...'"

the mobilization of soldiers, their leaving and—a year and a half later—their return after a humiliating defeat; but he has seen a great many wars in his lifetime and he has only one distinct memory of that one, since it resulted in a permanent change in his appearance.

He was walking one evening in 1905 with his parents when a discharged soldier approached and asked for money to get home. Czesław Neyman took out his purse and gave the man a coin.

"How much did you give him?" Kazimiera asked as the family continued their stroll.

"Five rubles."

"Five rubles!"

The incident remains vivid in Neyman's mind, for, as he walked along with his parents, he was holding both hands against his head. A passerby stopped and asked, "Why are you holding your head like that?"

"Because it hurts!"

The man drew him under the street lamp and looked carefully into his eyes.

"This boy should have glasses."

The Russo-Japanese War was followed almost immediately by a revolution in which Konstanty Neyman, the son of Czesław's brother Tytus, participated. Like his father and his uncles before him, "Kot" was arrested and sent to Siberia. His special fate must have been the subject of whispered concern in the Neyman household in Simferopol; but Jurek, loved and attended by parents and grandmother, shooting at birds with his friend Kazik, enjoying school and doing well, was not aware of any such discussions.

This secure childhood came to an end in the spring of 1906 when sometime around Jurek's twelfth birthday his father died of a heart attack.

1906
—
1912
Although Czesław Neyman had earned a comfortable living for his family during his lifetime, there was very little money after his death. A small pension or an insurance policy. Neyman does not know. The newly widowed Kazimiera acted promptly and with decision. Within a few months she had packed her son and her mother and moved them to Kharkov, where she had relatives on both sides of the family.

The city of Kharkov lies four hundred miles south of Moscow and east in an almost straight line from Kiev and Lvov, being about as far from Kiev as Kiev is from Lvov. It has never been a beautiful city—in no way comparable to Kiev of the Golden Domes, with which it is usually coupled, since both cities are important in the Ukraine. At the time the Neyman family arrived in the early summer of 1906, it had a population of somewhere around three hundred thousand and was the fifth largest city in Russia, a busy trading and industrial center.

Tytus Neyman, the father of the exiled Konstanty, had also died, leaving a largish piece of property in Kharkov to two daughters who lived in another town. It included a substantial residence which had been extended and turned into apartments and two small houses also occupied by tenants. Kazimiera Neyman took over the management of this property for her husband's relatives and established her small family in the spacious main floor apartment of the big house.

Neyman draws a little map of the area of Kharkov in which he lived. He describes the city as flat in comparison to Simferopol, but flat in a curious fashion, being cut by deep, wide washes. It was in one of these washes that the family's new home was located. A sluggish little river called the Netech—the "doesn't flow"—ran nearby. Across the river was the town market and open land. The important buildings of the town—the gymnasium which he would attend and the university, the Roman Catholic church and the "Polish House"—were on the high, flat tableland, the approach to which was very steep. How steep? He gives an example from later in his life in Kharkov. He was returning from somewhere on the high land and took a vehicle—"not a taxi but similar." When it reached a certain spot in the descent, the driver announced that he would go no farther. "So we discussed, and instead of twenty-five kopeks I paid him fifteen and walked the rest of the way. Steep," he says.

In Kharkov, as in Simferopol, he knew more than the other boys. It was pleasant to be able to speak to the teachers of French and German in those languages and to have their permission to read foreign novels in class. But the most memorable event of his first school days in Kharkov was a fight.

"One of the boys in the school, for some reason, he decided he wanted to become friends with me, and he adopted a very unusual method. He beat me up. Without any reason. Well, I tried to do my best, but it was most unexpected. I was beaten. No doubt. But then he came to visit me. It was again a great surprise. And then his mother came and said to my mother that she wanted her son to live with us and that she would pay so much."

The boy, Yevgeniy Butkos, "Zhenka," as Neyman would later know him, was the son of a government paymaster who worked some little distance from Kharkov. He thus could attend the gymnasium only if he boarded in town.

"So there was an agreement between them—his mother and my mother—and so we shared a room—the room was substantial."

In the winter Kharkov was much colder than Simferopol, the temperature falling sometimes to twenty degrees below zero. Livestock was slaughtered in November, and miscellaneous odds and ends of meat were combined with sauerkraut and spices and stashed away in a storehouse on the property. The result was a delicacy called *bigos*, which had to be guarded by a dog from November to March. Neyman recalls it with relish: "And I tell you about Christmas time it becomes very tasty. But it has to be eaten before March."

In the summer Kharkov was hot and unpleasant. Jurek was invited to Zhenka's (where the two boys went hunting with Zhenka's father) or to the homes of relatives in the country. Were these dachas? No, they were places where people lived all year, not just summer houses. There were many cousins. He remembers at one place at least a dozen boys "and a horse for every one of them."

One summer Kazimiera's brother—the father of Roman Lutosławski and an engineer with a railroad company in Odessa—obtained complimentary passes for Jurek and his mother so that they could go to visit another brother of his mother's who was an accountant in a town beyond the Volga, the name of which he has forgotten. It was quite a long journey, probably with many new experiences; but what Neyman remembers most vividly is that his little cousin Roman ("Romek") threw a box of chocolates out of the window of the train because he could not have one. Roman Lutosławski became a flyer and after the defeat of Poland in 1939 flew transport planes and ferried tactical aircraft for the RAF in North Africa and Italy.

(As far as I know, Roman Lutosławski is the only person still living who knew Neyman as a boy. Although he is six years younger, he had a brother, Czesław, who was exactly the same age as Neyman. What he writes me of his memories of "Jurek" and "Czeska": the two older boys' making him tickle their heels because they had read that it would put them to sleep—fighting so violently over the best place to fish that they fell in the river—disappearing from a party at their Aunt Helena's and so angering her that she informed their mothers they were both savages—all testify that Neyman was a quite normal boy.)

Another summer—Neyman thinks it was 1909—when all the banished uncles had returned, the Neyman family, its members numbering almost a hundred, held a grand reunion in Kiev, renting an excursion boat for the occasion. A picture of the golden domes reflected in the Dnieper leaps into the mind—the old uncles recounting their adventures and sufferings to the young people. But what Neyman remembers is that he was told by his uncles that it was now appropriate for him to kiss his girl cousins.

He does not think that his cousin Konstanty was at this reunion. Kot had been permitted quite a bit of freedom in his exile, which was in an area where there was a great deal of game. (Neyman remembers being present while his uncles discussed sending him a better gun than the one he had taken with him.) When he decided to escape, he did so quite easily by walking to a railroad station more distant than the one toward which most of the prisoners headed. Back in Kharkov, he gathered up his girl and fled to Paris. He lived there until his death in the early nineteen seventies. His eldest son, Jean-Louis de Neyman, was a member of the French *resistance* during the Second World War and was executed by the Nazis.

Neyman has in his possession a copy of Konstanty Neyman's academic record at the Kharkov Institute of Technology, from which he was expelled "without right to reenter" in 1904 when it was discovered that he was a member of the Polish Socialist Party of Joseph Pilsudski.

"This is something I am extremely proud of," he says as he spreads out the large sheets on which his cousin's courses and grades are recorded "for submission if and when he wishes to enter a foreign institution."

11

Is he proud because his cousin was such an ardent Polish patriot?

"No. Because he stood up for what he believed in. For this idea that that's the appropriate thing. Well—I have respect for that."

As a boy, Neyman says, he was not himself a revolutionary; but he shared the contempt for the czar which he feels was general in people of his social class, whether Polish or Russian. He recounts a story he heard while he was convalescing at his brother's home after a bout with typhoid fever in the winter preceding his fourteenth birthday. His brother showed him the card left by the head of the local gendarmes, a man who had distinguished himself in the suppression of the Revolution of 1905 and had applied to the czar for permission to change his name. Delicately, Neyman explains that the name meant "Seven Behinds"—"Well, 'behind' is not a very elegant term, but there are other less elegant terms to describe that same part of the anatomy and so—in this man's name—that term was the worst! After some period of time a reply came. On the man's application His Imperial Majesty had written in his own hand, 'Seven is too many, five is enough!' And so the man's name on the card which he left at my brother's on New Year's Day was 'Five Behinds.'"

That was a joke his brother told him?

"No. That was not a joke," Neyman says firmly. "His Imperial Majesty was stupid."

Neyman has pleasant memories of his school days in Kharkov. The Latin teacher required his three sons to converse with him in that language and the "senators" of the class—"I was a senator"—to write to each other in Latin during their summer vacation. There was also an excellent mathematics teacher who, Neyman thinks, may have been responsible for stimulating his own interest in that subject. What he says is "It is not impossible that the tendency towards mathematics was due to this teacher." But when asked how that was accomplished, he gives a little shrug. "Who knows? A boy becomes interested in a subject. Must be a good teacher." The physics teacher was mediocre. "So—no interest at that time in physics."

When asked where he stood in his class in the gymnasium, he will say only that "there were boys who did not have as much intellectual interest as I did." He explains that it was "kind of customary" for him to prepare his friends for examinations. Six or eight of them (including Zhenka) would gather at the house on Rogatinskiy Pereulok, tables would be pushed together in the big dining room, "and then we would get to work." It is clear from his account that he enjoyed these sessions. Although he had already begun to do a little tutoring in the summer to earn money, he received no pay—"It was just for fun." When, however, one of the boys who had expected to fail passed his examinations, he gave Neyman "something golden," inscribed "Thanks to my friend."

Religious instruction was given at the gymnasium according to individual preference. Students of the Russian Orthodox church received instruction from a teacher at the school. Polish students went across the street to the priest at the Roman Catholic church.

And Jewish students went to the synogogue?

"Presumably. But I don't remember any Jewish students."

Neyman was brought up, like most Polish children, in the Roman Catholic faith. He served as an altar boy and still remembers some of the Latin responses to the mass. He traces the beginnings of doubt to the priest who gave him religious instruction during his gymnasium days. For some occasion this man organized a picnic. In the course of the event he became "disgustingly drunk." Such behavior, in the young Neyman's opinion, did not speak well for the higher authority the priest was supposed to be representing on earth.

As the time for her son's graduation from the gymnasium approached, Kazimiera Neyman decided that it would be "appropriate" for him to see something of the world outside Russia. Although she had very little money, she arranged through an organization in Moscow that he join a student group making a journey by train through Europe in the summer of 1912. Neyman is still impressed by her vision in conceiving this plan—"it shows she was not provincial"—and her initiative in carrying it out.

Did he go to Poland on the trip?

"No," he says, "the train passed through Warsaw without stopping."

He recalls Vienna, Rome, Venice, Naples, Pompeii, Capri.

Upon his return to Kharkov, he entered the university. He had decided that he wanted to study mathematics, and his mother supported him in his decision.

"She didn't know any mathematics to speak of. She knew that such thing exists. All right. She was trying to do her best to help me develop in matters which she herself wasn't competent in. She had respect for intellectual activity."

Since he had rejected training for the profession which his father and brother had followed, did he plan to earn his living as a teacher—a professor perhaps—of mathematics?

"I do not remember thinking about what will I do with mathematics in the future."

1912
—
1914

Eighteen-year-old Jerzy Neyman (Yuri Czeslawovich at the Russian university) became a student at the University of Kharkov, now Maxim Gorki University, in the autumn of 1912.

The university, as he recalls it, consisted of a few buildings surrounding a central courtyard. There may have been some outlying structures which he doesn't remember but nothing at all like the "university city" which he visited half a century later when he was an exchange professor at Kiev. At the university he wore a still different uniform from the one he had worn as a gymnasium student. He describes it in detail. The color was grey "with a little blue in it." There was a cap with a ribbon to indicate the university—he gestures that the cap had a bill—and a blouse which buttoned high at the neck. Was it belted? He doesn't remember. Then there was a long skirted overcoat for winter. Did he wear boots? No. Black shoes with elastic high around the ankles to keep out rain and snow.

Asked if he would write to the university in Kharkov for a transcript of the courses he took during his days there, he says, "regretfully," no.

"In America you can do something like that, but in Russia—it would be very suspicious—somebody would wonder what you would want to do with it."

He says he has no objection to my writing, but he obviously thinks I will be wasting my time.

(As it turns out, after a number of unsuccessful efforts, both direct and indirect, I do receive a letter from Dr. E.V. Balla, the director of the Central Scientific Library at the University of Kharkov. He informs me that the university records were destroyed during the Second World War when Kharkov was one of the Russian cities that suffered most dreadfully at the hands of the Germans.)

Not surprisingly, the first member of the university faculty who impressed Neyman was a Pole, Antoni Przeborski, who lectured on the calculus. Przeborski had a mocking, irreverent way which the young man found very attractive. He had no personal contact with him. "It was not like it is here. You did not talk to professors." But he observed that Przeborski was a great reader, a frequent visitor to the university library.

"So I spied. What books does Professor Przeborski read? And I took out those same books."

This was the way he was introduced to *Penguin Island*. Like Przeborski he read it in French although Russian translations of foreign books were easily available at the time. He was immediately fascinated by the story of the rise of the Penguins' civilization (paralleling that of France itself) after their inadvertent baptism by a befuddled and halfblind priest. To a young Roman Catholic who had already begun to doubt his faith, such a scene as that in which God and His most learned clerics debate the validity of the baptism of birds was irresistible.

Neyman's other reading in his early university days was very much like that of many intellectual eighteen-year-olds. He remembers that in addition to *The Memoirs of Casanova*, to which he was also introduced by Przeborski, he and his friends read Oscar Wilde. He can't remember whether it was in Russian or Polish translation.

The Picture of Dorian Gray?

"Yeh yeh yeh yeh."

It is a delighted chortle—a response of surprise and pleasure which he slips into from time to time.

One day, as he and a Polish friend, Bronisław Mąkowski, were walking along a street in Kharkov, they saw a sign advertising a school of modern languages. Wouldn't it be nice, Mąkowski suggested, to be able to read Oscar Wilde in English?

"So we went there and started attending. The lady really had a talent of teaching. Within a matter of a very brief time—I remember it was brief, but whether it was weeks or months I can't tell—we were talking, really talking to each other in English! Then it happened somehow, somewhere, I met an Englishman, and I tried to speak to him in English. And then it appeared that he could not understand a word I said."

Did you understand what he said?

"Again very difficult. But then he told me—he may have said it in broken Russian—that he had a sister who taught English, and I suggest that you switch teachers if you want to learn to speak English."

Unfortunately the Englishman's sister lived some distance out of town, and after a few visits Neyman abandoned the lessons. He has never really had any further lessons in English. He likes to say that he learned his English as he learned his geography—from experience.

With the exception of Mąkowski and perhaps one or two other Poles, most of his friends at the university were Russians—boys with whom he had attended the gymnasium. There was, however, a contingent of Polish students who had come to Kharkov from Warsaw as a protest against the fact that professors at the University of Warsaw were appointed by the czar rather than elected by their colleagues as they were at universities in Russia. Once, Neyman remembers, he was told by one of these students that Joseph Pilsudski was going to talk in Kharkov. He went to the Polish House, which was across the street from the gymnasium which he had attended, and heard a man talk there about the prospects for Polish independence. Pilsudski, who was already training Polish sharpshooters in Lvov, was convinced that there would soon be a general European war. He foresaw—given a German and Austrian victory over Russia and a French victory over Germany—the possibility that Poland could reestablish herself as a nation. This was essentially the tenor of the talk which Neyman remembers hearing; but for some reason, which he cannot now explain, he never believed that the man he heard speak was actually Pilsudski.

"I was told it was Pilsudski," he says, "but I have doubts."

Although Neyman had entered the university with the intention of concentrating his efforts on mathematics, he shortly changed his mind. An interest in physics, which had not been aroused in the gymnasium, was stimulated by a report on Einstein's special theory of relativity which he heard at a meeting of the Mathematics Club.

"Inspired me, this," he says. "But I remember that an old professor of physics commented, now the speaker said that one observer sees this and another observer sees that. Well. All right. So the comment of the old professor of physics—he mumbled but everyone heard him—was that one observer knows how to take derivatives and the other one doesn't."

Interest in physics was also stimulated by the fact that the year before he entered the university a Nobel Prize had gone to Marie Curie—a Polish woman, he points out—for her discovery of radium and the study of its properties.

He approached one of the physics professors—not the man mentioned above but a younger man, D.A. Rojanskiy. Could he do some work in the physics laboratory?

Does he mean that he just wanted to do a little something in the laboratory?

"No no no no."

He really wanted to become a physicist?

"I really wanted to."

In the laboratory, however, he was singularly inept at handling equipment and performing experiments. Fuses blew, glass broke, simple mechanisms jammed in his hands.

He has, it seems, never become more dexterous. When I suggest that he can refill a stapler, he shakes his head and pushes it away.

"I used to think maybe just American-born could do it," he explains, "but then somebody suggested maybe only Poles *couldn't* do it, and then Mark Eudey—you know Mark Eudey [one of his early students at Berkeley]— suggested maybe only *one* Pole couldn't do it."

Fortunately an employee of the university whose job it was to care for and repair the laboratory equipment took a liking to Neyman and covered many of his mishaps.

"I used the electroscope of Curie. Very primitive. So it was this—radium emits gas, I've forgotten what the gas is called, and this gas decays. So the assumption was, probably still is, that the decay is exponential. I was told to obtain empirically the decay curve. So after much time I take measurements, one, two, three, and eventually get a point and so forth."

The professor was satisfied with the curve which Neyman found, but he himself was not. What he had actually noted, although neither he nor his professor recognized it at the time, was the effect of so-called "daughter atoms" which also emit radiation as they decay.

During Neyman's second year at the university a great opportunity for field work with the electroscope offered itself. There was an announcement that student volunteers were being solicited for a summer expedition to the Urianghai region of Mongolia. This area had been a part of the Chinese empire until recently when Russia had successfully encouraged a separatist movement. Now the czar's government was in the process of placing the country and its people under the protection of His Imperial Majesty. The purpose of the proposed expedition from Kharkov was to determine the suitability of the area for settlement and farming by Ukrainian peasants. Originally it was to have been a modest excursion by two university agronomists, but in the course of preparing for the journey one of them had chanced upon a book by an Englishman who had earlier explored the same region in search of the grave of Genghis Khan. The Englishman described the savagery of the natives in such a hair-raising fashion that it seemed prudent to enlarge the expedition. This could obviously be done most inexpensively by taking along a few students.

Several of Neyman's schoolmates were excited at the prospect of going to Mongolia for the summer. One young man, wealthier than most, offered to pay his own way: he could hunt and thus help feed his companions. Another, a medical student, pointed out that he could immunize the natives against small pox. Mąkowski, who wanted to become a writer, offered to keep the journal of the expedition. Neyman proposed that he would prevail upon the physics professor to let him borrow the electroscope—he would look for radium ore! He had other skills as well: "I could also hunt, and I could take photographs."

These four youths were accepted by the leaders of the expedition; and in June 1914 they set off from Kharkov by train for Krasnoyarsk. It was there that the Trans-Siberian Railroad crossed the Yenisey, one of the great rivers of the world, extending 2500 miles and draining more than a million square miles.

In Krasnoyarsk the little group from Kharkov boarded a steamer with the trappers, traders, miners, and assorted adventurers who swarmed into the area during the summer months. The ice floes which jammed the river in the late spring and early summer, sometimes actually backing up the powerful flow, were past; but the trip south—a short distance—took a week because they were going against the current, which still ran with spectacular swiftness. At Minusinsk, the leaders of the expedition purchased horses and supplies and hired a guide who could communicate with the natives in their own language. (There were quite a few escaped convicts in the area, and Neyman suspects that their guide was one of these.) With not a little trepidation, for they were all informed of the report about the savagery of the natives, the little group set off.

"We each had a horse and something, a broad sack here and a broad sack there, and that was all. Oh, there may have been one or two more horses which were carrying food but very little really. And so we depended on hunting."

The Yenisey south of Minusinsk is composed essentially of two rivers known as the Large and the Small Yenisey. It is Neyman's recollection that the assignment of the expedition was to investigate the situation at the origin of the Small Yenisey. It was rugged country and in the course of the journey over the mountains, not having yet used the electroscope, he dropped the instrument when his horse fell in a rushing mountain stream. Although he was able to retrieve it, the essential gold leaf through which it functioned had become separated from the rod to which it was attached; and he could not repair it. In spite of this disappointment, he distinguished himself by hunting and, in fact, shot an elk, the largest animal to be brought down. The kill was especially welcome because it took place at a time when the party had been without meat for a number of days. He also took photographs, but he no longer has any of them.

He remembers well how he felt his horse tremble when he and his companions came upon a woman matter-of-factly milking a yak in front of the circular tent of the chief of the first native tribe they encountered. He

17

trembled himself when their guide explained that they were all expected to leave their guns outside the chief's tent when they went in.

The chief offered them tea and sold them a sheep—for a few beads. Neyman is embarrassed as he tells the story.

"Hospitable, no question. No enmity at all. The Englishman—it was all lies!"

What was the ultimate decision of the expedition about the feasibility of transplanting Ukrainian peasants to the land of the Urianghai?

Neyman has no idea.

At the beginning of August 1914 he and his companions began the journey back to Minusinsk. Returning, they were able to travel by river on a large raft which carried thirty or forty people and their horses. Among the returnees was a party of Scandinavian scientists.

"I was surprised to see how they behaved. During the night, we would come out from this raft and, well, all right, let's build a fire, eat something, whatever. But when they came out, they had some sacks and they collected rocks in these sacks to take back with them to Sweden or wherever to analyze. They were real scholars. We were essentially kids."

When they reached Minusinsk, they sold their horses, paid off their guide, and boarded the steamer for Krasnoyarsk. The journey upstream at the beginning of the summer had taken a week; downstream it took a day.

At Krasnoyarsk, where they would catch the train for Kharkov, the inhabitants greeted them with astounding news.

Germany and "Australia" were at war!

Australia?

Yes! And since "Australia" was so close, Russia was bound to be involved very soon!

It was in this garbled fashion that Neyman and his companions learned of the beginning of the First World War.

1978   Neyman lives by himself in a five-room white stucco house with a red tile roof in the Berkeley hills, where the majority of the professors of the University of California have their homes. A small gate opens into a garden bursting with color in the early Berkeley spring. The first rhododendron, with its bright red flowers, is just coming into bloom by the front door. Camelias and azaleas are already in full force. Spring bulbs and cinerarias, scattered at the whim of some past gardener, contribute to an impression of cheerful disorder.

The front door is open.

"People smoke in here," he complains, coming to the door with the inevitable cigarette in his hand, "and I am determined that this house will not smell."

The living room, like all the other rooms, is furnished in the Danish style. There is a large asymmetrical desk in front of the big windows that overlook the glistening expanse of San Francisco bay; but, as in his office, the curtains are drawn. The glare of daylight has bothered him since he was operated on for cataracts.

Over the fireplace is a painting which he says he mildly likes, modern but not completely abstract, bought because, he explains, something had to be hung over the fireplace. The pictures he really likes are the copies of Holbein portrait drawings which hang in his bedroom.

Some Polish trinkets people have given him and a few snapshots are on the mantel of the fireplace. A photograph of the portrait of Griffith C. Evans which hangs in the library of Evans Hall is the most prominent among these. It was Evans who was responsible for bringing Neyman to Berkeley in 1938.

Everything in the house is orderly except the big desk, which is covered with papers, books, a pad of pale green graph paper, a long magnifying glass and a big round one, several red and white packages of Marlboros, and a large ash tray, half full.

Neyman has been separated for a quarter of a century or so from his wife, Olga, whom he married in the spring of 1920 in Kharkov. Their only son, Michael, who is forty-two in 1978, tells me, "Since I was seventeen, the three of us have had our own places. We get along much better now. Sometimes when I come up from Hollywood [where he has lived for many years on what he calls the fringes of the film industry] we go to a movie together." He says that his father, whom he describes as a "workaholic," reminds him of the professor in Ingmar Bergman's *Wild Strawberries* who, through his absorption in the intellectual life, has cut himself off from his family and human love and suffering; but he concedes that such a comparison is maybe not totally fair.

Mrs. Olga Neyman lives a dozen or so blocks farther up the hill from Neyman in the old family home on Euclid Avenue, which she and Neyman bought a few years after they came to Berkeley. She is five years younger than her husband but neither so well nor so active. She reminds one of an old gypsy. Her hair, bobbed in the style of the nineteen twenties, is still dark and her manner lively. In the old days she was known to her Polish in-laws as "Muszka"—the little fly. She speaks in a spirited fashion, repeating words three or four times in a row for emphasis and interspersing her descriptions with such expressions as "Good Lord!", "My my!", or "Wow!"

The marriage of the Neymans was always stormy, according to Mike Neyman—the clashings of incompatible Russian and Polish temperaments.

"Mr. Neyman," his wife says with a still even heavier accent than that of her husband, "was a great mathematician. Mathematics mathematics mathematics mathematics. That was it. But somehow he had enough time to have girl friends. Mr. Neyman was more romantic than I was. I was busy busy busy all my life. I always always always—since I was a little girl—I wanted to draw and do sculptures and so on. This was my real love. Love love love was art."

Neyman goes to his wife's for dinner on Easter. They own their two houses together. She is the primary beneficiary of his life insurance policy. He files an income tax return for the two of them but pays all the tax himself since, as he explains, he has the greater income.

This year he gets a refund, and he is delighted.

"I made a mistake of one digit in addition," he explains.

But that is not very much of a mistake.

"Yes," he says happily, "but it was in the fourth place!"

He has paid the tax, but he considers that half the refund should go to his wife. He asks to be driven up the hill so that he can give it to her.

In front of his old home he gets slowly out of the car and makes his way to the front door through the ragged garden with the two great trees that have long ago outgrown the space planned for them.

He has said that he will be just a minute, but he does not return for some time. He is apologetic as he climbs back into the car.

"She is lonely," he says, "so I stayed and talked for a while."

A few blocks down the hill, after a period of silence, he says something in Russian and, asked to translate, repeats slowly:

The time will come—
Old age tending toward silence—
And then at that time
Possibly you will understand,
However it will be too late for love.

1914
—
1917

In the autumn of 1914 in Kharkov—far from the capital and from the front—Neyman, rejected by the military because of his poor eyesight, settled down to what quickly became normal life at the university. It was a life different but not very different from what had been normal in the past—at least as he remembers.

In spite of his failure to find radium in the land of the Urianghai, or even to make use of the electroscope, he persisted in his determination to become an experimental physicist.

"Physics with hands," he explains. "That was an expression used. Not official. *Physics with hands* and then *Physics without hands.*"

His own ability to handle equipment had not improved. On one occasion he was instructed to manufacture a glass tube with two smaller tubes coming out at the side. Failure followed failure. At last, never having asked for assistance, he had used up most of the glass supply in the laboratory. This was a serious matter, since, because of the war, additional glass was impossible to obtain. Even his friend, the mechanic in charge of equipment and supplies, could not cover for him this time. Professor Rojanskiy had to be called.

"And I remember his face when he looked. The disappointment. He had hoped that I could do it, but then he saw this pile of glass…"

Perhaps, Rojanskiy suggested, Neyman was not really suited for "physics with hands."

It was after this incident that Professor C.K. Russyan, a Polish professor whose lectures on the theory of functions he was attending, spoke to him about a new integral from France—quite different from the classical integral of Riemann. Because such prerequisites as set theory were not offered at the university, Russyan had not been able to treat the Lebesgue integral in his lectures. He now suggested that Neyman investigate it on his own and take it as the subject of a paper to be entered for one of the university's gold medals.

Neyman looks in his library for Henri Lebesgue's *Leçons sur l'intégration et la recherche des fonctions primitives* and quickly locates it.

"*Conceptual mathematics*," he says with satisfaction. "Not manipulative."

How does he distinguish between *conceptual* and *manipulative* mathematics?

"Very difficult to divide," he replies, thoughtfully making a cutting motion with the index finger of his right hand. "Let me give you an example. So people became at one time aware of velocity and also aware that velocity changes so that there is acceleration and deceleration. So they produced concept, mathematical concept of velocity, which is derivative. And then there was the question. Suppose we know the formula for the derivative. All right. Now here is the particular case. How do we compute it? That's manipulative. And then the reverse thing. Suppose we have a formula which represents, let us say, derivative. How do we get the function of which this would be the derivative? That would be integral. And that you have in Lebesgue directly. *Recherche des fonctions primitives*. The search, the looking for primitive functions, the ones which are differentiated. O.K. So that's conceptual."

It is a distinction upon which he places much emphasis. He may praise a colleague for his great ability in manipulative mathematics—frankly admitting his own inferiority ("An error in the fourth place, you remember.") —but it is clear that what he values is conceptual mathematics.

From the first page, the *Leçons* enthralled him. When he is asked for other recollections of this period of his life, he squirms a little in his chair, embarrassed by his inability to produce any.

"The war," he apologizes. "Wiped out everything. And then there was Mr. Lebesgue…"

But it seems to have been "Mr. Lebesgue" who wiped out everything, even the war, as far as Neyman was concerned. The year 1914–15, which saw defeat after defeat for the Russian army on its western front, was a year of intense self-education for the young man. His first task was to familiarize himself with set theory, and he found that subject fascinating. Especially intriguing was Ernst Zermelo's axiom of choice. He remembers how he walked the icy streets of Kharkov trying to explain the concept to his friend Leo Hirschvald.

"He was not a creative person," Neyman says of Hirschvald. "He was a type of people who enjoy understanding delicate things in mathematics and learn and can explain them. And I liked to communicate! And I remember walking with him late at night, close to morning, in the streets discussing this which I wanted to understand, the hypothesis of Zermelo. And suddenly we felt the smell of freshly cooked bread, and then we felt hunger, and so, sniffing, we located the bakery. It was closed so we knocked to get in. And so we got some milk and some freshly cooked bread, and I remember the pleasure of eating it."

In addition to set theory, Neyman had to familiarize himself with the state of integration before Lebesgue. Not only did he have to learn the French and German historical background, which was laid out quite fully in the *Leçons;* but he also had to investigate the development in Russia, not treated.

What was most exciting in the book was contained in the last chapter. There Lebesgue had set out the conditions which the integral should satisfy and had showed that these conditions led to the theory of measure and measurable functions. Professor Russyan wanted Neyman to present these results and to follow them with applications of his own. This was not an easy task, since, as a later writer, J.C. Burkill has commented, "[in the last chapter] Lebesgue slipped into inaccuracy more often than one would have expected of so fine a mathematician and so lucid a writer...."

In May and June 1915, while Neyman was absorbed with sets, measure and integration, defeat was turning into disaster for the Russian army and by the end of June its front had been cut in two. But in Kharkov, through war and later through revolution—at least as Neyman recalls—"summer was always summer." He spent the vacation months of 1915 coaching in the country. His pupil was the son of a Jew who in spite of his religion was permitted by the government to own property because he produced something of commercial value with the land. Neyman believes this was canned fruit or preserves, for he remembers extensive orchards and a little factory. Many young people—"including a number of pretty girls"—came and went during the summer. Among these was a younger boy, a relative of the owner, named Eugene Rabinowitch. He later married one of the pretty girls and emigrated to the United States, where he became a well known physicist and the founder of *The Bulletin of the Atomic Scientists.*

At this point in his account of his life Neyman announces that he can produce "a document." He goes to his study and brings out two much handled notebooks which contain the manuscript of his paper on the Lebesgue integral. Since a manuscript for a gold medal had to be submitted anonymously, the title page bears an identifying motto instead of the author's name, which was written on a card and placed in a sealed envelope bearing the same motto: *Per aspera ad astra.*

"Through difficulties to stars!"

He handles the old notebooks tenderly.

"Substantial," he says of them with satisfaction. Together they contain 530 pages of closely written Russian. With a little surprise he notes that there is also a dedicatory line in Polish. He reads it aloud and translates it into English

"To the memory of my father I dedicate this work."

He has forgotten that he dedicated the paper to his father.

"And then there is a place. Lutowka. On the day of the seventh of July 1915."

He points out that a great part of the paper is not in his handwriting. Although he had been required to take penmanship during his gymnasium years, then as now he had a very low opinion of his handwriting and he paid a fellow student to copy his paper for him.

"It is very nice handwriting," he says, gently turning the pages. "Isn't it?"

Unfortunately, after he had had the paper so beautifully copied, he found many things which he wanted to change or add. He was always to be an inveterate reviser of manuscripts; and there is a joke with his secretary about the time when there were eleven versions of a certain page 11 in his office. In the Lebesgue paper, where he wanted to make changes, he cut a piece of paper that exactly matched the section to be corrected and pasted it over the original. In some places, where his correction did not exceed the length of the original, he wrote it on the pasted paper. In other places he wrote on the opposite page.

Although the date in the notebooks is July 7, 1915, he did considerably more work on the paper during the following academic year. He also attended lectures on several other aspects of mathematics. Among these were lectures on probability theory by a docent named S.N. Bernstein, one of the great mathematicians of the first half of the twentieth century.

"Not a professor," Neyman points out. *"Docent."*

Bernstein was in his mid-thirties then, a bookish-looking man with an impressively thick reddish moustache. (Neyman indicates with his hands up to his face how the moustache hung down on both sides of Bernstein's mouth.) A biobibliography in Russian lists a doctor's degree from the University of Paris in 1908 and another from Kharkov in 1912.

Why would two degrees be necessary?

"Well, probably—Jewish," Neyman offers, embarrassed.

He explains that there was a quota for Jewish students at the university and that Jews—unlike Poles—were not permitted to become professors.

"However, Przeborski and Russyan are today forgotten. But Bernstein— he is not forgotten!"

As a student, Neyman says, he recognized Bernstein's mathematical stature and was impressed by the fact that he devoted himself to big problems. ("I have always tried to emulate him in this.") But he says also that he was not particularly interested in probability theory, even as presented by Bernstein.

"I was interested in Mr. Lebesgue—in measure and integration."

He makes the statement a little apologetically, for he is not unaware of the irony. In less than two decades A.N. Kolmogorov—in 1915–16 still a gymnasium student in Russia—would make explicit the fundamental connection between probability theory and the theory of measure; and in

1962, Neyman himself, returning to Russia as a guest professor at Kiev, would lecture on the topic: "If Measure Theory Did Not Exist, It Would Have to Be Developed to Treat Probability Theory."

At the time that Neyman was hearing Bernstein's lectures, the theory still steered a far from consistent course between probability as a mathematical abstraction of frequencies (a view which went back to the work of Jacob Bernoulli) and probability as a measure of intensity of belief (a view identified with the name of Thomas Bayes, an English clergyman).

The basic distinction between the two approaches turns on the events or statements to which each is willing to assign probability. "Frequentists" restrict their attention to those which are capable of a large number of repetitions, either in reality or in principle, while "Bayesians" are willing to assign probability to *any* event or statement, even though it may be unique. In those days, this could have been the probability of a revolution in 1917.

(Many years later, Neyman, by then identifying himself quite completely with the frequentist point of view, described "a particular attack on the Bayes front for which I personally have great respect...[which] I learned of in Bernstein's lectures in 1915 or 1916." Here he was referring to a theorem of Bernstein's (to come up later in his own work) which provides interesting information on the behavior of the Bayesian and frequentist approaches in situations where large numbers of observations are made.)

In addition to attending Bernstein's lectures on probability theory, Neyman—as a member of the Presidium of the Mathematics Club—also helped to prepare them for mimeographed publication.

The Presidium was a small group of students, elected by their fellows, who had the duty and the honor—"No pay!"—of mimeographing, sorting, and binding the lecture notes of the mathematical faculty. They had a room in which to work, and this became a kind of clubroom where they could gossip and argue and play endless games of chess. Sometimes members of the faculty dropped by to check on their work. It may have been on just such an occasion, Neyman thinks, that Bernstein suggested to him that he read a book called *The Grammar of Science* by an Englishman named Karl Pearson.

There was only one copy of the book—a Russian translation of the second (1900) edition—in the library of the university, and very soon Neyman and his friends were passing it excitedly from hand to hand. They found even the chapter and section titles heady: "The Need of the Present," "The Ignorance of Science," "The Term Knowledge Meaningless if Applied to Unthinkable Things." The strong, confident voice from London seemed to speak directly to them.

Many years later, after Karl Pearson was dead, Neyman recalled in a review of a new edition of the *Grammar* the discussions which the reading of the book had provoked in Kharkov:

"We were a group of young men who had lost our belief in orthodox religion, not from any sort of reasoning, but because of the stupidity of our priests. [But] we were not freed from dogmatism and were prepared in fact to believe in authority, so far as it was not religious.

"The reading of *The Grammar of Science*...was striking because...it attacked in an uncompromising manner all sorts of authorities....At the first reading it was this aspect that struck us. What could it mean? We had been unused to this tone in any scientific book. Was the work 'de la blague' [something of a hoax] and the author a 'canaille' [scoundrel] on a grand scale...? But our teacher, Bernstein, had recommended the book; we must read it again."

One of those with whom Neyman discussed the challenging new ideas in the *Grammar* in the spring and early summer of 1916 was Otto Struve, a tall youth, several years younger than he, who had just recently come to the university from the gymnasium. Struve was the son of Ludwig Struve, the professor of astronomy, and the great grandson of Wilhelm Struve, the founder of the Pulkovo Observatory, the first in Russia. Eager to get into combat, Struve did not return to the university in the fall of 1916 but applied instead to the elite St. Petersburg military academy. A high-ranking officer, on the lookout for tall men for a regiment favored by the czar, intervened on his behalf and he was accepted in spite of the fact that he wore glasses.

That same fall Neyman submittted his notebooks on the integral of Lebesgue, and in December—"either before or immediately after Christmas"—received the long awaited announcement of the winners of Gold Medals. Stuck between the pages of the notebooks is a yellowed offprint from the *Zapiski*, or "Transactions," of the university entitled "Report on the 'Medal Paper' on the Subject of 'The definite integral of Lebesgue' Submitted Under the Motto 'Per Aspera ad Astra.'"

"Only one piece of work was submitted on this subject," Professor Russyan, the author of the report, stated, "...and it appears to have been written by Yuri Czeslawovich Neyman."

Russyan's use of the word *appears* amuses Neyman.

"He knew all along that I had written it because he had suggested the subject to me."

After summarizing Neyman's work in detail, Russyan concluded, "...the author has completed all the requirements [and] has manifested a degree of independence in this theory, which requires fine logical delicacy. This is shown by his detailed proofs of theorems of relevant authors who have given only brief indications of how their theorems might be proved. [He has also] given his own theorem on necessary and sufficient conditions for the summability of unbounded functions. All of which brings me to the determination that this paper deserves a Gold Medal." (The translation is Neyman's.)

Did he really receive a *gold* medal for his paper?

"No. This was wartime, and so I got money instead. One hundred and fifty rubles, I believe."

What did he do with the money?

"I used it to buy a railroad ticket to St. Petersburg. The war was going badly. There had been a speech in the duma—that was legislative body. Was it incompetence or was it—," he searches for the word,"—*treason. Was it incompetence or was it treason?* I wanted to try to get into the officers' training school.''

Since his friend Struve had been accepted, "in spite of some question, I believe, about his eyes," Neyman had high hopes for himself; but he was not tall enough to catch the eye of a recruiter. Again rejected by the military, he returned to Kharkov and his studies.

1917
—
1919
At the beginning of 1917 the three-century rule of the Romanovs was proceeding inexorably to its bloody end. The czar had taken personal command of the army at the front. The czarina, representing him in the capital, had surrounded herself with a bizarre band of advisers, most of them chosen for her by her confidant, the monk Rasputin, who had just been murdered. The duma had not been convened for more than three months. The conduct of the war was heatedly debated. In the capital, workers struck and regiments mutinied.

Was there revolutionary activity at the university in Kharkov?

"Of course! But it was hidden. It was too dangerous to show."

Once Professor D.M. Sintsov came to the room where the members of the Presidium worked.

"It seemed that some time ago a young man came to him, told him that he was a student and that he belongs to a revolutionary party. This was forbidden, and so he is afraid he is going to be arrested, could Professor Sintsov keep him in his house, kind of hiding, you know. What political attitude of Sintsov was I don't know. He was not a good lecturer, but he was well liked by the students. Somehow he impressed us by width of horizon. So he found it impossible to deny a fellow to stay in his house who is a student and is afraid of being arrested. But then it lasted long. So he came to us and said, 'Now you are Presidium. Can you help me?' And so forth. We understood the attitude of the professor. We never heard that he was revolutionary minded. So, yes, we had some money. So we paid for a ticket for this guy to go some other place."

Soon there came a time when all meetings of students out-of-doors were explicitly forbidden by the government.

"It was explained that more than two, it is already a meeting; and Cossacks, that's cavalry essentially, were to be used to disperse. And this is what happened. My mother saw to it that I feel warm, and so I have an overcoat with sheepskin. Thick. And this is relevant to the story because after maybe lectures I was walking in the street with two colleagues, and then suddenly a bunch of these Cossacks came on horseback, and since we were three—a meeting, you know—we were beaten with their whips. So I received a hit on the back. But because I had my coat I got only a blue mark. Not much. That was the beginning."

By the end of February 1917 a provisional government with Prince Lvov as prime minister was in charge in Petrograd* and the czar and his family were under arrest. This is what Neyman thinks of as "the February Revolution" to distinguish it from "the October Revolution" by the Bolsheviks later that year.

"It is my impression," he says, "that the change in government met with general approval and a substantial degree of enthusiasm."

He spent the summer of 1917, as was his custom, coaching the son of a wealthy landowner in the country. Life had by no means settled down with the end of the czardom. His mail often arrived already opened; and, receiving a letter from a colleague discussing some mathematical question, he found scrawled on the envelope—"What fools!" One day, riding with the overseer of the estate, he saw crude signs set up by the peasants who farmed the land as tenants: "After the Revolution this land will belong to the People!" The overseer, "a little bit drunk," jumped off his horse and, pulling up a row of corn growing along the boundary of the peasants' fields, shouted angrily: "We'll see about that!"

The war continued—Russia's involvement was one of the principal subjects of conflict in the capital between the provisional government and the Petrograd soviet. One day Neyman rode over to a nearby cavalry post and again volunteered his services. This time he was told to try the infantry.

"I was just a little bit offended."

In July, A.F. Kerenski, who was an officer of the Petrograd soviet as well as an officer of the provisional government, became prime minister. That same month separatists in the Ukraine proclaimed an autonomous Ukrainian republic with its capital in Kiev. National minorities such as Poles, Germans, and Jews—as well as "Russians"—were promised freedom of cultural and national development, a freedom which had been denied to Ukrainians under the czar. When Neyman returned to Kharkov at the end of the summer, he and his mother went to the appropriate government office and formally declared themselves to be Poles. That fall, since quotas for Jewish students had been abolished, there was a great increase in the enrollment at institutions of higher learning. Also, Bernstein became a professor.

"I remember a young man, Jewish, making a speech. I can still hear his exact words." He repeats them in Russian and then translates into English. "'Now things will be different! Now we will be doctors! Now we will be engineers!'"

By September 1917 Neyman had concluded his undergraduate studies in pure mathematics. On Russyan's recommendation he was awarded a government stipend to permit him to remain at the university and prepare for an academic career. He was also given a position at the Kharkov Institute of Technology as lecturer and assistant to Przeborski, a member of that faculty as well as a member of the faculty of the university. He supervised the

---

*St. Petersburg had been renamed Petrograd in 1914 at the beginning of the war.

"practice sessions" in analytic geometry and the introduction to analysis for Przeborski and lectured himself on elementary mathematics.

It was the custom in Russian universities, he tells me, for a candidate for an advanced degree to lay out the course of study on which he would ultimately be examined. This was then subject to the approval of the members of his examining committee, who also assigned him a topic of their choice on which he was to prepare himself in a comparatively brief period of time. ("I later tried to introduce this system at Berkeley," he inserts, "and to a degree I was successful.") The topics which Neyman personally chose were all areas of pure mathematics related to his interest in set theory and the work of Lebesgue: theory of functions of a real variable, theory of functions of a complex variable, and theory of elliptic, algebraic, and abelian functions.

The academic year 1917–18 had scarcely begun when, with Petrograd menaced by the Germans, Kerenski and his ministers announced their intention of transferring the seat of the government to Moscow. The proposal furnished the Petrograd soviet with an excuse for forming a military revolutionary committee to protect the capital.

In Kharkov, Neyman, walking along a country road with a friend, was stopped by a peasant woman who ran up from the fields to slip a note into his hand.

The note said simply, "The sappers are ready."

He had no idea for whom the note was intended or why he was mistaken for that person.

How did the Bolshevik revolution come to Kharkov?

He cannot say exactly. It was difficult to know what was happening. There were no longer any newspapers. Instead enormous broadsheets appeared on vacant walls.

"You would see a large group of people gathered around, obviously interested, and one person speaking, the only one who could read."

Soldiers, armed with every weapon they were able to carry, swarmed home from the front. The rich fled, leaving behind their empty residences, businesses, and factories. The largest church in town disappeared over night, in its place a pile of rubble.

"As a progressive academic Professor D.M. Sintsov warmly welcomed the October revolution," A.F. Lapko and L.A. Lyusternik have reported in their history of Soviet mathematics. "From the very beginning he took an active part in the work of Kharkov University, which had in fact become a national university. The training of staff for national economy and culture was the more important task. A special lack of staff was felt during the first years of Soviet power...."

Neyman listens to this quotation, surprised and doubtful.

"I don't think so," he says. "Sintsov was not the kind of person that people listen to."

His own memories of events between 1917 and 1919 are disconnected, and the chronology which follows is an approximation attached to known historical facts of the period.

In February 1918 the Ukrainian Nationalist Government concluded a separate peace with Germany and the other Central Powers. The day after the peace was signed, the Bolsheviks occupied Kiev, the nationalist capital. The Germans marched into the Ukraine to support the beleaguered government with which they had recently signed the Treaty of Brest-Litovsk. Neyman stood in the crowd lining the streets of Kharkov and cheered the arrival of the smart stepping, spike helmeted Germans in their long skirted winter coats, each bayonetted rifle held exactly at the same angle over each right shoulder.

He was *glad* to see the Germans come?

"Of course! They represented the return of law and order!"

The Germans, however, soon became a burden upon the population, first taking over and then ultimately replacing the government which they had come ostensibly to support. When they left, following the collapse of Germany in the fall of 1918, they were cheered away.

By that time there were two nationalist governments in the Ukraine, one struggling with the Poles, who were moving in to free Lvov, and the other battling the Bolsheviks, who had reoccupied Kiev as soon as the Germans left. The situation in Kharkov was not quite so chaotic as the situation in other parts of the Ukraine, for it was soon under the complete control of the local soviet.

That fall, life at the university and at the technical institute was quite different from what it had been in the past. The degrees of master and doctor and the title of assistant, along with all the rights and privileges of those degrees and titles, had been abolished. The division of the teaching staff into full and associate professors, lecturers and assistants was no longer in existence. All persons carrying out independent teaching in the university were to have the same title of professor. The requirement of gymnasium degrees for admission to the university had also been abolished.

A prominent member of the official committee which took over educational reform for the Bolsheviks was V.I. Meshleuk, the oldest son of the Latin teacher whom Neyman had so admired at the gymnasium. In some way, which Neyman does not now remember, Meshleuk managed to involve him in this activity, too. His first task was to interview young people who wanted to enter the university and to help decide which ones would be able to handle the work with the assistance of a little extra preparation.

"This included special courses. On a different level from regular courses. We have here at Berkeley—again I participated in it—such courses to bring in essentially blacks. 'Special Opportunity Something Something.' Well, so exactly similar. Now, all right, this is a reasonable approach. A variety of people came. I don't remember quite, but Khrushchev may have been one of them. May have been. Whether Khrushchev tried to get into the University of Kharkov, I personally am not sure. But we had people from Donets—like him—who came."

In addition to the classes he had been conducting at the technical institute, Neyman had been teaching mathematics and physics a few hours a day at a Polish high school which had recently been established in Kharkov—no easy task, since he had learned all scientific terms in Russian. Now he began also to teach special classes in elementary mathematics for the new, ill-prepared young people who were flocking to the university.

He remembers he would start out in the morning with a rucksack on his back and go from place to place where various classes were held. Sometimes he could barely see his students in the unheated, dimly lighted classrooms and the rag he used to wipe the blackboard would freeze in its bucket. After each class he would get his "pay"—on occasion a ruble or two (although according to the Education Reform Act teachers were not supposed to receive money), more often a bit of butter, a little coffee, a cup of flour, a potato.

On one occasion he received a communication from the Cheka (at that time merely the revolutionary committee, only later the secret police known as "The Red Terror"). It informed him that he was assigned to teach five peasants to read and write—"and I was told that their not coming to me to learn would be no excuse for my not teaching them! So they came. There were five—four women and one man. I taught them in my room at my house. Teaching reading was not so hard. The difficulty came in writing. So you have a hand marked with hard work, not very clean, and you have to take this hand and guide it around to make the letters. Still there was an obvious willingness and interest. They wanted to learn."

He approved very much then, and approves now, of the Bolshevik educational reform. It was one of the good things.

"An effort was being made," he says and then repeats the expression. He likes efforts being made.

With all his teaching activity and the many hardships of the times, was he doing any scientific work?

"Well, I tried," he says with a little helpless gesture of his hands. "I was interested in this work of Lebesgue, and I did a little bit, proving theorems of my own."

Two handwritten vitas, prepared by him in 1923 and 1925, respectively, mention as well "working in 1920 under the guidance of Professor Bernstein on the application of the theory of probability to experimental problems of agriculture" and "study in the theory of mathematical statistics with S. Bernstein from 1917 to 1921." But Neyman shakes his head. He does not recall any formal study. "A few conversations maybe."

Later I show him a letter (written in 1960) in which he describes how Bernstein used to urge him to perform sampling experiments. The young Neyman, according to this letter, "felt revolted," arguing that if the sampling experiment agreed with the theory, nothing was gained; if the sampling experiment contradicted the theory, that meant only that the performance of the experiment was wrong, and who cared?

"I wrote that?"

He is pleased that some documentation for the statement in the vitas has been found, but he maintains that he has no recollection of working on mathematical statistics with Bernstein. He is also surprised to learn that during these years Bernstein, whom he admires as "a highly theoretical mathematician," published papers with obvious connections to applications. (One dealt with the relation of the wind force to the total weight of crops produced in the Kharkov province between 1913 and 1918.)

For the most part, during the years of war, revolution and later civil war, scientific life ceased almost entirely in Kharkov. The *Zapiski* of the university was no longer published, and the meetings of the Mathematical Society were abandoned, not to be resumed on a regular basis until 1925. No member of the mathematics faculty published anything in 1919 or 1920.

"Existence," as Neyman puts it, was the primary concern. Peasants demanded exorbitant prices for farm products. The best way to get food was simply to go to the country and offer them something which they could not produce themselves. One such thing was matches, and a number of chemistry students began to manufacture these. They were not very good matches. "The peasants knew this so you had to show them that the matches would light. You could learn the way." Neyman illustrates exactly how he used to hold a match to make it light and laughs delightedly because he remembers so well.

It was in the course of selling matches for food that he was first arrested.

"It was forbidden, you see. What people in the Second World War called black market."

He was taken by the police with a number of other guilty parties to a building that had been the residence of a governmental official in pre-revolutionary days.

"In the basement there were little cots, tight against one another, where you were supposed to sleep. Under the cot you kept your things. And next to me was a Red Army general. Yes. And he was accused of something. I think it was stealing things. The first night or the second night, but soon after I was deposited there, the door opened and they called names one at a time. 'Bring your things out.' We were all wakened up, and we heard what happened behind the door. 'Take your clothes off!' 'Clothes off?' 'Yes!' 'And shoes too?' 'Yes!' And then—walking walking walking walking—and a couple of shots. Door opens and another name is called. And so this general on the cot next to me was called and shot. Well, I tell you, it wasn't pleasant."

He is still impressed, however, that the discipline applied to ordinary people like himself was applied even more harshly to a general.

After a few days, as he recalls, he was released to return to teaching.

Lack of heat, as well as lack of food, was a problem for the professors as it was for everyone in the city. On one occasion—"may have been winter 1918–19"—the authorities questioned Przeborski, who was then rector of the university, "Why so many deaths among professors?" He explained the effect of lack of heat on old bones.

"So the government allowed professors to cut down trees in the park so that they could have fire—it was called University Park—but then the militia, or whatever they were, had not been informed. So they arrested everyone."

Neyman, with others, was in prison again.

"But this was just an incidental thing. The real prison experience was later. There were several. Difficult to count."

During the year 1919 the civil war between Reds and Whites raged over Russia as a whole. Neyman's sympathies were with the Whites, who represented the remnants of the bourgeoisie holding out against the Bolsheviks. Fighting with them was his cousin and youthful chum, Czesław Lutosławski, and also his friend Otto Struve.

In Kharkov, although there was no actual fighting (as far as Neyman recalls), the hardships of everyday life continued: cold and darkness in winter, hunger in the city amidst plenty in the country, the patrolling Cheka, roaming bands of anarchists, epidemics—a general feeling of impending disaster.

Life was not all bleak. The government made a mansion deserted by the rich available to citizens for approved entertainments, and Neyman's friend V.L. Goncharov obtained permission to stage an operetta he had written. Neyman can still sing many of the songs. "How nice it is to be the Commissar/Of things needed to live/Soap, matches/Sugar, and salt!" He sings in Russian and then translates into English. One song that he remembers is about Professor Sintsov, "the progressive academic."

"Before the Revolution he earned 3,000 rubles a year/And he was able to buy stocks./After the Revolution he earns 300,000 rubles a year/And he has to sell his pants."

Neyman chuckles and tells how he sold his pants, too.

In spite of such cheerful interludes, by 1919, after two and a half years of overwork and privation, he was beginning to show signs of incipient tuberculosis. In the late summer the doctor who had cared for him since he was a boy ordered him south for a rest before classes began again.

It is at this point that a second witness—Olga Solodovnikova—enters the account of Neyman's life.

1919
—
1921

On a train bound for the Crimea in the very early autumn of 1919, Neyman met two young girls. One was fair haired French Natalie, the other black haired Russian Olga. In spite of their parents' disapproval, the girls had taken the money they had earned at various jobs and, during a lull in the fighting between Reds and Whites, had set off for Balaklava, a resort on the western coast of the Crimean peninsula.

Neyman struck up an acquaintance with them on the train. When they said they were going to Balaklava, he said how lucky—Balaklava was also where he was going! The three sat together and shared the books they had brought along for the journey and exchanged comments on what they observed from the window of the train. By the time they reached their destination, they were friends.

At Balaklava the girls talked a landowner into renting them a cottage on the edge of the cliffs that overlooked the rocks and the sea. It consisted of a little kitchen and two rooms connected by a covered passage. Since all three young people got along so splendidly, Neyman persuaded the girls to let him occupy the second room of their cottage for a share of the rent.

For some reason—"for some reason" is all he will say in explanation—he found himself more attracted to the dark, lively Russian girl—Olga Solodovnikova, or "Lola"—who sketched and swam and gloried in the ocean and the cliffs with such passion. Her father was a doctor, the author of a standard medical text; and her mother, the sister of M.P. Artzybasheff, a well known novelist and dramatist—his *Sanin* had been a sensation in 1909 because of its open treatment of sex. (A cousin, "who practically lived at the house," was Boris Artzybasheff, later a famous illustrator in the United States who did many of the early covers of *Time*.) Lola had earned money for the trip to the Crimea by working as a nurse in her father's hospital and by drawing huge portraits of the revolutionary leaders to be carried in parades.

"I did so many posters, enormous big posters, I was telling people that I can close my eyes and still sketch Lenin or Trotsky or somebody," she explains to me. "It wasn't completely true, but I did them so many times. Trotsky had quite a face, you know. Lenin was a better man and a better brain, but Trotsky had a wonderful face."

The holiday in the Crimea was perfect. Every day there was an exciting expedition to the shore or along the cliffs. Then, just before Neyman had to leave, Natalie fell ill with typhoid. Fortunately Lola was able to find a doctor in Balaklava who had been a student of her father's. He arranged for the sick girl to be admitted to the hospital, and reassured her friends. Neyman returned to Kharkov—classes were taking up again—and Lola remained behind until a relative could come and stay with Natalie.

By the time Lola was able to leave for Kharkov, the White army, which had been regrouping in the Crimea under General A.I. Denikin, had begun its advance toward Moscow. Her train moved slowly through the battlefield. "Horrible horrible horrible horrible" is all she can say now of the experience. When she finally arrived in Kharkov, she found that she could not get transportation to Akhtyrka, the town where her parents lived. Having no place else to go, she presented herself at Rogatinskiy Pereulok.

Neyman was delighted to see her again. His mother—it seemed to Lola—was not so delighted.

"She was regretful," Neyman concedes, "but she tried not to show it."

By the beginning of November the White advance, of which Lola had been a witness on her return to Kharkov, was within two hundred miles of Moscow. Another part of the White army was threatening Petrograd. Victory seemed imminent. Then, suddenly, the Whites were thrown back on both fronts.

Lola stayed several weeks at the Neymans' house. Then her parents separated, and her mother moved to Kharkov. "So then I moved to her place, and Mrs. Neyman was very pleased. She thought that this is the end of the story, but it was not the end of the story." Now "every day every day" Neyman was coming to the home of Mrs. Solodovnikova to see Lola.

"And then, well, he told me, you are going to be my wife, and that's it." This was November 4, 1919.*

The proposed marriage was complicated by the fact that she was Russian Orthodox while he was, nominally at least, Roman Catholic and wanted to be married in that church to please his mother. Since Lola and her family did not feel strongly about their religious ties, she docilely agreed to take instruction from the Polish priest. It was not long though before she rebelled. In the mornings she was a student at the Kharkov Academy of Arts and Crafts and in the afternoons she worked in the hospital as a nurse. She simply could not spend time studying dull things like religion! Neyman had a long talk with the priest, who ultimately agreed to marry the couple without Lola's becoming a member of the church. His only condition was that both young people sign a statement that any children of their marriage would be raised in no faith other than the Roman Catholic. Neyman was extremely careful about the wording of the promise they made.

While the Polish-Russian romance was thus progressing toward the altar, the political situation was changing once again. Simon Petlyura, holding together the remnants of the army of the once proudly proclaimed Ukrainian republic, had gone to Warsaw to meet with Pilsudski, by then the head of the newly formed Republic of Poland and the commander-in-chief of its army. A few weeks before the date set for the wedding, these two leaders signed a treaty of alliance. Almost immediately a Polish offensive began in the Ukraine and Polish forces occupied Kiev. By May 4, 1920, when Jerzy Neyman and Olga Solodovnikova were married, a state of war existed between Russia and Poland. Ten days later Neyman was arrested and imprisoned.

In 1978 I talk with Mrs. Neyman. Her account of her meeting and developing romance with Neyman varies only in some minor details from his. She, for instance, tells of friends he was initially visiting in Balaklava while he insists, "I knew no one in Balaklava." But there is a substantial difference between his account of his imprisonment following their marriage and hers: they are completely at odds.

Neyman's recollection is as follows:

"I was arrested after the Riga Peace was signed [between Russia and Poland at the end of the Russian-Polish War]. One of the paragraphs was that there would be an exchange of nationals, prisoners or whatever; and so then people were arrested, including Przeborski, who was rector of the university."

In other words, the Russians wanted to have some Poles on hand to exchange for Russians?

"Yes."

The only difficulty with this account, taken alone, is that the date of his

---

*This and subsequent dates are given according to the Gregorian calendar, which was adopted in Russia in 1918.

arrest—May 20, 1920 (given in a questionnaire filled out in 1948 and presumably correct)—falls in the middle of the Russian-Polish War and not at the end. An armistice agreement was not signed until October 11, 1920; and the peace treaty itself, not until March 18, 1921. It appears that Neyman must have been arrested because he was a Pole and Poland was at war with Russia.

Olga Neyman's account of the reason for her husband's arrest almost immediately after their marriage is the following:

"Why was he arrested? Because one of the mathematical professors, with whom Mr. Neyman worked, had a beautiful daughter who was a ballet dancer and this ballet dancing girl got involved very much with a Polish spy. He was there in Kharkov, a servant of the Polish government, to discover all kinds of things and doing it fine; and this girl, daughter of a mathematician, she was extremely pretty, wonderful wonderful girl, and so this Polish spy got in love with this girl, and the girl was providing him with some secret information, because she was beautiful, she was a dancer, and she would talk to people and get some information, and give this information to this boy. And then our Russian chiefs, the Bolsheviks, they discovered what was happening. This girl was arrested, and Mr. Neyman, he was very friendly with the girl's father—this girl's father was a mathematician. Well, this father mathematician was arrested, Przeborski was also arrested, and Mr. Neyman was arrested."

Neyman listens, bemused, as I recount his wife's version of his arrest. He has no idea who the mathematician she mentions could be. He goes over the mathematicians he was close to. Przeborski had no daughters. Russyan had a daughter, but she married a Bolshevik, much to her father's sorrow. Bernstein? Neyman remembers only that Bernstein had a son who perished in the Siege of Leningrad during the Second World War. There were, of course, still other mathematicians at Kharkov, and—in a letter written in 1966—Neyman did mention that "Przeborski and another mathematics professor" were arrested with him.

Olga Neyman (according to her account) went directly to Christian Rakovsky, the governor of the Ukraine, who as a young medical student and revolutionary refugee had been befriended by her doctor father.

"I was very angry and I went to this Rakovsky, to his office, and his secretary went to Rakovsky and came out and said, 'Go away—Mr. Rakovsky is very busy and he will not discuss this with you.' And I was just frightfully angry and I, oh, threw something on the floor and I yelled. 'When he needed help, my family pulled him out, and now he doesn't even want to see me.' So I went out and slammed the door; and then when I was still going down down down down the stairs, his secretary ran after me and said, 'Stop stop, Mr. Rakovsky will see you!'

"Well, he listened to me and he said, 'I DO remember that your family was extremely nice to me. I DO remember that your father got me a job and all that. I DO. But I CANNOT release Mr. Neyman, because he was connected with this mathematician whose daughter was connected with this spy.' But then I yelled at him that Mr. Neyman was connected with professor because of mathematics and not because of spying. Rakovsky said, 'Yes yes yes yes. All right. It is quite possible. You go back home and wait. I will send a commission to investigate. Then if Mr. Neyman is not guilty of spying, and so on, we will release him.'"

Quite pleased with herself, she went immediately to the prison where her husband was being held.

"I was on the outside—walking and walking and walking and looking— and then—Mr. Neyman's face at the window! And I shouted, 'Oh oh oh, YOU WILL BE RELEASED!' And then guards came and took me to the police station and arrested me, too. Well, the place was frightfully filthy. I looked around and said, 'I am going to stay here?' And the guard looked at me and said, 'Sure, you are going to stay.' And then he went out and came back with a bucket of hot soapy water. 'If you find it dirty, wash it!' 'All right! I will!' So I washed the room beautiful."

When Mrs. Neyman's account is communicated to Neyman, he responds again with a shake of his head: "No memory."

I point out that Christian Rakovsky did in fact exist and was in fact the governor of the Ukraine at the time—a graduate doctor in medicine, arrested and imprisoned in various places during the World War, and released by the Russians in 1917. Neyman shakes his head again after listening to this summary and repeats: "No memory at all." The following week, however, having thought about the matter, he concedes that he does recall "something about scrubbing floors"—but the mathematician and the beautiful ballet dancer and the Polish spy? No.

"Mr. Anatole France," he says and smiles at the reminder. "The testimony of the second witness. Always contradictory and nonreconcilable."

The police released Olga Neyman the same day that she was arrested, but Neyman was held in prison for a number of weeks. On one occasion, in the nineteen thirties, he mentioned in a letter to a colleague that it was six weeks.

Outside the walls of the prison, his fate and that of the other incarcerated Poles was being determined on the battlefield.

"I was just in desperation," Mrs. Neyman says, "but Mr. Neyman was not. He had an awful lot of spirit, you know, as a young man. Good lord! And I occasionally had permission to see him and to talk to him with some other people around. He was always always always big smiles and jokes!"

In fact, as Neyman recalls, he did rather enjoy his stay in prison. This time he was not afraid of being shot, and he was fascinated by the characters and experiences of some of the criminals who had also been arrested in the roundup of Poles. One of these, a petty thief, taught him to smoke.

When he was permitted to receive packages from the outside, he asked his wife to bring the second volume of Emile Picard's *Cours d'analyse,* which he had been studying when he was arrested. Because it was written in French and France was supporting Poland against Russia, his guards would not let him have it. He indicates the place in his bookshelves where there is a gap between Volume I of the Picard and Volume III.

"The guards used the paper to make cigarettes," he explains, adding with satisfaction, "but it was too good and didn't make very good cigarettes."

Przeborski was the first to be released from prison—"for hard labor at the university." Neyman was released a short time later for the same duty.

Since there was a housing shortage in Kharkov, he and Lola had continued to live in his mother's house after their marriage. Later Lola's friend Natalie, who had also married, moved in with her husband, sleeping in what had been the dining room. It was a big house with big rooms, and after a while Neyman's brother, his German wife and their son joined the household. Karol Neyman had decided to go to Poland as soon as possible, selecting as his destination that part of the new republic which had previously been under German rule.

"Poor Mrs. Neyman," Olga Neyman says. "Mother of these two young men. She was completely disgusted because the older brother married a German girl and the younger brother married a Russian girl!"

Indomitable, however, Kazimiera Neyman began to teach her young grandson Polish.

In the fall of 1920 Neyman was at last ready to take his examinations for the master's degree. Apparently these examinations continued to be given even though the degree of master, like other academic degrees, had been officially abolished. In addition to being examined on the various aspects of the theory of functions which he himself had selected for study, he was questioned by his examiners on physics and applied mechanics. He produces the copy of Appel and Dautheville which he used to prepare for this special assignment and points out that while it is a summary of an earlier three-volume work it is still a very substantial volume.

After he had successfully passed his examinations, he became a lecturer at the university. In his 1923 vita he stated that during 1920–21 he conducted exercises in higher algebra, integration, and the theory of sets and supervised a seminar on the theory of functions of a real variable in addition to teaching courses in elementary mathematics.

Also according to this vita, he was approached in 1920 by a Professor Iegoroff of the agricultural department for assistance with the statistical analysis of some agricultural experiments which Iegoroff had been conducting over the past three years. Neyman says that he remembers nothing about this contact, although Iegoroff and his experiments are specifically referred to in a paper he later wrote. In the 1923 vita it is also stated that during 1920 he gave a talk before the university's agricultural department on the application of probability theory to agricultural experimentation.

"So—must be fact, but no memory."

Early in the spring of 1921—this may have been at the time when the Treaty of Riga was actually signed—Neyman heard that he was going to be arrested again. The warning came to him from an old friend by then a Bolshevik—the man who had been responsible for the equipment in the physics laboratory and had so often covered for him in his ill-starred days as a student of experimental physics.

He promptly gathered together a few possessions and took a train to a small village not far from Kharkov where a Russian uncle, I.P. Osipov, a professor of chemistry who was married to his mother's sister Helena, had a country house. He had no money, but he obtained food from the peasants by offering to teach their children. It amuses him that the subjects the peasants most wanted their children to learn were French and German.

When he was not teaching, he spent his time trying to write an account of the events he had lived through in the years just passed—the two revolutions of 1917 and the subsequent civil war.

"But many things happen, and you have to select selections, and so you make selections and then you read it and the general impression is absolutely unrealistic." His first attempt sounded to him like White propaganda. "And so then I tried to correct it, and then again it was a selection too one-sided." It sounded now like Red propaganda.

"How shall I say? I was very unfavorable toward the government of His Imperial Majesty. So, all right, there was a revolution of people who were educated. That fitted. Oh, there were details, this and that, but in general it fitted with my ideas. And then there was this unavoidable—I saw it!—revolution of the mob! And you cannot expect these people who are illiterate to behave in a manner to which you yourself have become accustomed to behaving. And so allowances have to be made."

So one thing he wrote represented the White point of view and the other represented the Red point of view, and neither of them represented the real situation?

"Not quite. Whether it was the real situation—that is still different. But neither of them represented my own ideas of what the situation was."

Some years later, he read a long story about the same period, written by A.N. Tolstoi and entitled, in English, "The Adventures of Nevzorov, or Ibikus."

"This Nevzorov is a man who was a minor employee in a trading company. Yes, he must have been literate, but he was definitely not a member of the same category to which I belonged. No intellectual interest, but interest in beer and cards—interest in becoming rich. Now so—the revolution. And this revolution has several phases. One was the first revolution in February, and Mr. Kerenski symbolizes it; and the second was the revolution in October—a revolution essentially by soldiers who ran away from the front, millions of them, carrying with them the guns they had and could get. Then Mr. Lenin came to Russia from Switzerland or somewhere, and he tried to put intellectualism into that revolution. But it was not easy. It was not really Lenin's revolution. It was a revolution of the mob. Now, all right, in this interface between the mob—call it that for brevity—and Lenin and so on, there was this guy Nevzorov but also there were others—like me. I was interested in Mr. Lebesgue. But then—question of survival. Here is my mother, my grandmother, and later my wife. I am not identifying myself as Mr. Nevzorov. But the description of what happened—I feel that this is something that Tolstoi did excellently while I was trying, twice, and failed."

He does not remember when or how he received word that it was safe for him to return to Kharkov, but he knows that he was back in the city in the early summer of 1921.

The Bolsheviks had rid their country of the czar and his family, the provisional "Kerenski" government, the Allied interventionists, the Germans, the Whites, and Ukrainian separatists. The Poles were about to go. Now, however, there was a new enemy—famine. Lola was still earning some money by making political posters, and at Neyman's suggestion she made a poster urging the people to work to defeat "Czar Hunger."

As always, plague accompanied famine. Walking on the street in Kharkov one evening, Neyman and Lola met a friend of her family from Akhtyrka. They stopped to chat and were told that her father had died recently of typhoid contracted while he tended victims of the epidemic. A little later Lola herself also came down with the disease in spite of her many years of exposure.

Karol Neyman and his family had by that time established themselves in Poland, and arrangements had already been made for Neyman, his mother, and his grandmother—as well as Lola—to go to that country as part of the exchange of nationals agreed upon between Russia and Poland at Riga. Lola insisted that the others leave as planned and that she would follow.

How were they able to get money for train tickets?

"Tickets?" Neyman smiles at the naïveté of the American. "We didn't have to have tickets. We were shipped in empty railroad cars to the Polish border. No seats. People sat on their bundles."

The trip must have been leisurely, if such a word can be used, for on at least one occasion Neyman left the train to return to Kharkov for another glimpse of Lola, getting back in time to continue the journey with his mother and his grandmother.

It is at this point in our discussion of his early life that Neyman corrects me.

"You keep referring to my *return* to Poland," he objects. "I never *returned* to Poland. I *went* to Poland for the first time in 1921."

1978   Neyman's day begins at six with a cup of strong tea sipped during an hour-long bath. He does his thinking and planning in the tub, and he likes to explain that Monday is always an especially busy day for his secretary because there have been three baths since Friday.

What language does he think in?

"It depends on what I am thinking about."

When you are thinking about your work?

"Then English."

He fixes his breakfast, sometimes taking from the refrigerator some cold *bayalda*, a version of the French ratatouille, which his mother learned to make from Bulgarians in the mixed population of Bendery and with which Betty Scott keeps him supplied. Fruit, toast, sometimes a cup of good Brazilian coffee, the beans sent to him by a former student in São Paulo, two vitamin tablets—C and B-complex—complete the breakfast. Before he leaves the kitchen, he packs a lunch.

Afterwards he works at his desk until it is time to go to the university. Since he can no longer drive, he has arranged that in return for the use of his car a young woman named Marilyn Hill—an illustrator for the Statistical Laboratory—will run errands and chauffeur him to and from school. She has her own car, but she makes the agreement out of friendship. When it is almost time for Marilyn, he exchanges his sandals for shoes and puts on a tie. He repacks his briefcase, checks to make sure that he has three packages of cigarettes and his lighter, puts on his overcoat—he leaves it behind only on those rare days when Betty Scott phones and tells him that it is warm enough for him to go without it—gets his dark blue beret from his closet.

Marilyn drops him off in front of Evans Hall, and he takes the elevator to his office on the fourth floor.

During the spring of 1978 his secretary is Tere Sanabria, a young woman originally from Cuba. She has been working for Neyman for a year. His expectations about what can be accomplished—and how quickly—have always been high; and although she likes and admires him, she finds the tensions of working for him very great. "Professor Neyman has just no sense of reality!"

The two of them begin the morning by going over the mail. Neyman is as eager as a child to see what has arrived. Then he gives Tere the work which he has done at home.

After these matters have been arranged, he goes into his office with a freshly typed copy of the paper he is currently working on, and begins to revise it.

The phone rings frequently. Tere answers and switches the call to him, remaining on the line to take notes.

By noon the neatly typed manuscript which she gave him in the morning has been cut in pieces and a number of insertions, written in very black ink on pale green graph paper, have been taped in. Neyman leaves it on his desk. He is not through. But it is time for lunch.

Unless he has some off-campus visitor, he eats in Evans Hall with other faculty members and graduate students who bring their lunches. It is a source of regret to him that not everyone in the statistics department eats lunch together anymore.

Half a dozen men and one or two young women are already around the long table when he arrives. Today most of them are reading newspapers. Neyman does not remember the names of some of the younger people so he does not introduce them to me but announces, "There is a rule here that everybody is presumed to know everybody else."

He places two paper towels on top of each other so that the fold of one is vertical and of the other horizontal.

"That is your plate."

One towel is enough.

"No. Keep two."

He prepares a similar plate for himself, pours two mugs of coffee—he has brought the mugs from his office—and begins to extract from a large plastic bag the lunch he has prepared for himself and Betty Scott: three hard-boiled eggs, a piece of liverwurst in waxed paper, a piece of salami unwrapped, an assortment of yellow apples and bananas.

"This could be a meeting of the Scandinavian Statistical Society," he says, indicating the young people in the room as he peels a hard-boiled egg. "They are all Norwegians."

It is pointed out to him that the young man at the end of the table is a Swede.

"With the exception of one Swede," he agrees.

Betty Scott has still not arrived although she said, when he phoned her office, that she would be down in two minutes. He fusses that Betty does not eat properly, that she is always dieting. He adds that she doesn't like yellow apples.

Perhaps she might like red apples?

"But I like yellow apples."

Eventually she arrives with Lucien Le Cam, a French-born professor who eats regularly with Neyman. He is a tall man with a cheerful face and very dark eyebrows which make almost perfect semicircles over his lively eyes. An informal meeting between him and Neyman takes place at the corner of the table as they discuss nominating Laurent Schwartz for membership in the International Statistical Institute. Neyman explains to me in an aside that five members of the ISI must nominate a new member, and one must be from his own country. Le Cam, who got his Ph.D. at Berkeley in 1952 and who has taught there ever since, is still a French citizen as is Schwartz. So that requirement is met.

Neyman asks Le Cam if he has spoken to C.R. Rao of the Indian Statistical Institute who, like Schwartz, is a current visitor in the department. Le Cam says that Rao has indicated that he is willing to join in the nomination.

"So that is three," Neyman says with satisfaction as he places a slice of liverwurst on a slice of apple. "Le Cam, French; Rao, Indian; and I am an American."

It is now 12:30 and time for his class. He carefully folds his papers, deposits them in the wastebasket, rinses out the coffee mugs and gives them to Betty Scott to take back upstairs to his office.

Waiting for the elevator to carry him to the ground floor of Evans Hall, where his class meets, he comments on the peculiar characteristic of the French which keeps them French even to the extent of maintaining their French citizenship long after they are permanently residing in a country other than France.

He does not understand.

"I want to put down roots," he says, "roots in the university here—where my life is."

1921
—
1923
The newly formed Republic of Poland in which Neyman found himself in the summer of 1921 had been patched together from the provinces of three empires and lay uneasily surrounded by ancient enemies. Elections of the legislative body provided by the constitution had not yet been held, and Pilsudski still served as provisional president; but already, in Warsaw, an exciting new mathematical school had come into being. Its journal, the first issues of which had appeared in 1919, announced its interest in the foundations of mathematics— *Fundamenta Mathematicae.*

As soon as Neyman had delivered his mother and his grandmother to the home of his brother in Bydgoszcz in northern Poland, he set off for Warsaw to seek out Wacław Sierpiński, the leader of the new generation of Polish mathematicians and one of the founders of *Fundamenta.*

How had he heard of Sierpiński so quickly?

He smiles at the question.

At the beginning of the war, Sierpiński, a Pole but a professor at Lvov, then a part of the Austrian empire, had been interned by the Russians as an Austrian national. When the mathematicians in Moscow had heard about his situation, they had arranged for him to be permitted to come to the university there. All during the war he had lived and worked with them, even writing some joint papers with N.N. Luzin, the head of the Moscow school.

"How did I hear about him? I just heard. The name was obviously Polish, and he was working in the same direction I was. Foundations, set theory, and integration."

In Warsaw, though, Neyman learned to his disappointment that Sierpiński was away from the university during the summer.

"But then by walking and listening and smelling—Sierpiński! And he lives *there.* All right. So let's go. I think I paid several visits actually before I found him at home."

He had brought with him from Russia the notebooks which contained his work on the Lebesgue integral and also five theorems "filling in some gaps in Lebesgue" which he had discovered and proved during the intervening period of war and revolution. He remembers distinctly that he had five theorems which he wanted to show to Sierpiński. In time, these and the work for the Gold Medal have merged in his mind, and he is surprised to find that the theorems are not contained in the notebooks which he handles with such affection.

Sierpiński was friendly; but after each of the first four theorems, he went to his bookshelves and took down a volume. "That theorem has already been proved," he told his young visitor. After the fifth and last theorem, he paused, thought a moment. (Neyman mimes.) "Ah! That theorem is new!"

Sierpiński suggested that the work be written up in French and submitted to *Fundamenta Mathematicae.* He was generally encouraging as to Neyman's scientific future. He could not make promises, but he was certain that in the fall a place would be found at one of the universities, most probably Lwow, which—returned to Poland under the Treaty of Versailles—had resumed the

Polish spelling of its name. Neyman, nevertheless, went back to his brother's house in Bydgoszcz depressed and discouraged. Talking to Sierpiński, he had become aware of how far the "conceptual" subjects that interested him had developed since he had been introduced to them at the beginning of the war. Even the vocabulary had changed. It seemed impossible that he would ever be able to catch up.

Lola had arrived from Russia, and the two of them were living at the home of his brother. Karol had a large house, which currently lodged his own family, his mother, his grandmother, and his wife's mother in addition to his younger brother and his sister-in-law. There was still plenty of room, but Lola did not feel welcome. "His mother thought he had left me behind in Russia, and she was not glad to see me again!" Neyman himself considered that it was "not appropriate" for him and his wife to live on the bounty of his brother. He felt that he must find some employment immediately.

But hadn't Sierpiński virtually promised him a job in the fall at a Polish university? July. August. September. That wasn't very long to wait and see.

"It seemed a long time then."

His brother suggested that he should apply for a job as a statistician at the National Agricultural Institute, which was located in Bydgoszcz; and he went to see the director, Kazimierz Bassalik. There was no statistical unit at the institute, but Bassalik was interested in forming one. Neyman's credentials from Kharkov, while modest, were satisfactory for a beginning. He asked Neyman to give a demonstration of his proficiency in French and German and the name of a Polish mathematician who could recommend him. Neyman gave the name of Sierpiński. Bassalik, who in addition to being the director of the institute was also a professor at the University of Warsaw, was satisfied.

The historian of *Penguin Island* would not be surprised at certain contradictions between my account and Neyman's of the beginning of his career as a statistician. As Neyman tells it, his brother's suggestion was made only because he had attended Bernstein's lectures on the theory of probability while he was a student in Kharkov; but sometime after he told me the story, I came across a reference in a paper he wrote at Bydgoszcz in 1922 to "results which Professor Iegoroff communicated to me two years ago in Kharkov." It occurred to me then, for the first time, that Neyman might have had more contact with statistical theory and its application to problems of agricultural experimentation than he remembers himself as having when he applied for his first job as a statistician. A reference (in a later application for a Rockefeller Fellowship) to having studied the theory of statistics with S.N. Bernstein from 1917 to 1921 gave further support to this conjecture; but when I commented to this effect to Professor Scott, she found the idea of Neyman's having had any contact with statistics while he was still in Russia contrary to everything he had ever told her about his early life. Perhaps, she suggested, applying for a fellowship in mathematical statistics, he had quite naturally inflated his contact with Bernstein. Ultimately the question was settled by a statement in a 1923 vita, in Neyman's own hand: not only had he studied the applications of mathematical statistics with Bernstein but his interest had been sufficiently well known that an agronomist (Professor Iegoroff) had consulted him in regard to agricultural experiments he was conducting—as a result Neyman had given the talk, mentioned earlier, before the agricultural department of the university in Kharkov. Even with the vita in his hand, Neyman still has no recollection of the study of statistics with Bernstein nor of the contact with Iegoroff. He truly thinks of himself as having become a statistician entirely by accident and economic necessity: "My brother suggested, so I went."

The Polish town of Bydgoszcz, where he was to live and work in his first job, had been the German town of Bromberg. It was located on a tributary of the Vistula in an area of farm and forest lands. Although the soil was poor, standards of cultivation were high. The Agricultural Institute itself was a relic of German rule. Since almost two-thirds of the land in Poland was devoted to agriculture, it was a large operation employing several hundred people in its various departments and carrying on extensive agricultural experiments at its model farm.

Neyman's official title was "senior statistical assistant," although, he points out, there was no one for him to assist. He received only a small salary but also, importantly for him, living quarters for himself and his wife in the building of the institute. Lola took various jobs. At one time she was painting designs on dishes and at another, assembling toys. (There was considerable hostility toward Russians, and sometimes a worker would refuse to sit next to her.) Neyman's hours at the institute were flexible, and he was able to augment their little income by teaching mathematics a few hours a day at a local high school, the function of which was to prepare enlisted men who had distinguished themselves in the recent war with Russia for officers' training.

In his first months at the institute, Neyman set about educating himself in statistics and agricultural experimentation. Unfortunately, when the Germans had left, they had taken almost everything of value with them. The materials remaining in the institute library were inadequate even for his modest purposes. Sometime in the winter of 1921–22 he received permission and funds from Bassalik to go to Berlin to purchase statistical books and journals. The defeated German capital was in the throes of ideological strife and exponentially spiralling inflation. Neyman had no contact with any of the mathematicians there, not even Richard von Mises, the head of the Institute of Applied Mathematics at the University of Berlin, who had recently published his first attempt at a general formulation of the foundations of probability theory—part of the same modern movement in which Bernstein had been working when Neyman was his student in Kharkov.

He returned to Bydgoszcz loaded down with reading material. (This did not include copies of *Biometrika*, the English statistical journal founded by the author of *The Grammar of Science*.) Since his teaching at the high school had been interrupted by his journey to Berlin, he abandoned it and devoted himself entirely to educating himself in statistical theory and its relation to agricultural experimentation.

After he had been at the institute some nine months, there was a large gathering in Bydgoszcz of agricultural specialists from all over Poland, at which he delivered a paper on the application of correlation theory to agricultural research. In the program he is listed as Jerzy Spława-Neyman. This is the name under which all his early papers were published. The word *Spława* refers to the Neyman family's coat of arms. It is not a title, he explains, but "a sign of nobility." According to family tradition, during the reign of one of the last kings of Poland, a Neyman, presumably a merchant ("Perhaps Jewish, who knows?"), managed to float a raft carrying much needed supplies over the water to the monarch, who was besieged at a confluence of rivers. Thus succored, the king triumphed and rewarded his benefactor and his descendants with the right to bear arms on the coat of which was to be represented their specific service to the king. *Spława*, Neyman explains, comes from the Polish verb *spławiać* meaning "to float." The actual arms, which he never saw until he came to Poland, show the rivers "and some other decorations."

When, referring to *Spława-Neyman*, I describe him as "assuming" the name after he came to Poland, he objects.

"*Assume* implies a little more initiative on my part than there was. I became suddenly confronted with this *Spława-Neyman*."

A few years later he dropped the *Spława*, which he describes as "more of a written thing"; and today he is embarrassed that he ever used it.

By the time that Neyman had been at the Agricultural Institute in Bydgoszcz for a little more than a year, he had written three long papers and two short articles. The first of these appeared the following year in *Roczniki Nauk Rolniczych*, a journal of agricultural science. A continuation of the work was published later that same year in *Miesięcznik Statystyczny*, the journal of the Central Office of Statistics. Both papers were in Polish but were followed by lengthy summaries in French: "Sur les applications de la théorie des probabilités aux expériences agricoles" and "Essai d'application de la statistique mathématique à la résolution de quelques problèmes agricoles."

Neyman has always deprecated the statistical works which he produced in Bydgoszcz, saying that if there is any merit in them, it is not in the few formulas giving various mathematical expectations but in the construction of a probabilistic model of agricultural trials which, at that time, was a novelty. On one occasion, when someone perceived him as anticipating the English statistician R.A. Fisher in the use of randomization, he objected strenuously:

"…I treated *theoretically* an unrestrictedly randomized agricultural experiment and the randomization was considered as a prerequisite to probabilistic treatment of the results. This is not the same as the recognition that without randomization an experiment has little value irrespective of the subsequent treatment. The latter point is due to Fisher and I consider it as one of the most valuable of Fisher's achievements."

By the time Neyman had completed these papers at Bydgoszcz, he was beginning to have confidence in his own ideas regarding the statistical treatment of agricultural experiments, even though they differed from the accepted methodology to which he had been introduced by Edmund Załęski, a professor at the university in Krakow and the recognized Polish authority. He remembers that he went to Bassalik and explained that he felt the "conceptual" foundation of the methodology of Załęski was not quite right.

·"Are you sure?"

"Well, no."

"Better wait then."

The time spent in Bydgoszcz was not unhappy. Neyman was appreciative of the opportunities Bassalik gave him. He participated in meetings connected with the work of the institute and helped with the editing of its journal, being the only one of the sub-editors who received some remuneration at the end of the year. He made one very good friend, Czesław Klott, an employee of the chemical division. Klott was for Neyman a different kind of Pole from himself and the Polish friends he had had in Kharkov—a young nobleman—"in fact, a baron"—he had grown up in "Poland Proper" and so had real roots in the country. Still today, from time to time, he carefully sends checks to Klott's surviving sister and her daughter.

Despite the satisfactions of his work in Bydgoszcz, Neyman was increasingly eager to get to Warsaw. The "conceptual" researches of Sierpiński and his students seemed infinitely more attractive than the applications of the theory of probability to agricultural experimentation. He felt, as he had felt when he was at the technical institute in Kharkov, that a university was where he belonged. It is not clear how much, if any, pressure he himself exerted to bring about the change which took place in his situation after a little more than a year in Bydgoszcz; but at the beginning of December 1922 he went to the capital to take charge of equipment and observations ("Section P") at the State Meteorological Institute.

Neyman has nothing good to say about this new job except that it took him to Warsaw and provided him and Lola with a place to live in a little installation on the Vistula—"really a hut." He recalls one of his duties with particular distaste. For some scientific project in which he himself had very little confidence he had to take measurements connected with the amount of sunshine each day. He especially disliked the handling of the instruments and always tried to get Lola to do that part of the work. Since when the sun did not shine he could not take measurements, he composed a little song in Latin, calling upon cirrus and cirrus-stratus to cover the face of sol. He says he sang it each morning; and, since the weather in Warsaw was not especially good, the incantation often proved effective.

Other than this employment, there are few established facts about the year 1922–23 in Neyman's life. A date that does stand out in his memory is February 19, 1923, the 450th anniversary of the birth of Copernicus. On that occasion he took part in a commemorative program in the opera house in Bydgoszcz. In his talk, he remembers, he emphasized Copernicus's achievement in overcoming "the routine of thought"—"another idea I got from Karl Pearson and *Grammar of Science!*"

He has no record of this talk although at the time, he tells me, it was reported in a local newspaper. Fifty years later, however, on the occasion of the 500th anniversary of Copernicus's birth, he returned to the same theme, writing in *The Heritage of Copernicus:*

"Quite apart from external difficulties connected with attitudes of the existing establishments [in science and religion], Copernicus faced and solved a great 'internal' difficulty resulting from the omnipresent psychological phenomenon of 'routine of thought'.... Along with other domains of human behavior and activity, research is also affected by routines of thought and, particularly, of premises. Here again many routines are very useful by providing important economy of mental effort. But there are exceptions. It does happen that some premises of our thought, acquired through traditional learning in schools and universities, have no other backing than tradition. And the longer the tradition of a commonly accepted premise, the more difficult it is, psychologically, to notice its routineness and to question its validity."

(During our conversations Neyman cites the expression "routine of thought" so frequently—in quotation marks—that I turn to a copy of the 1900 edition of *The Grammar of Science,* which is the one he read in Russian translation in Kharkov in 1916, to find out what Pearson himself has to say on the subject. Karl Pearson indeed comes out firmly against dogma. "If the reader questions whether there is still war between science and dogma, I must reply that there always will be as long as knowledge is opposed to ignorance...." But I cannot find anywhere in the *Grammar* the expression that Neyman quotes so often. Pearson does, however, refer frequently to the "routine of perception" as the basis for the formulation of scientific laws. When I recount my difficulty, Neyman says, "You mean I may have modified the idea a little bit? Possible.")

In Warsaw he continued to work over papers he had written in Bydgoszcz. Two of these had been accepted for publication: one, the long paper which appeared that year in *Roczniki Nauk Rolniczych;* the other, a short paper sent to *Ziemianin,* or "Country Gentleman," as Neyman translates it. The publication of the latter paper was being held up because the press had no mathematical characters.

In the spring of 1923 he received word that the paper presenting the fifth set-theoretical theorem he had brought from Kharkov had been accepted by *Fundamenta Mathematicae.* The news was not so pleasing as it might have been, since Sierpiński, dissatisfied with Neyman's French, had completely rewritten the paper. Although Neyman often speaks of having an "emotional attachment" to certain pieces of his work—and these include, above all, the notebooks on the integral of Lebesgue—he says that he has no emotional attachment whatsoever to the work published in *Fundamenta* under the title "Sur un théorème métrique concernant les ensembles fermés."

It appeared in 1923—his first published paper. He was twenty-nine years old.

"Documents" from the first thirty years of Neyman's life are, as has been noted, virtually nonexistent. All the more welcome, therefore, is the discovery of several handwritten sheets of notes from 1923. These are stuck in the back of one of his notebooks on the Lebesgue integral. He is surprised to find them there.

"I am describing that I intend to write something or other," he says, glancing curiously at the old papers. They are dated *Carnival*—the days of festivity preceding the forty fasting days of Lent—but then *Carnival* has been firmly crossed out and *March* inserted.

"'The present work falls into three different parts according to its contents,'" he reads aloud, translating the Polish into English as he goes. "'Part One is concerned with principles which justify the use of abstract mathematical theory in studies of natural phenomena, especially in the domain of agricultural experimentation.'"

I comment that it is interesting how in the first years of his career what was to be a lifelong interest in the application of theory to practical problems had already surfaced.

"Yes," he agrees. "Generated by Karl Pearson's *Grammar of Science* suggested to me by Bernstein in Kharkov. Yes, spirit but no action—yet."

He smiles then as he translates the next sentence.

"'The thoughts in the first part came to my mind through Karl Pearson's *Grammar of Science....*'"

He says that he was most impressed by the idea, which he met for the first time in the *Grammar*, that science is *description*, not *explanation*. A scientific theory is merely a *model* which happens to fit the observed facts more or less perfectly than another model, or perhaps equally well, the essential point being that there is no single model which is the only possible one for a given set of phenomena.

"The difference between reality and theory, yes, that was very impressive to me. The most impressive, of course, was the realization that geometry—I learned the euclidean geometry—that's one thing, but real two-dimensional space is a different thing. That was—at present it's *trivial*." He gives an embarrassed little laugh. "But then—"

Then it was a real shocker?

"Yes."

He reads bits and pieces of the old notes aloud to me.

"'In *The Grammar of Science* the author notes the absence of a real substitute for the term "true value" of some measurable characteristic. It seems to me that the concept of "true value" is frequently very useful—to the point of appearing indispensable—and in consequence I make an effort to produce a definition that will appeal to intuition and will allow for the application of the theory of probability. Part I ends with familiar formulas for calculating "probable errors." Also I emphasize the falsehood of the frequently heard assertion that the applicability of probability theory to agricultural field trials requires that the yields from particular plots follow the law of Gauss.'"

He takes up Part II, which is devoted to the problem of precision of field trials.

"'Using the method of mathematical expectations I make an effort to solve the problem of the dependence of the expected precision of the experiment on the number of plots in the fields and the number of replications. As far as I know, this problem has not been properly treated thus far.'" In an aside he explains: "Starting with Fisher, this and that, it's old stuff, but at the time it wasn't." He slowly moves his magnifying glass back and forth across the page. "'Subsequently I study the problem whether precision could be increased if you took into account the non-homogeneity of the field, the manner of seeding, and the thing that was sown the year before and also in years past.'"

He stops to tell me that he is not making a direct translation.

"'An effort is made to explain why on occasion the correlation coefficient is negative. On occasion those fields that are better one year incline to be worse the next. There are some signs of periodicity.'—These things later developed into something called uniformity trials on which I gave papers in London."

He continues.

"'Part III contains the method of evaluating agricultural field trials using the method of least squares of Markov. This chapter is likely to be difficult for non-mathematicians, but I make efforts to illustrate assertions by numerical examples which are based on practical problems. I expect it to be useful because in the literature the concept of Standard Error does not appear to be quite clear to real agriculturists.'

"Then I explain that I am not an agriculturist, and I don't claim any competence in agricultural things. Another apology. 'The conditions of my life prevented me from learning the literature, which I presume contains many very valuable works. Since I am not in a position to provide a complete bibliography, I am giving no references whatsoever.'"

He laughs at this last statement and puts down the paper.

"A Part IV was contemplated but then crossed out."

He is not quite sure about the purpose of these notes. Things in them appear in both earlier and later papers. Perhaps he was playing with the idea of a long work, for he considered the notes important enough to save.

"You remember," he says—"Bassalik told me, 'Better wait.'"

In Warsaw, Neyman quickly established contact with the mathematicians at the university. His old friend and teacher, Przeborski, had also come to Warsaw from Kharkov and was lecturing on theoretical mechanics. Undoubtedly it was through him that Neyman obtained a position in the fall of 1923 as an assistant at the university. In addition, probably through the good offices of the agriculturist Załęski, he became a special lecturer in mathematics and statistics at the Central College of Agriculture—*Szkoła Główna Gospodarstwa Wiejskiego*, or the SGGW.

Thus, in the fall of 1923, six months before his thirtieth birthday, he was back at last at the university, where he felt he belonged. But even with his new academic duties, he continued as head of "Section P" at the State Meteorological Instititute. He also began to prepare himself for the examinations required of a candidate for the doctor's degree in mathematics.

Poland was still politically unsettled. Pilsudski had refused to stand for president and a friend elected in his place had been assassinated almost immediately. For Neyman, as for almost everyone he knew, life was hard. An exception was an acquaintance who had made a great deal of money locating old railroad locomotives and arranging for their sale as scrap. Neyman remembers standing wistfully before a display in a shop window and seeing in the glass between himself and the books which he coveted the reflection of this same man quoting to him in Latin: *Even though there is a lack of power, there is a laudable intention.*

As an assistant at the university, Neyman taught the prerequisite courses for Przeborski's lectures on theoretical mechanics. At the SGGW, which was located in an old apartment building fifteen or twenty minutes on foot from the university, he lectured on differential calculus and the theory of statistics. A few years later, describing this period in his life to a friend, he remarked that he had delivered twenty-five hours of lectures a week.

The plunge of the Polish mark continued to follow that of the German mark, to which it was pegged. Neyman and the other assistants and docents went early in the morning on the days when salaries were paid and stood in line in the dark until the university controller's office opened. (Sierpiński and the other professors strolled in later, secure in the knowledge that a place at the head of the line would be yielded to them.) As soon as he received his salary, Neyman literally ran to the bank to exchange his Polish marks for American dollars. He remembers getting six dollars in exchange. During the following month, as he and Lola needed money to buy food and other necessities, he would go to the bank and cash a single dollar at a time.

In spite of the lack of money and the "disgusting" circumstances—"really slums"—in which it was necessary for him and his wife to live, he was happy to be back in academic life. He recalls with particular pleasure the weekly meetings of the Polish Mathematical Society, which were held in the building called the Palace of Staszic after a nobleman who had done much to sustain Polish culture during the time that Poland was under the control of other nations. A wide corridor outside the meeting room was lined with long tables on which were laid out the most recent copies of mathematical journals. Each week he eagerly took up the current *Comptes Rendus* of the Paris Academy to follow what he described to himself as "the tennis game" between French and Russian mathematicians, particularly Paul Lévy and N.N. Luzin, as they developed and extended the ideas that interested him so much.

The other students working for their Ph.D's at the university were, most of them, several years younger than Neyman. One of these was Alfred Tajtelbaum-Tarski, the son of a Warsaw banker and later a colleague of Neyman's in the mathematics department at the University of California. Another was Antoni Zygmund, later a professor at the University of Chicago.

Neyman and Tarski were never close; but a large photograph of Zygmund sits on the file cabinet in Neyman's office next to the portrait of Copernicus. He points to it and says emphatically, "My very cordial friend!"

Zygmund, who was born in Warsaw, is seven years younger than Neyman. Both of his parents were peasants, his mother a woman who never learned to read. Neyman was even then attracted to intellectuals from a background that he saw as "disadvantaged." In addition, even in his youth, he shared political sympathies with Zygmund. "Leftish but not left" is the way Zygmund describes them.

"Mathematically we were not close," he says of Neyman and himself. "He was in probability. I was in analysis. So mathematically we were not close, but somehow—possibly it was human sympathies which made us close, especially in the Jewish matters. Both he and I felt very strongly injustice."

The two young men saw each other frequently at lectures and at meetings, gathered with others at the Cafe Karaś across the street from the Palace of Staszic and the great statue of Copernicus, a duplicate of which now stands in Chicago. But they were not the friends then that they are now.

"I liked him," Zygmund says, "but, you see, seven years' difference at that age is great."

What was Neyman like in 1923?

The soft-spoken Zygmund, in his office at the University of Chicago, is silent, trying to reach back over more than half a century.

"I wouldn't say he was sure of himself," he says at last, "but he could insist on things. He had very strong opinions. Especially in the political area. He was not aggressive, but he was uncompromising." He is silent again for a moment and then repeats, "He did not compromise."

In a departure from common practice, Neyman did not produce a thesis specifically for the doctor's degree. He already had several substantial papers in print, and his bibliography lists the first published paper from the Bydgoszcz group as his doctoral thesis. He says that in fact his degree was granted on the basis of *both* the long papers mentioned in the last chapter.

The load of teaching, attending lectures, and preparing for his examinations was heavy. At the beginning of March 1924, according to an official document, he requested that his employment at the Meteorological Institute be terminated to give him time to study for his Ph.D. examinations.

For his examinations—there were traditionally two—he was required to appear formally dressed in a cutaway and striped trousers. Later he would lecture at the university in these same clothes. At the *Rigorosum Minor* (in philosophy) a professor of the history of philosophy, whose name he has forgotten, asked him to give his opinion of a certain Greek philosopher, the name also forgotten.

"Well, I must say that I smiled at him, 'Sorry, I never heard of the gentleman,' and the professor threw up his hands like this!" (Neyman illustrates.)

The other examining professor for the *Rigorosum Minor* was Tadeusz Kotarbiński, whose specialty was the philosophy of science. He asked

Neyman about the ideas of Ernst Mach, which he had studied and which had close connections with the ideas expressed by Karl Pearson in *The Grammar of Science;* and Neyman had no trouble answering these questions. The dean, who was also present, voted in favor of him.

His examiners for the *Rigorosum Major* (in mathematics) were Sierpiński, Przeborski, and Stefan Mazurkiewicz, one of the founders of *Fundamenta Mathematicae* but a mathematician whose interest extended to probability theory.

Why was he not examined in statistics?

"There was no one in Poland to examine me. I was *sui generis.*"

So then whose student was he? Who was his "Doctor Father"?

"I am the student of Lebesgue," he replies firmly.

In 1924, when Neyman finally obtained his doctor's degree, he was thirty years old—an age which is today described by such researchers as Daniel Levinson as an age of transition, "strongly colored by the imminence of Settling Down and the need to form a life structure through which one's youthful dreams and values can be realized." It is not clear from Neyman's recollections how he saw his future at the time. His first priority was still "existence," supporting himself and Lola, who was again studying art.

His teaching schedule in 1924–25 was even heavier than it had been the year before. In addition to lecturing at the University of Warsaw and at the SGGW, he went to Krakow several days a week to give lectures on statistics at the university there—work again probably obtained through the agriculturist Załęski. He describes how he used to catch a train after his lectures in Warsaw on Thursday afternoon, lecture in Krakow on Friday and Saturday, then go back to Warsaw on Sunday evening, sleeping on the train.

He tried to give good lectures and to give personal help to his students, many of whom—as a result of the democratization of life in the new Polish republic—came from worker and peasant backgrounds. At the SGGW he also had a number of women in his classes since—it amuses him to explain— the Department of Horticulture had been created to provide an appropriate subject for the daughters of the well-to-do to study when, after the war, higher education was opened to women.

The rector at the SGGW was sympathetic with Neyman's efforts to establish a modest mathematics library and the rudiments of a statistical laboratory, but there was little time for research. From his bibliography it appears that he produced only one piece of work after he left Bydgoszcz: a paper on "The problem of value and price from a statistical point of view."

He participated in the weekly meetings of the mathematical society and gave several talks on statistical subjects. He also continued to attend Sierpiński's lectures on the foundations of mathematics.

"You see, this subject, the foundations of mathematics—*fundamenta mathematicae*— has an advantage, had at the time at least, that you didn't have to have prerequisites. You started from the foundations. You had to start from the beginning, however, from the first lecture. If you missed any, then you were lost."

He remembers Sierpiński in his cutaway and striped trousers—"the quality and price much higher than mine"—standing thoughtfully before a formula on the blackboard and speaking so softly that the students could hardly hear him. When somebody would finally get up the courage to ask that he repeat something, he would not turn around but would wave with the hand holding the chalk: "Please, don't disturb me." He also remembers feeling sometimes a little wistful when he saw Zygmund, a docent by that time, discussing "conceptual mathematics" with professors or students. As Neyman has always seen his career, through the vagaries of fate—the scientific isolation in Kharkov from 1917 to 1921 and the necessity of supporting himself and his wife in Bydgoszcz—he became a statistician rather than a mathematician.

At that time, in his opinion, one of the few places in Poland where any respectable statistical work was being done was the Central Office of Statistics. He is inclined to think that it may have been there that he was able to read *Biometrika* as it came out. Through it he became acquainted with the work of a man who signed his papers "Student." These were of special interest because, while statisticians in the past had dealt entirely with large samples, Student dealt in his papers with small samples of the type that arise in agricultural and other experiments, and had worked out methods of obtaining from these useful statistical information.

From the evidence of Neyman's early published papers it appears that he knew at that time very little about the statistical work of Karl Pearson. His thinking, however, continued to be heavily influenced by Pearson's *Grammar;* and, according to Olga Neyman, he talked of nothing but "Karl Pearson Karl Pearson Karl Pearson!" Neyman, nevertheless, insists that he had nothing to do with creating the circumstances which were shortly to result in his going to London to study with Pearson.

"These two guys—Sierpiński and Bassalik—came one day and said to me, 'You know, you're doing all this statistical stuff and nobody in Poland knows if it is good or bad. So we are going to send you to London for a year to study with Karl Pearson, because you say he is the greatest statistician in the world. And if he will publish something of yours in his journal'—that was *Biometrika*—'then we'll know you're O.K. Otherwise—don't come back.'"

By that time, in a desperate reform move on the part of the government, a new unit of currency—the zloty—had been pegged to the French franc. The financial crisis seemed to be over. Neyman would receive a Polish Government Fellowship that would provide him with the equivalent of twenty-five English pounds a month. Twenty-five pounds at that time was the equivalent of $125. He and Lola felt as if they had come into a fortune. They agreed to split the money down the middle. *She* would go to Paris to study art for a year, and *he* would go to London to study statistics with Karl Pearson!

1925
—
1926

In September 1925, outfitted in the cutaway and pinstriped trousers in which he was accustomed to lecture in Warsaw, Neyman presented himself at the Porter's Gate of University College, on Gower Street in Bloomsbury, and inquired where he could find Professor Karl Pearson. It happened that W.S. Gosset, of the Guinness brewery in Dublin, was leaving the college at that moment; and he stopped to inform Neyman that one could not see "the Professor"— the Professor was at lunch. Then, since by his dress, manner, and speech Neyman was obviously a foreign visitor, Gosset suggested that they lunch together around the corner. This was Neyman's first meeting with the man who signed his statistical papers "Student."

Gosset, or Student, was an Oxford-trained chemist who had become a statistician as a result of circumstance. According to the memoir which Neyman wrote at the time of Gosset's death, one of his superiors at the brewery had read a book on probability theory and "[had] thought there might be some money in it." Be that as it may, the nature of brewing, to which scientific methods were just beginning to be applied at the turn of the century, was such that it showed up very quickly the deficiencies of contemporary statistical theory when applied to the necessarily few replications of an experiment at a brewery. Gosset became interested in the statistical treatment of brewing problems and, when he had been working as a brewer for five years, submitted a report on "The Application of the 'Law of Errors' to the Work of the Brewery." Shortly afterwards he met Karl Pearson and then began to attend Pearson's lectures at University College. In his spare time he worked on the problem of small samples:—"a greater toil than I expected, but I think absolutely necessary if the Brewers are to get all the possible benefit from statistical processes." The result of this study was "Student's t-statistic" for testing a normal mean. One story has it that when the work was published Gosset signed himself "Student" because the brewery did not want it known that one of its employees was writing scientific papers. According to another, Karl Pearson suggested using the pseudonym in preference to his having to designate the author of a paper in *Biometrika* as a brewer.

What did Neyman and Student say to each other over lunch? Neyman lifts his shoulders helplessly. He thinks that they must have eaten pretty much in silence. Already, in the few days he had been in London, there had been a contretemps with language when, inquiring where he could find a woman to do his laundry, he had substituted a form of the French *laver* for the English *wash* and had appeared, to his horrified landlord, to ask for "a woman to love."

After lunch with Gosset, Neyman returned to University College, where he was admitted to the presence of Karl Pearson. The author of *The Grammar of Science* was sixty-eight—a commanding figure, very tall and straight, his features and high forehead testifying to a handsome skull under the ruddy, tightly drawn skin—a fact of which Pearson himself would not have been unaware, since one of his many great interests was craniometry.

There was no question but that he was at that time in absolute charge of the department at University College and still the dominating figure in English statistics—although frequently and increasingly challenged by the gifted but contentious statistician and geneticist R.A. Fisher of the Experimental Station at Rothamsted—a man some thirty years younger. Pearson told Neyman firmly that he would not talk to him until the term began in October but that in the meantime he would be welcome to use the library of the Department of Applied Statistics.

In the ensuing month Neyman came each day to the college and read, one after another, the statistical papers of Karl Pearson, almost all of which had been published in *Biometrika,* the journal which Pearson had founded with W.F.R. Weldon and Francis Galton when the Royal Society had refused to take papers in which mathematics was combined with biology. In spite of the fact that Pearson had spread himself over several fields—he had already been admitted to the practice of law when he accepted a professorship at University College—his papers on statistics would have made a substantial bibliography for any statistician.

When the term began, Neyman found that he was one of seven or eight post-graduate students, some of whom, like himself, already had their Ph.D. degrees. There was only a single Englishman. The others were Americans with the exception of Neyman and Katsutaro Yasukawa, a Japanese. They all did their work in one room with two rows of tables, each table being shared by a pair of students. Neyman sat beside Yasukawa, and a fragile friendship began to develop between them.

"He was very punctual, and I tried to be punctual, and so we would go to the lecture together, and then kind of an encounter, close encounter, would occur. He would say something, and I would say something. But he didn't understand what I said, and I didn't understand what he said."

There was a tall, reserved young man in the laboratory approximately Neyman's own age. He gave a few lectures and assisted students by helping them to work out examples and by demonstrating to them the use of the Brunsviga, a hand-operated calculating machine. This was the Professor's only son, Egon. On one occasion, quite early in the term, Neyman commented to Egon Pearson that Professor Pearson's mathematics was not quite the kind of "conceptual" mathematics which he had expected to find at University College.

"Whether I knew he was his son or not, I don't remember. But I thought being young he would feel the same way I did. His reaction—he didn't take offense, didn't scold me or anything, no quarrel, but it was obvious that he disapproved of what I said."

Neyman had no way of knowing that Egon Pearson was then at a turning point in his relationship with his father—a psychological crisis precipitated, at least in part, by R.A. Fisher and his rough attacks during the past few years on the scientific work of the elder Pearson.

"...I had to go through the painful stage of realizing that K.P. could be wrong," he was to explain to Neyman much later in their friendship. "...and

I was torn between conflicting emotions: a. finding it difficult to understand R.A.F., b. hating him for his attacks on my paternal 'god,' c. realizing that in some things at least he was right."

There is no evidence that when Neyman came to London in the fall of 1925, he was familiar with the work or even the name of Fisher. Like Karl Pearson himself, Fisher was both a geneticist and a statistician and thus—it might have been thought—the natural heir to Pearson's mantle. In fact, in 1919, recognizing Fisher's obvious gifts, Pearson had offered him a position in his laboratory at University College—but under what Fisher's daughter-biographer, Joan Fisher Box, has described as "castrating" conditions: "to teach and to publish only what Pearson himself approved." Fisher had refused the position and had taken instead a job as the only statistician at the small experimental station in Rothamsted. The next year, Pearson had returned a paper submitted to him by Fisher with the comment that "under the present printing and financial conditions, I am regretfully compelled to exclude all that I think erroneous...." Fisher is said to have "vowed" never again to submit a paper to Pearson, and he never did. The year following, he published, in the *Journal of the Royal Statistical Society*, a trenchant criticism of the application of chi-square,* a statistical test which Pearson had introduced at the beginning of the century, telling the older man in effect, as Dr. Box writes, "that he did not understand the primary principles of its application, that he had misled his followers and spread his own confusion, even in cases where he managed to get the right answer...."

In 1925, the year that Neyman came to London, Fisher's *Statistical Methods for Research Workers*, in which he continued the attack on Pearson and chi-square, had just appeared.

While Egon Pearson was struggling to understand Fisher's new ideas—he was not alone in finding them difficult—Neyman was concentrating on the work of Karl Pearson.

The Professor suggested and supervised almost all the research of the students. It was his custom, twice a day, to come into the room where they were working and talk with them individually. One day, when he arrived at the desk which Neyman shared with Yasukawa, he remarked that he would like to see reprints of any papers which Neyman had published. Neyman replied that the reprints, which he had sent to England before he left Poland, were already in the Professor's possession. Two were in French and the others in Polish with French and German summaries.

"So he took all these papers, read them, studied them, this and that, and said, 'This one is going to appear, but here is a mistake in it.'"

The mistake Pearson referred to, according to Neyman, was the following:

"Karl Pearson deduced a number of distributions, Pearson curves, so called. And so in one of these papers which I brought from Poland there is a note saying that the variance and the mean are independent for just one of the Karl Pearson curves. A paper I wrote while I was still in Poland. Somewhere

---

*Throughout the text the few mathematical symbols which appear have been written out.

at the end it says that. Mean and variance are independent for one of the distributions but not for the others. And at that time it seemed that Karl Pearson did not understand the difference between independence and lack of correlation. He was talking to me in this room with the desks. People whom I hardly knew. It was the first weeks essentially. And I tried with my inadequate English to explain to him."

In the midst of Neyman's efforts to make himself understood, Pearson said angrily, "That may be true in Poland, Mr. Neyman, but it is not true here!" and strode out of the room.

"The suggestion being that he had missed something, you see—something which he obviously hated to think—and then that it would be communicated to these people."

In the embarrassed silence which followed Pearson's outburst, Neyman could think of nothing to do but pick up his things and leave. He stayed away from the college for several days, trying to decide how he could possibly retrieve his situation.

"First of all, the scene. And then, in addition to the scene, would be the consequence. He had intended to publish my things, not new things, but republishing things I had published in Poland. So I could say to those guys—Sierpiński and Bassalik—'See, I'm O.K.!'"

He decided to approach some other member of the staff at University College and try to explain the point in dispute—then maybe that person could explain it tactfully and in more correct English to Professor Pearson. But whom should he approach? He thought of young "Mr. Pearson" but he seemed "too committed" to the Professor, as did most of the staff, to take someone else's side in a situation where the Professor was wrong. He turned instead to J.O. Irwin.

Irwin, who is retired and living in Switzerland, remembers the misunderstanding as Neyman recounts it, including Karl Pearson's "That may be true in Poland, Mr. Neyman…"

"It concerned a point which was new to us at the time, but which Neyman had explained to me, and where I thought he was perfectly right. Karl Pearson thought he was wrong and they seem to have had a discussion about it—heated at any rate on K.P.'s side.… However, I managed to get the point across to Egon, who in the course of a few days convinced his father."

During this time Neyman continued to stay away from the college.

"How I got the message from Irwin I don't remember. And so Karl Pearson came in. He came, opened the door, looked. 'Oh, Dr. Neyman is there. Just a minute.' And so he left. Brought the paper and said, 'Well, let's put it slightly different.' And so it was published."

Although this first work of Neyman's in English was later to be described by Oscar Kempthorne as "surely a landmark paper," it does not include the problem of correlation between plots and its effect on the precision of an experiment, which Neyman had dealt with in the original Polish paper, but treats only the simplest type of random sampling and the technical details of deriving the necessary formulas. In a footnote Pearson explains his reprinting a paper already published on the grounds that English biometricians may not have seen the original and "several important corrections" have been made. These, however, appear to be essentially changes in notation.

After the incident with the Professor, Neyman felt that there was a change in Egon Pearson's attitude toward him.

Sometime during the winter Neyman invited the younger Pearson and Yasukawa to have dinner with him. In the course of the meal a reasonable amount of alcohol flowed, and by the time the coffee came around the Polish Neyman and the Japanese Yasukawa had discovered that they could understand one another. It was "a miracle" in Neyman's words—"and then we conversed. Egon was there. He was smiling."

Except for a few such incidents, the first term at University College was a time of isolation, hard work, and—very shortly—seriously straitened financial circumstances. The drastic manner in which the Polish government had carried out its reform of the currency had resulted in economic disaster. Neyman's grant, which he was dividing with Lola, was suddenly worth $25.00 a month to each of them instead of the $62.50 of a few months before. There was also the probability—he was afraid it was a certainty—that the grant would be cut off altogether at the end of the year. In desperation he applied at the Polish embassy for assistance. A friendly official gave him the job of carrying diplomatic dispatches once a week between London and Paris. The financial arrangement for the transport of these provided first class fare and accommodations for the courier. By going third class and staying with Lola, Neyman was able to pocket the difference.

Sierpiński also came to his assistance, as did Stanisław Michalski, a patron of science, the founder of Mądralin, a retreat for Polish scientists outside Warsaw. The two men joined in proposing Neyman for a fellowship of the Rockefeller Foundation's International Education Board.

Karl Pearson recommended him as "an exceedingly able mathematician, much interested in modern statistics, [who] has had to struggle in Poland with many difficulties, being largely out of the main current of modern statistical development....He...is just such a one as the Rockefeller endowment would do real service to, and in the future could itself earn credit for having assisted in his earlier struggles for wider knowledge, and to increase the bounds of existing knowledge...."

Neyman's application and the associated correspondence now in the archives of the Rockefeller Foundation provide a statement of his scientific goals and interests at that time.

"I would like to follow the guidance of Professor K. Pearson," he explained, "and to investigate the properties of the coefficient of variation (to find its moments, its frequency distribution in various cases, etc.). Afterwards I intend to proceed [in] my occupation with applications of statistical methods to some problems connected with agricultural experimentation."

In response to the question, "Why do you wish to carry on the study or investigation above named?", he replied:

"The coefficient of variation is a very important statistical character of populations, is very often used in many branches of science, but its theory is as well as unknown.

"The methodology of agricultural experimentation has recently very largely developed, but the accuracy of experiments is often useless as there are no or only few correct statistical methods of calculation of results. As the progress in agriculture is connected with the progress of experimentation, it seems to be of great importance to work off the statistical methods suitable to application to the agricultural problems."

The International Education Board moved promptly. Its chief concern seems to have been that Neyman's Polish sponsors guarantee that he would have a position at the end of his fellowship. Five days after the expiration of the grant from the Polish government, Neyman learned that he would be receiving $182 a month for living expenses for the next eight or nine months with additional payment for tuition fees and return travel from London to Warsaw. He had asked for only $150; but Augustus Trowbridge, the director of the International Education Board, had recommended the additional amount, "[noting] the tendency in all these Polish fellowships to rather grossly underestimate the needs...."

Overnight the Neymans' almost worthless Polish zlotys had been replaced by solid American dollars. With the unexpected largesse they began excitedly to plan a holiday in Italy at Easter. Until then, Neyman continued to concentrate on investigations to determine the moments of sample moments, both for infinite and for finite populations. He had a second paper accepted for publication in *Biometrika*, this one developing earlier ideas of Karl Pearson on the theory of non-linear regression.

By spring Neyman had become sufficiently familiar with the name and work of Fisher to express a desire to Student to meet with that statistician. On March 22 Gosset, who at that time still managed to keep a friendly foot at Gower Street as well as at Rothamsted, wrote to "warn" Fisher, who had recently become the director of the experimental station, that "one Dr. Neyman" was eager to visit him.

"He is fonder of algebra than correlation tables and is the only person except yourself I have heard talk about maximum likelyhood as if he enjoyed it."

According to Gosset, Neyman was especially eager to receive reprints of Fisher's papers because, since Fisher published in a number of different journals, these would be hard for him to trace and obtain in Poland. In spite of his expressed interest in meeting Fisher, Neyman did not follow up on Gosset's letter of introduction until almost four months later. They met for the first time on July 20, 1926. Neyman is frankly surprised by the date. He is under the impression that he did not meet Fisher until considerably later. He has memories of several visits to Rothamsted, and he can't be sure which was the first. But one stands out particularly.

It was a Saturday or Sunday afternoon with tea being served to guests on the lawn.

"Yes, there was Mrs. Fisher. Yes, there were many children. Yes, there were various other people who talked. Mrs. Fisher was a silent person. She and the children seemed isolated somehow."

He remembers a big house and in back a large tree. Under the tree was a pram which contained a small baby who squalled continuously and to whom neither Fisher nor his wife and children paid the slightest attention. During the course of the afternoon, one of the daughters came up to her father and asked permission to join the discussion. Fisher shook his head, and she went away. Neyman was curious.

"Why don't you let her?"

"Because she was wrong," Fisher replied, "and she would not admit it to me."

In the spring of 1926, at the same time that Neyman was being introduced to Fisher by way of a letter from Student, Egon Pearson was composing a letter of his own to Student—a letter which would change both Pearson's and Neyman's professional lives and have its effects on Fisher, too.

Pearson had decided that if he was going to be a statistician he was going to have to break with his father's ideas and construct his own statistical philosophy. In retrospect, he describes what he wanted to do as "bridging the gap" between "Mark I statistics"—a shorthand expression* he uses for the statistics of K.P., which was based on large samples obtained from natural populations—and the statistics of Student and Fisher, which treated small samples obtained in controlled experiments—"Mark II statistics."

In both Mark I and Mark II statistics, statistical hypotheses were commonly accepted or rejected according to various rules based on the calculation of intuitively chosen test statistics, or functions of the data. If the test statistic exceeded a certain critical value, the hypothesis was rejected. Since in any given situation a number of appropriate test statistics could be constructed, a statistician had to rely on his own judgment in choosing which statistic to use in any particular testing situation.

The question which Pearson had begun to ponder was the following:

Was there some general principle or principles, still appealing to intuition, by which the statistician might be guided in choosing?

It seemed to him that the justification for the use of the standard test statistics was most questionable in the new small sample theory which had been developed first by Student and later, more mathematically, by Fisher. One day at the beginning of April 1926, down "in the middle of small samples," wandering among apple plots at East Malling, where a cousin was director of the fruit station, he was "suddenly smitten," as he later expressed it, with a "doubt" about the justification for using Student's ratio (now known in the form of the t-statistic) to test a normal mean.

Young Pearson did not find Gosset, who dropped by University College with a rucksack on his back, the least bit intimidating; and he took his doubt directly to him in a letter written on April 7, 1926. Exactly what interpretation could be put upon Student's test?

---

*During the Second World War successive models of military hardware were designated as Mark I, Mark II, and so on.

"I have not really thought of the matter much before," he explained in his letter, "but as it is a stepping stone from which so much small sample theory (particularly of Fisher's) starts, I feel the whole thing rather important...."

He did not immediately mail his letter to Gosset, but instead went off to Italy to enjoy the sun and gaze upon the pictures of the great Italian masters.

There had begun, by that spring, to be considerably more social contact between Neyman and Pearson; and although Pearson traveled to Italy alone, there were (as Neyman puts it) "encounters" there between him and the Neymans during the holiday. The three visited various tourist attractions together; and, in a later letter of Neyman's, there is a suggestion that Pearson and Lola both sketched some scene and agreed to exchange their pictures when they were finished. Neyman's third paper in *Biometrika* expressed his "warm thanks to Dr E.S. Pearson for his help in the English redaktion..."

That spring, in London, the two men attended a concert together.

"May have been Kreisler playing," Pearson recalls. "Jerzy got up from his seat, clapped his hands, and shouted 'Bravo!' He looked so foreign with the little moustache and the pince-nez. People looked around. I was embarrassed, but I didn't take it against him."

The friendship, tentative though it was, was still a little surprising. Pearson was an introverted young man who felt inferior for a number of reasons. He had grown up in the great shadow of K.P., "lovingly protected" in his childhood and kept out of the war in his youth. At Cambridge he had felt cut off from classmates of his own generation, all veterans of the conflict. He suspected that K.P. was disappointed in him, for he had not gone on to his second mathematics "Tripos" but had taken his degree on the basis of work he had done during the war. After joining the staff of K.P.'s laboratory, he had continued to live at home and to have almost all his social contacts with relatives. He spent his summers sailing with a cousin and her husband. His best friend was another cousin.

Neyman's stories of his life experiences fascinated Pearson as much as the settled order of the life of the English, so different from the struggle for existence in Poland and Russia, fascinated Neyman. On May 6, when Pearson mailed Student the letter which he had written a month before, England was in the middle of a General Strike involving more than three million trade-union workers. To Neyman's East European eyes it was an astoundingly decorous strike. He was especially impressed when his landlady, hearing that the government was urging people not to hoard, took back to the grocer half the food she had just purchased. In Poland or Russia—he was sure—such an announcement would have produced exactly the opposite effect.

The English General Strike ended on May 12—on the same day that Pilsudski marched into Warsaw at the head of the army. The English mails went through. Gosset received Pearson's letter and responded promptly with two letters written on successive days.

Pearson's question required "profound study," he wrote in his first letter, but "you can't expect Student to surrender...without a struggle." In his second letter, in the course of responding at length to the question, he pointed out that the only valid reason for rejecting any statistical hypothesis, no matter how unlikely, is that some alternative hypothesis explains the observed events with a greater degree of probability.

In such a case, Student concluded, "you will be very much more inclined to consider that the original hypothesis is not true."

Here, explicitly stated on the fourth page of Student's letter to Pearson, was the essence of what was to become the Neyman-Pearson theory of hypothesis testing. Like so many original ideas—after they have been accepted—it appears obvious to the layman as well as to the statistician. To appreciate its originality, one must go back to the time of Student's letter.

Statisticians of the first third of the twentieth century talked simply about testing a statistical hypothesis. They did not think, at least not explicitly, about the alternatives to the hypothesis under consideration. They had to have these vaguely in mind; but they did not put them on the table, as it were, and say—as Student said to Pearson—here is what we have to look at before we can make up our minds.

Receiving Gosset's letter, Pearson realized that he was "on" to something; but he felt inadequate as a mathematician to deal with the ideas he saw developing. Although he was always to give Gosset credit for his suggestion—"the germ of that idea which has formed the basis of all the later researches of Neyman and myself"—he did not consider turning to Student for mathematical help. The brewer from Dublin was even less of a mathematician than he was. In fact, Student had once forwarded the report of some experimental work to Karl Pearson with the comment that he hoped K.P. would find it interesting "though its chief merit to the likes of me (that there are no mathematics in it) will hardly commend it to you."

No,—Student was not the man to provide the kind of help Pearson felt that he needed for the mathematical formulation of his exciting new ideas. Instead he thought of his father's post-doctoral student from Poland.

"Neyman struck me as just the right man."

He admits he doesn't know quite why. It may have been because Neyman was so "fresh" to statistics. The other English statisticians would all have had preferences between Mark I and Mark II statistics. But Neyman—"I think K.P. meant little more to him than *The Grammar of Science*; he had had hardly any contact with Student, and none really with Fisher."

Instinctively, as Pearson sees it, he chose, "a foreigner whom I liked."

Sometime in late May or early June he invited Neyman down for a weekend at his father's summer place, the Old School House at Cold Harbour on Leith Hill in Surrey. He thinks that Neyman talked a great deal about his early life and also about Pilsudski; but he also thinks that at that time he may have spoken to Neyman, "mildly," about the problem of statistical inference that he had been puzzling over. He is no longer sure whether he also suggested at the same time the possibility of their working together on the problem. Later letters seem to indicate that in fact he did. But such collaboration would have to be carried on almost entirely by mail, for Neyman was already planning to obtain an extension of his Rockefeller

Fellowship for a year's study in Paris—the city of light, from which, as he knew from childhood, "emanated all things attractive."

1978   Neyman's class meets in a small room on the ground floor of Evans Hall.

What is the title of the course?

"Something this and that." He dismisses the question. "I teach usually now about what I have done myself."

The half dozen students are already in the room when he enters. David, a young man from Liverpool who has recently had a malignant mole removed from his arm, has got rid of the cast since the last meeting of the class; and Neyman begins by asking him how his hand feels. David says he is limited only in bending the hand backwards, and he illustrates its mobility by making various motions.

After this brief discussion, which takes place at the door, Neyman proceeds slowly to the front of the room, talking about the work of the day.

"Now, today, the question is *What properties should a test criterion have to make us happy?*"

Having reached the front of the room, he ignores the lectern and stands a little to one side of it, one hand in his pocket, his fingers curled over a package of Marlboros, which shortly emerges. He looks around at his students.

"Now who smiles?"

All immediately straighten their faces.

"I think maybe you smile a little bit?" he says to the only woman in the class.

"I'm not sure."

"Well. All right. There is a rule. *Ladies first.*"

A young woman wearing faded jeans comes to the blackboard, obviously not looking forward to what lies before her.

Neyman sits down in one of the chairs in the middle of the front row, his left arm, with a cigarette in hand, hanging over the back of the chair. His voice is strong and quite audible even though his back is to the students. He guides the young woman, whose name is Linda, through the material, somewhat in the manner of a Polish Plato leading a slave through a proof of the theorem of Pythagoras.

Linda is only a little more knowledgeable than Plato's slave. Having come from operations research rather than statistics, she does not possess as much mathematical background as the other members of the class.

Customarily—Neyman explains later—he likes to take the poorest students first to find out how much they understand.

He begins by instructing Linda what to put on the board. Then come questions. So what is wrong here? What can we do? What do we need? She almost never is able to answer, partly it seems because she does not understand what he wants her to say and so she guesses. The other students make suggestions.

At one point Neyman demands, "Now who is responsible for this distribution?" She does not know, but someone in the class offers the name of Cauchy, which is what Neyman wants.

"Cauchy. Cauchy was a very nice gentleman," he tells the class. "But his distribution *has—no—moments.*"

The demonstration continues, Linda still being led step by step by Neyman.

"Do not cross your little *x*. It may be mistaken for a capital *X*. Capitals are for variables, small letters are for constants." "Put in *theta* there. Now do you need it somewhere else?"

Announcing emphatically what is obviously a familiar rule to the class— *Laziness is a prerequisite to progress,* he asks, "Can we shorten this long equation?"

This time, on her own, Linda writes "*P =*" and follows the equals sign with a large part of the equation.

"Have I tortured her long enough?" Neyman asks the class.

"No!" chorus the young men.

"You have just lost your beers," Linda tells them as Neyman moves from Cauchy to Taylor.

"This lady is not quite comfortable with Mr. Taylor," he observes finally and, to her relief, sends her back to her seat.

"What about David?" he asks of the young man with the arm recently removed from the cast. "Are you able to write?"

David says his hand is all right. It emerges white and thin from his long-sleeved T-shirt. He shyly passes it over his fuzzy beard. As David goes to the blackboard, Neyman remarks that he can shave now that his arm is out of the cast.

David writes on the board with confidence, but he has to switch the chalk to his left hand as he moves ahead of the equation. He is obviously much better prepared than the young woman who preceded him, but Neyman is not quite satisfied.

"You are jumping a little bit. Let's keep that root and then see—"

David's continuing difficulties with the physical act of writing concern Neyman, and very shortly he sends him back to his seat.

"What about Luis? Does he smile?"

Luis, a tall Venezuelan with lively eyes, black hair and a black moustache, comes eagerly to the board and begins to write, explaining what he is doing in emphatic, heavily accented English. Sometimes he runs ahead of Neyman and is firmly reined in.

The students are becoming interested. Dr. Odoom, a big soft-spoken visitor from Ghana, older than the others, follows the work on the board closely, Luis's Latin-accented English being interrupted from time to time by Odoom's English, perfect except for a slight African lilt. Two other students, Steve and Ken, enter into the discussion.

At one point everybody in the class, except Linda and David, is talking about the equation on the blackboard. Odoom gets up to show Luis what he thinks should be done.

Neyman sits smoking and quietly watching them for a while. Then he takes over again, and Odoom returns to his seat.

At a quarter past two Neyman glances at his watch.

"Well. All right. I must say that we have not advanced much today."

As Neyman gets up to leave, Odoom joins Luis again at the blackboard. He is followed by Ken. The three of them argue about the equation.

Neyman closes the door of the classroom behind him, a gleam of pleasure in his eyes.

"You see," he says, "they are *interested*."

Later he tells me that Linda did not appear at the next meeting of the class.

"But I sent word to her through one of the other students to come back, and I think she will."

(She does.)

1926 — 1927

Scientifically and emotionally, even before the end of the academic year 1925–26, Neyman had already "left" London. University College had been a disappointment. He had come to what he had expected to be the frontier of mathematical statistics—to Karl Pearson and *Biometrika* and the Biometric Laboratory—and he had found that there was very little mathematics there. While in London, he had continued to follow "the tennis game" between French and Russian mathematicians in the *Comptes Rendus* of the Paris Academy, and he had noted that the papers in the game were always presented by Emile Borel, Professor of Probability and Physics at the University of Paris. He sent a short paper on a question in probability theory to Borel. This was presented by Borel at the academy meeting of June 21; another paper was presented at the meeting of July 12.

At the end of the term, Neyman went to Paris to join Lola.

"So. All right. Paris. Borel. Let's listen."

The name of Borel was already part of the established vocabulary of mathematics—the initial B a conventional symbol familiar in many different contexts. Since 1898 when, in his thesis, Borel had taken the first significant step on the path that led from set theory to the modern theory of functions, he had been simultaneously research mathematician, teacher, editor and author, both of serious mathematical works and of such popular little books as *L'aviation,* written with his friend Paul Painlevé, and *Le hasard,* a small book on chance which heralded a change in interest on his part from function theory to the theory of probability. The year before Neyman's arrival in Paris, when Painlevé had been premier of France, Borel had served in his cabinet as minister of the navy. He continued to bear on his cards (as he would for the rest of his life) the inscription "Ancien Ministre de la Marine."

After one of Borel's lectures, Neyman went to the front of the amphitheatre in the Sorbonne and, introducing himself, expressed his desire for a personal interview. He recalls Borel's response as follows:

"You are probably under the impression that our relationships with people who attend our courses are similar here as elsewhere. I am sorry. It is not the case. Yes, it would be a pleasure to talk to you, but it would be more convenient if you would come this summer to Brittany where I will be during vacation."

Thus instructed, Neyman traveled to Brittany in August; but the interview turned out to be, from his point of view, quite unsatisfactory. On holiday Borel was more interested in swimming than in talking about mathematics. He was an excellent swimmer, and Neyman found he could not keep up with him.

After his meeting with Borel, Neyman had a long interview with Dr. Ralph Tisdale, the assistant to Trowbridge, at the Paris headquarters of the International Education Board. Presumably Tisdale asked him questions similar to those which he had been asked earlier on the application form which he had completed for his study in London:—to describe in detail the study or investigation he wished to carry on and to explain his reasons for his interest in it. The answers to these questions would provide an indication of the direction of Neyman's thinking at this point in his career, but unfortunately there is no record of them. It is clear, however, from the two papers presented to the Paris Academy that he had turned away from the problems of statistics to which he had devoted himself at University College and toward the theory of probability. According to a letter which Trowbridge wrote to the Executive Board of the Rockefeller Foundation in New York, Neyman had "a very definite continuation of his problem in mind."

"His researches [have] carried him into a field in which he believes that further work with Professor Borel and his associates in Paris would be extremely advantageous.... Professor Borel, Professor Lévy, Professor Galbrant [Galbrun?] and others are willing to aid [him] in his general field."

What was "the very definite continuation of his problem"?

Neyman shrugs helplessly.

He had asked for a year's extension, but Trowbridge had recommended instead an extension of six months. This would take him almost to the end of classes in Paris the following summer.

"Inasmuch as his stipend is a bit higher than ordinarily might be necessary in Paris, he can manage with this to finance himself for a year if he so desires."

In the fall of 1926, Neyman settled down with Lola in the little hotel where she lived. The Sorbonne was just up the hill. Nearby was the Collège de France.

He had admired the English and the settled order of their lives; but he had been lonely in London, cut off to a great extent from communication by his "abominable" English and by being so conspicuously "a foreigner." In Paris, of course, he was still a foreigner and still looked upon with condescension. When he reported that his wife had had an ermine wrap (which she had brought with her from Kharkov) taken from their rooms, the gendarme asked where they lived and, when given the address, said, "What can you expect—only Poles live there!" Although he found, and still finds, the French "difficult to approach," he was no longer alone as he had been in London. Lola was there, passionately studying and painting. Konstanty Neyman (now de Neyman), the cousin Kot who had escaped from Siberian exile after the Revolution of 1905, was living in a nearby suburb with his wife and three children. In addition, there was in the French capital at that time a contingent of Russian mathematicians who had been shepherded to the city of light by Luzin, one of the leading players on the Russian side of the "tennis game" which Neyman had been following in the *Comptes Rendus*.

"France is for me Paris," he likes to say, "and Paris is for me the Rive Gauche."

He murmurs a few lines of poetry. It is the dedication to a play by Edmond Rostand which he learned in his youth in Russian translation and repeats in English:

"Dedicated to those whose laziness involves a universe of ideas/ Whose thoughts are too delicate to be useful to people."

He explains that these lines recall the Left Bank to him.

"Yes, the intellectuals who sit there, probably over a cup of coffee or a liqueur, and discuss this and that without caring whether it will feed them. Essentially intellectuals with this intellectualism overwhelming."

Life in Paris was enchanting. Cafés open to the street. Shops and stalls along the Seine. Strange new pictures being displayed. A little night club, run by students in the cellar of a bistro where there were once dungeons: *Les Oubliettes Rouges*. He takes pleasure in recalling the name. But most impressive to him was the stimulation which he received from the mathematicians in Paris, at that time an especially remarkable and original lot.

He continued to attend the lectures of Borel on probability theory; but although it was for these that he had come to Paris, he does not have much to say about them. For a description of Borel's lectures, I have had to rely on the English mathematician E.F. Collingwood, who heard them at approximately the same time as Neyman. According to Collingwood, Borel's performance at the lectern was as distingué as his appearance. He moved deliberately in front of the long blackboard, returning only now and then to glance at the halfsheet of paper which was all that convention allowed him for notes. His manner was "magisterial" but his style "vivid, even racy." Many years later, after Borel's death in 1956, Collingwood was to write that "phrases which fell from his lips...remain with me until this day."

There was also to be a phrase of Borel's which was to remain with Neyman. In fact, he would later attribute to it—and thus to Borel rather than to Student—the inspiration for the joint work on which he was shortly to embark with Egon Pearson. But at the time, although he had come to Paris primarily to attend the lectures of Borel, he found himself more attracted by the lectures of another mathematician—at the Collège de France.

"The Collège de France, it's somewhat like the Institute for Advanced Study. Some selected people give lectures there. Not school courses like at the Sorbonne but modern developments. No examination, no registration, nothing. It's open to everybody. And there, on occasion, Lebesgue lectured—and Lebesgue for me was 1916, and this was 1926. Ah, just a decade!"

Henri Lebesgue did not have the distinguished appearance of Borel; but the eyes behind his spectacles were startlingly penetrating, and anything he turned his mind to (as I was told by Hans Lewy, a Rockefeller Fellow in Paris a few years after Neyman) seemed to have been "illuminated from another source."

Neyman found Lebesgue's lectures as absorbing as he had found the *Leçons*. In addition, he remembers with pleasure, they were very amusing: "full of smiles and jokes."

As in the case of Borel, he had only one contact with Lebesgue outside the lecture hall. This occurred at the first meeting of the traditional twice weekly seminar conducted since 1913 by Jacques Hadamard, the doyen of the mathematicians in Paris. Hadamard's seminar was a high class affair, attended by many professors, including on occasion both Borel and Lebesgue; but younger people like Neyman were also welcomed. The initial meeting, when papers on advanced developments in mathematics were handed out to the seminar participants for later reports, was held at teatime in Hadamard's home.

"Now how did I hear about this? I don't know. But I went. And there I had for the unique time essentially a social encounter with Lebesgue. And then he was kind of reserved. He was asked to present a writing of Sierpiński. And I remember he looked at it, frowned, and said, 'Oui, if I can understand it.' And so he presented it. That was later. 'Yes. I suppose this is right.' He tried to be objective, but I don't believe that at that time he smiled. But when he described his own things, you know, he did smile and he also made jokes."

Neyman's enjoyment of Lebesgue's lectures was intensified by the fact that Luzin and his young Russian mathematicians also attended. These included an old friend from Kharkov days—that same Goncharov who had written the operetta poking fun at Professor Sintsov—but also some more mathematically distinguished followers of Luzin, such as Nina Bari, M.A. Lavrentev, and D.D. Menshov. After Lebesgue's lectures Neyman and some of the Russians, most often Lavrentev and Menshov, never Luzin himself, who was considerably older, would gather in one of the little discussion rooms off the main reading room of the Sorbonne and struggle to understand what they had heard.

Did they ever discuss politics?

"Politics somehow I have no memories of discussing. They were interested in Lebesgue and these analytic sets. I was interested. They felt the need of contacts with what happens, and so I was with them."

Neyman attended the lectures of a number of other French mathematicians including Paul Lévy, a man a little older than himself, highly gifted. But the names which he conjures up from his memories of his study in Paris are always the same—Borel and Lebesgue. He admires these two tremendously and considers that both have had great influence on his own work. He seems not to have been aware that there was a certain amount of ill feeling between them. In this, Lebesgue was apparently the more unyielding. According to a story circulating during the period (although Neyman seems never to have heard it), Borel on one occasion attempted to mend matters. "Without doubt," he conceded, "I have some faults. But who does not? Such a one, does he not?"—"Yes," agreed Lebesgue.—"And such another one?"—"Assuredly."—"Then you also, perhaps you have some faults?"—"I do not believe so," answered Lebesgue.

When, in 1978, I show Neyman copies of obituaries of Borel and Lebesgue, he is disturbed to find references to a controversy between them, described as "teilweise heftig" by the *Mathematical Encylopedia*. He goes to the French mathematician Laurent Schwartz, who is visiting in Berkeley at the time, and asks, "Is it true?" "Yes," Schwartz tells him. "There was a controversy, *teilweise heftig.*"

Later Neyman says reflectively to me, "Human nature is a funny thing. It bothers me for some reason to learn that there was an argument between Borel and Lebesgue. 'Somewhat violent' controversy can only mean that one man calls another man liar." He shakes his head sadly.

Since leaving University College, Neyman had had no contact with Egon Pearson, who had spent the summer sailing off the northwest coast of Scotland with his cousin and her husband. That autumn, while Neyman was being stimulated by the lectures of Borel and Lebesgue, as well as other mathematicians in Paris, Egon was having a new professional experience in London. The Professor was recuperating from an operation for the removal of cataracts; and, for the first time since he had come to work in his father's laboratory, the younger Pearson was being permitted to deliver the first and second year lectures on statistics. He had not abandoned his concern with statistical inference; but, what with his new duties in the Biometric Laboratory and a serious illness on the part of his mother, he felt that he could not go to Paris in December as he had hoped. Only in late November did he even find time to compile a long series of notes on his ideas, a dozen pages or more, and mail them to Neyman.

Pearson's letter and the accompanying notes have not survived; but from Neyman's response, dated December 9, 1926, it is clear that these put forward, right at the beginning of the collaboration, a number of ideas which have since become familiar concepts of the Neyman-Pearson theory of hypothesis testing. It is also clear that Pearson directly or indirectly quoted Student's statement that "if there is any alternative hypothesis which will explain the occurrence of the sample with a more reasonable probability...you will be very much more inclined to consider that the original hypothesis is not true."

Neyman's answering letter begins in medias res with the following paragraph, which recognizes immediately the importance of what Student had written but at the same time misinterprets it:

"I think that to have the possibility of testing it is necessary to adopte such a principle as Students, but it seems to me also that this principle is equivalent to the principle leading to the inverse probabilities and that it must be generalised. I think we must have curage enough to adopt this principle, because without it, every test seems to be impossible."

The eight-page handwritten letter contains no personal news or comments about Neyman's experiences in Paris but concentrates on discussion of the ideas that Pearson has put forward in his notes. Although well aware of his difficulties with the English language, he plunges intrepidly ahead, unselfconsciously thinking aloud, crossing out and inserting words without regard for the appearance of his letter.

The only personal reference occurs at the end:

"We are awfully sorry that the probability of your visit is diminishing. But still we hope to see you and spend joyfully some time together."

In the notes Pearson had sent, he had proposed the likelihood ratio as the criterion for deciding between tests of significance for statistical hypotheses. This was an idea which was to be basic to the Neyman-Pearson theory; and although Gosset had not explicitly suggested the use of the principle, the choice was a natural jump from his comments. In the later published work of Neyman and Pearson, the likelihood principle is attributed to Fisher, who had introduced it, although very casually and somewhat vaguely, in his theory of estimation. In Neyman's early correspondence with Pearson, however, he always refers to it as "your principle."

Neyman's attitude toward the likelihood ratio is quite different from Pearson's. The latter, with apparently a somewhat proprietary air, seems to have accepted it as a satisfactory basis for his thinking about tests. Neyman, on the other hand, keeps searching—although in a confused way—for a better justification for the principle.

Finally, Pearson expresses some irritation.

"Why do you think that I am trying to prove that your ideas are worse than any others?" Neyman responds. "I say only that they are applicable in certain cases in which I think they are the best [ones]. In other cases...only, I would only apply other methods....I think that everybody may believe in the test he likes, but I would like to make you believe with me that in some cases...the decision must depend up on chi-square test [which at another point in his letters he refers to as 'Prof. Pearson's beautiful chi-square test'] and not upon yours."

The first group of letters between Neyman and Pearson comes rather abruptly to an end after a discussion by Neyman of "two categories of things in our polemics"—those that can and those that cannot be proved.

Pearson's reply seems to have been satisfactory.

"I absolutely agree with your last letter. No objections at all!" Neyman tells him, adding modestly in regard to the contents of his own earlier letter: "It is nothing new and I lerned it in professor Pearson's 'Grammar of Science' and perhaps in Kant's 'Criticism.'"

That spring, presumably between the London terms, Egon Pearson arrived in Paris.

Fifty years later, sitting among his books and papers in his room at the Pendean Home, he writes of the visit to the Neymans with still fresh delight.

"I stayed perhaps 10 days with them. Statistically, our thoughts developed. But another attraction (because of Lola), I [had] discovered *art* besides *mathematics....*Click, friendship and cooperation were established. No choice. Luck!

"We drank coffee and yoghurt on the Left Bank. Lola could not talk English, only Russian and Polish. French too, I guess, but my French was very poor. But we tried to communicate. Statistics in our talks must have been combined with art, just what I needed...."

There is no record of the statistical conversations that took place; however, since Neyman and Pearson were not to meet again until the summer of 1928, by which time their first joint paper would have been published, they must have sketched out at least the rough plan of their first paper during Pearson's visit. From the few letters which survive between this meeting in 1927 and the next in 1928, it appears that they divided the work, each man taking a section to draft.

"I suppose Jerzy must soon have realized that I was no mathematician," Pearson says. "Only he never told me. Once or twice later he encouraged me by saying, 'No, you can solve this yourself, get on with it!' At the time I thought he did not realize my mathematical weakness, but perhaps he did....I never knew! I was too introverted I suppose to ask him. But I did not feel inferior...."

This kind of friendship was new to young Pearson.

1978   Neyman has never seen his early letters to Pearson since he wrote them some fifty years ago; but when copies arrive in Berkeley from London at the end of 1978, he glances at the packet with only casual interest. Urged to look at the first letter, he takes it and begins to read aloud, inserting from time to time a comment, most often a correction of his English grammar or spelling. After a page or two, he hands it back.

"Musings," he says.

It is clear that he is not interested in going over his old letters, historically significant to mathematical statistics though they may be. I ask him if he would have any objection to my showing them to Erich Lehmann, one of his early students at Berkeley who has grown up, quite literally, with the Neyman-Pearson theory of hypothesis testing and has written the definitive book on the subject.

"It's a free country," he tells me. "No objection."

I ask Lehmann if he will read the letters and give me his reaction, and he agrees. To his surprise, he finds the task difficult. These early ideas of Neyman's are far from the point of view to which he is accustomed and are grounded in a historical situation with which he is not very familiar.

"There is a lot of fumbling, no really clear logic," he says. "The confusion involves a variety of issues, such as whether the hypothesis is simple or composite, the relationship between the value of a test statistic and the probability of observing so extreme a value, the difference between fairly specific or very general alternatives, and—of course—the role of prior probability, the Bayesian approach, which comes up immediately in the first paragraph of the first letter when Neyman concludes, wrongly, that Student has suggested assigning numerical values to the alternative hypotheses. For quite a while I thought it was mainly my fault that I could not catch what Neyman was trying to say, but the second page of Letter No. 6 finally convinced me that the concept of clearly defined alternatives to the hypothesis, which Student in his letter had correctly identified as the key to the puzzle, had not yet been absorbed by Neyman. He simply did not see the problem at that time."

Lehmann is not alone in this evaluation. In after years Neyman himself was to write: "I am afraid that I was not very helpful to Egon in my reply. Frankly, prior to his letter and also for a period thereafter, I did not notice the existence of the problems that bothered him." Even in his early letters, at one point he exclaims, "It seems to me what I have lastly written...is quite incomprehensible."

In the opinion of Lehmann, the very early letters from Neyman to Pearson contain an interesting positive insight in regard to the alternative approaches to probability—the frequentist and the Bayesian.

"As you know, hypothesis testing is basically concerned with the question whether the data furnish sufficient evidence against a hypothesis so that it should be abandoned. In Letter No. 1 and particularly clearly in Letter No. 3, Neyman points out that whether the evidence will be considered sufficient depends upon the strength of one's prior belief in the hypothesis. If, for example, a belief in extrasensory perception runs counter to your view of the world, you will ask for overwhelming evidence before you accept it. This is one aspect of Neyman's example in the letters of one's differing attitudes toward a run of bad hands in a game of cards with a friend whom he trusts as compared to a similar run in a game with a stranger who he thinks may be out to cheat him. Unfortunately this issue also gets confused in the example with various others. The very interesting sentence relating to this subjective element, as opposed to the completely objective, is the following in Letter No. 3 where Neyman says, 'When I have written about probabilities a priori I did not suggest that it is necessary to appreciate them by numbers.' In other words, to assign numerical values to them as the Bayesians do.

"This is my view," Lehmann continues, "and I would think it is that of most Neyman-Pearsonians today. There is a subjective element in the theory which expresses itself in the choice of significance level you are going to require; but it is qualitative rather than quantitative. While it is logically not very satisfying and rather pragmatic, I think this reflects the way our minds work better than the more extreme positions of either denying any subjective element in statistics or insisting upon its complete quantification. Unfortunately this idea, which Neyman expresses in Letter No. 3, was later jettisoned by Neyman and Pearson; and its absence has provided a source of criticism of the theory."

He goes to his bookshelves for a copy of the *Joint Statistical Papers* and opens the volume to the first paper (1928). In retrospect, he explains, this paper and the one that immediately followed represent a convenient way station ("even if it is a little bit out of the way") at which the advances made so far have been consolidated and strength gathered for the second stage of the journey.

"Pearson seems to have reached this way station (class of alternatives, two types of error, likelihood ratio principle) in 1926, after receiving the letter from Gosset, and to have communicated his ideas to Neyman in November of that year. Neyman's 1926–27 letters in response suggest that he did not see the picture very clearly. I would be very interested in knowing what happened between January 1927, when Neyman obviously didn't understand what Pearson was talking about, and the time when the 1928 paper was written. But there are essentially no letters for that period."

I am again reminded of the history of the Penguins:

"However that may be," Lehmann continues, "there can really be no doubt about the fact that the basic ideas of the 1928 paper were communicated to Neyman by Pearson and that—fortunately!—from the beginning Neyman did not find the likelihood ratio principle as compelling as Pearson did...."

Why *fortunately?*

"You'll see!"

1927
—
1928

The Rockefeller people were concerned, as they had been from the beginning of his fellowship, that Neyman might not have a position when he returned to Poland. In the latter part of March, as his stipend was expiring, they sent him to Warsaw at their expense to make arrangements for his future.

In Paris, attending the lectures of Lebesgue, he had felt again that this was real mathematics worth studying.

"...were it not for Egon Pearson," he wrote on a later occasion, "I would have probably drifted to my earlier passion for sets, measure and integration, and returned to Poland as a faithful member of the Warsaw school and a steady contributor to *Fundamenta Mathematicae.*"

It appears, however, that in spite of the interest in mathematics which had been revived by Lebesgue's lectures, such a drift was no longer possible in 1927. As a result of the efforts of Leokadia Borucka-Ubysz, who was responsible for an exhibit of Neyman memorabilia at the symposium held in Warsaw to commemorate his eightieth birthday, there are several documents which illuminate the situation which confronted him in Poland that March.

Since he had not yet "habilitated," he was not licensed to teach as an official member of the faculty, and the dean of the Department of Horticulture at the SGGW could offer him only the opportunity to give special lectures on mathematics and statistics, as he had before he went to London. This proposal presented one problem. As Neyman reminded the dean, when he had been employed previously at the SGGW, he had been paid for only nine months of the year and then rather modestly.

"This circumstance was the cause of continued uncertainty about the summer months and forced me to look for additional ways of earning, which did not favor research."

He and Lola had not lived so frugally during the past year as Trowbridge had thought they might. In addition, Lola now planned to stay on for a while in Paris, where she was soon to have her first show of paintings, sponsored by Raymond Duncan, the brother of Isadora. She would earn some money by painting Duncan's designs on fabric and she might even sell some of her paintings (as in fact she did)—but there would still be essentially two establishments for Neyman to support for at least a while, and with no money coming in during the summer.

Having originally gone abroad on a fellowship of the Polish government, he felt a responsibility to return to the Polish academic world. He was appreciative of the dean's past support of his efforts to establish a statistical unit and a little mathematics library at the SGGW. But he simply had to find better paying work.

It happened that his old acquaintance from Bydgoszcz days, Professor Załęski, had connections with K. Buszczynski & Sons, a firm in Krakow which dealt in pedigree seed culture and shipped seeds of sugar beets to such fabled places as "Utah" and "Wyoming." Załęski could arrange a job for him with that firm. His hours would be flexible enough to permit him to give lectures on statistics at the university and ultimately to habilitate in Krakow. He decided to accept the offer of K. Buszczynski & Sons; but he stipulated that if efforts to provide him with means to live and study in Warsaw were successful, he would like to have the opportunity to return there.

The dean at the SGGW promptly called a meeting of the Department of Horticulture and, as he later informed the appropriate minister, it was agreed that Neyman should be appointed an assistant professor of statistics as of May 1. The minister responded by pointing out that, since there *was* no official position in statistics at the SGGW, no such appointment could be made. Neyman could give lectures as he had in the past; and if the college had exhausted its quota of hours for such lectures, the minister would approve enough additional hours to make Neyman's return in the fall possible. But that was all.

Neyman went back to Paris to wind up his affairs and present his report to Hadamard's seminar (on two German papers on set theory).

"The answer for your last mathematical letter I shall send you from Krakow," he wrote to Pearson.

The friendship between the two men had developed greatly since Pearson's visit, and Neyman added in a postscript: "My wife and myself we find that your letters are much more kind than my [ones] which are rather 'dry.' Please do'nt think that the same is with our feelings. To write a kind letter one must have not only the good feelings but also a sufficient knowledge of the language."

Settled in Krakow in the apartment of Professor Załęski, Neyman found himself less concerned about mathematics than about an unpleasant situation which had developed in connection with two of his papers in *Biometrika*. Two members of the Royal Statistical Society—Leon Isserlis and Major Greenwood—had written in the society's journal that Neyman and another young worker in Karl Pearson's laboratory had "unduly depreciated and ignored" the work of the late Prof. Alexander Tchouproff of the Technical Institute of Petrograd. The Tchouproff work referred to had been published in January 1923 in the new English-language journal *Metron*, edited by the wide-ranging Italian statistician, Corrado Gini.

"I have got the reprint of Isserlis's and Greenwood's paper. It is indeed very ugly!" he wrote to Egon. "What is [striking] is that if dr Greenwood wanted really to defend Tchouproff it would be the easiest way to tell me or write me two words—he [Greenwood] was a teacher of mine and I attended to his lectures—of course it would be no what French people call tapage [an uproar]. I think I shall write some thing of that sort in the J.R.S.S.? Would you? Will your correct my paper? I am afrayed that as I do not know quite the English customs and circumstances I can make some stupide step—my temperament will not help me in not doing it."

He was embarrassed to be giving so much trouble.

"You are silent now—write me some good words!"

Karl Pearson considered the remarks of Isserlis and Greenwood directed as much against himself as against Neyman and Frank Church, the other young worker named. He proceeded to reply at length in the pages of *Biometrika*:

"One of these gentlemen [Isserlis and Greenwood] is a conjoint editor of *Metron* and must know full well that it does not appear on the date it is stated to be issued....In 1923 *Metron* was three to four months behind....Dr. Splawa-Neyman's paper was written in 1922 and sent to the editor of the Polish journal *La Revue Mensuelle de Statistique*...who published it in 1923. It would be thus impossible for Neyman to have seen Tchouproff's paper in *Metron*. It would be just as reasonable to accuse Tchouproff of disregarding Neyman, as Neyman of disregarding Tchouproff!"

The elder Pearson then proceeded to devote four pages to a defense of young workers and of youth, assuring his readers that he as an editor by no means regretted publishing Neyman's paper (the publishing of a paper already published having been another matter taken up by Isserlis and Greenwood).

"Only the future can demonstrate whether it is not more profitable for the progress of science to encourage the young who have worked and are working under very disadvantageous conditions, than to apply the bludgeon to promising scientific recruits on the ground—in this case illusory—that they are wanting in respect to the hierarchs."

By the time Pearson's defense appeared in *Biometrika,* Neyman had received the first part of the joint paper that he and Egon were preparing. The letter in which he acknowledges the receipt of the manuscript ("a manuscript is a very precious thing") is his last surviving letter to Pearson until after the publication of their first two papers in 1928.

What was Neyman doing during this time?

I am not the only person who has noted that although he is a great raconteur, it is very difficult to get him to talk about his early days in Poland. When even direct questions elicit few specific facts, I suggest he may have blanked out that period of his career.

"Could be," he concedes after a moment or two. "You see, there was this guy. One paper published in *Fundamenta.* Rewritten by Sierpiński..."

Professor Borucka-Ubysz has been able to locate only one document pertaining to Neyman's activities during the academic year 1927–28. On August 21, 1928, the Minister of Religion and Public Instruction—Neyman smiles wryly at the combination as he translates—confirmed that Jerzy Neyman had habilitated at the University of Warsaw on June 26 and was officially a docent of that faculty.

Neyman does not know whether he also habilitated in Krakow as planned; but he is quite sure that in the fall he began, again, to commute between Krakow and Warsaw, where he continued to be, as before, a special lecturer in statistics at the SGGW. Among his students in Krakow he recalls particularly a young man named Oskar Lange "with a mind of very high quality." The two became friendly, Neyman visiting Lange on occasion at his apartment and being surprised to learn that despite the obvious German origin of the family name the Langes had been "patriotic Poles" for centuries.

At least some of the work which Neyman did in connection with his employment by K. Buszczynski & Sons is documented in his published papers. Poland was one of the principal sugar-beet growing countries of the world, and there was a lively interest among landowners in raising beets with a high content of sugar as well as a high crop yield. During 1928 Neyman published two papers in the journal of Polish sugar-beet growers on methods of estimating the commercial value of beets. He had some trouble convincing the editor that there was a problem to be dealt with. "I had to document for him the fact that claims of excellence were on occasion overstated." He sees what he was doing in this work as the kind of interdisciplinary effort which he has always endeavored to make: the "willing and interested" statistician working with the similarly inclined expert in some other field.

He also wrote two much longer papers, one not published until 1929, on "The theoretical basis of different methods of testing cereals." (In Poland, he explains, sugar beets are classified as a cereal.) These papers appeared first in *Wiadomości Matematyczne*, a professional journal, and then in a series of pamphlets issued by K. Buszczynski & Sons. Since the latter were sent to customers abroad, including those in Utah and Wyoming, they were in English.

In the first of these papers Neyman treated briefly the method of Student and more fully that of his colleague Załęski. The latter, according to Neyman, was even less of a mathematician than Student but, like Student, a man who tried to apply common sense to the interpretation of agricultural experimentation. For some time (as has been mentioned) Neyman had recognized that the conceptual foundation of Załęski's methodology left something to be desired, and he now proceeded to reformulate it. He prefaced what he had done with the statement: "As the mathematical theory of [Załęski's] method has not yet been published, we do so now and show the hypotheses under which the application of the same is correct."

"When one works with people, one doesn't jump on them and say, 'You fool!'" he explains. "Załęski was no fool."

All during the academic year 1927–28, although documentary details are generally lacking, Neyman was being pushed very hard with his teaching, the weekly commute between Warsaw and Krakow, and the work for K. Buszczynski & Sons. He decided he had to withdraw from the collaboration with Egon Pearson, and a note from him to this effect is appended to the published version of their first joint paper:—

"I feel it necessary to make a brief comment on the authorship of this paper. Its origin was a matter of close co-operation, both personal and by letter, and the ground covered included the general ideas and the illustration of these by sampling from a normal population....Later I was much occupied with other work, and therefore unable to co-operate."

("I am very particular about not giving credit," Neyman says when I ask him about this disclaimer. When his sentence is straightened around, it appears that what he means to say is that he is very particular not to take credit for something which he has not done.)

Although Neyman had written in his appended note that he had had to withdraw from the collaboration, he obviously did not intend his withdrawal to be permanent. The first joint paper by its title—"On the use and interpretation of certain test criteria for purposes of statistical inference. Part I."—promised a "Part II."

1928
—
1929
Hypothesis testing, as set out in the first paper which Neyman and Pearson published together, is concerned with deciding whether a given statistical hypothesis is true. Mathematically—in the language of the authors—this presents itself as the question: is a particular sample likely to have been drawn randomly from a certain population or class of populations? Typically a test of such a hypothesis is expressed in terms of a test statistic, the hypothesis being rejected as false when the statistic turns out to be so large as to be implausible if the hypothesis were true. In a given situation many different test statistics can be devised. The question which first concerned Pearson, and then Neyman, was—why pick one such test statistic—as, for example, Student's t—rather than another?

In their first paper, which appeared in *Biometrika* in July 1928, the authors pointed out that the question had to be answered in terms of the alternatives to the hypothesis a statistician was willing to entertain. In the simplest case, either the sample must have been drawn randomly from the population specified by the hypothesis or it must have been drawn from some other specified population. Since it is usually impossible to distinguish between populations with certainty—no matter what test is adopted—two sources of error are possible. Sometimes when the hypothesis is rejected, the sample will in fact have been drawn from the specified population and the hypothesis thus falsely rejected. At other times when the hypothesis is accepted, the sample will have been drawn from a population other than the one specified and the hypothesis falsely accepted.

In their first paper Neyman and Pearson pointed out that the first kind of error (false rejection) could be controlled in such a way that the probability of rejecting the hypothesis when true would be very small. The second kind of error—accepting a false hypothesis—was, they said, "more difficult to control" but "both of these aspects of the problem must be taken into account."

The introduction to the first joint paper concluded with what was a crucial sentence:

"It is indeed obvious, upon a little consideration, that the mere fact that a particular sample may be expected to occur very rarely in sampling from [the suspected population] would not in itself justify the rejection of the hypothesis that it had been so drawn, if there were no other more probable hypothesis conceivable."

The authors then introduced the likelihood ratio—the ratio of the maximum probability of the observations under the alternatives to the corresponding probability under the hypothesis—which Pearson had proposed in his first letter to Neyman. Applying this principle to some specific examples, they showed that in a number of important cases, such as the case of "Student's problem," it picked out the same test statistic which had been selected in the past by statisticians on purely intuitive grounds.

By the time the first joint paper was in proof, Neyman was better established in his professional career. Although his habilitation had not yet been accepted by the SGGW, during the coming academic year there would be no more commuting between Warsaw and Krakow. His work with K. Buszczynski & Sons had provided him with extra money. He thought of travel, and he and Egon arranged to meet in Brittany in August and work on a sequel to the first joint paper, the promised "Part II."

At the beginning of the summer, on his own, Neyman took up again the problem of hypothesis testing, going over the work of some of the probabilists whose lectures he had attended in Paris. He has written since then that this collation of the writings of the French probabilists with Pearson's letters contributed importantly to his understanding of the issues which Egon had raised.

Before he left for Brittany he drafted a paper that he planned to present—in English—at the International Congress of Mathematicians to be held in September in Bologna. In it he reviewed the history of the problem of the application of the theory of probability to testing hypotheses and attributed to the English clergyman Thomas Bayes the first attempt at its solution. Bayes's contribution—a theorem not made public until after his death—provided a mathematical technique by which the statistician who has, a priori, assigned probabilites to values of unknown parameters of his problem on the basis of the degree of his belief in these values, can then, a posteriori, adjust these values to bring them in line with his data. This approach ("inverse probability") was one which, in Egon Pearson's opinion, had been forever discredited by Fisher in his 1922 paper in the *Philosophical Transactions* of the Royal Society.

Neyman now pointed out in his paper that the solution by Bayes's theorem had been "very seriously attacked since it needs for its application in practice a knowledge of the a priori probability law, which can only in quite exceptional cases be deduced from the conditions of the particular problem under consideration." He then brought up several interesting theorems which show that if one starts with some prior distribution—say, a preconceived notion that something in nature is more likely than something else— and takes a large number of observations, he will find that his prior distribution washes out; i.e., as the sample size becomes large, the posterior distribution becomes independent of prior conceptions about the result. One of these theorems was the Bernstein theorem with which he had become familiar while a student in Kharkov. In his paper he proved two other theorems of the same type which he mistakenly thought were not known. Although he found them interesting, he felt they were "of more doubtful value in practice," since they illustrated "the uncertainty that must always be associated with the method of inverse probabilities." When the method of Bayes's theorem was thus thrown in doubt, there seemed "to have been accepted very generally a new principle," which he formulated essentially as follows:

"If the observed event has a character, which from the point of view of the hypothesis considered is sufficiently improbable, then the hypothesis itself is not plausible."

Clearly not every character of an observed event is suitable for testing hypotheses. As Neyman pointed out, both Emile Borel and Paul Lévy had stated that such characters should be from some point of view *remarquable*, or "striking." He then proceeded to show, in a discussion of the joint work to date, that he and Egon Pearson had "refined" Borel's principle "with the only modification, that the notion of a 'remarquable' character [or event] is now defined." In short, Borel's rather vague notion of a "remarquable" character or event was made more precise when it was agreed to consider what today are called the alternatives to the hypothesis. This was, of course, the first crucial step in the Neyman-Pearson theory; and from this time forward Neyman was always to credit Borel—rather than Student—with having stimulated himself and Pearson into taking it.

The remark by Borel which Neyman always cites as particularly inspiring "to Egon S. Pearson and myself in our effort to build a frequentist theory of testing hypotheses" is a statement to the effect that the criterion for testing a statistical hypothesis should be a function of the observations *en quelque sorte remarquable*. He has referred so often to this statement that his French colleague, Le Cam, has tried to locate it in *Le hasard*, which Neyman always gives as its source.

"I found the paragraph which refers to 'remarquable' things," Le Cam tells me, taking up his own copy of *Le hasard*. "One should be very wary, Borel says, of the tendency that one has to consider as remarkable a circumstance that one did not make precise before the experiment, because the number of such circumstances that can appear remarkable from several points of view is very large. But I cannot find anywhere a reference to a function *en quelque sorte remarquable*."

I suggest that we see if Neyman himself can find the passage. We take Le Cam's book to Neyman's office and place it on the desk in front of him. He frowns a little as he leafs through it.

"I would have thought it was somewhat thicker," he says finally.

He turns to the title page. It is the 1948 edition. When the 1920 edition, which he would have known, is located and brought to him, it is indeed a somewhat thicker book; however, the passage which Le Cam has pointed out to me is the same in both versions. Although Neyman searches a little, he cannot find the sentence which has remained in his mind all these years.

"But the fact that I cannot find it does not mean that it is not there," he says as he hands the book back.

Later Le Cam says to me:

"It is in a way very puzzling. Of course I am not familiar with everything Borel ever wrote on the subject, but I certainly don't remember such a statement as the one which Neyman cites. In fact, what Borel says in *Le hasard* is actually something quite different. Now what Paul Lévy says on the same subject is a little closer to what Neyman says. He explains the requirement of 'remarquable' as simple and having a very low probability under the hypothesis. But he is sort of vague, too. The implication which Neyman saw in the sentence which he quotes is a much deeper one."

"Do you think that Neyman may have 'modified a little bit' what he found in Borel and Lévy?"

"Maybe."

Certainly Neyman must have had his paper for Bologna in hand when he and Lola traveled to Brittany in August 1928 to meet Egon Pearson again for the first time in more than a year; but Pearson does not remember his quoting the phrase from Borel until much later in the collaboration, and he is sure it never played a role in his own thinking. What he does still remember is that one day the two of them borrowed bicycles and rode into Concarneau. There was a little fishing harbor "full of lovely boats, painted in various colors and having colored sails." They each bought a pair of the colored trousers that the fishermen wore ("Mine were terra cotta!") and, wearing these over their shorts, bicycled triumphantly back to Lola. He and his friend "sat often at a table on the beach, drank coffee, talked statistics....Lola sketched. She was a good swimmer, but I doubt if 'Jurek' ever went in the water. He preferred to sit and smoke."

During the holiday Neyman and Pearson began to draft their "Part II." The principal goal of this paper was to show that the chi-square test, which was already the standard test on intuitive grounds in a number of situations, would also be picked out by the application of the likelihood ratio principle.

For Pearson, giving up his customary summer of sailing with his cousin and her husband to join Neyman represented a commitment to his career in statistics which he had not made before. There had also been something of a break with his father. Maria Sharpe Pearson had died that spring; and K.P., mourning his wife of almost forty years, had gone with his daughters to the Black Forest in Germany. He would have liked his son to go, too. "But I would not."

At the end of the fortnight in Brittany, the friends set off in different directions. Neyman headed for Bologna, blithely confident that his knowledge of French and Latin would take him through Italy. Pearson, who had fallen ill before the end of the holiday, traveled slowly up the Loire to meet his cousin and best friend, George Sharpe. Lola returned to Paris rather than to Warsaw, for she still spent much of her time there.

No letters, or at least none that survive, passed from Neyman to Pearson after they left Brittany. Upon Pearson's return to London, he finished drafting "Part II" of the joint work on hypothesis testing and passed it on to his father. It appeared in *Biometrika* in December 1928. Neyman, returning to Warsaw from Bologna, found that during his absence his mother, Kazimiera Neyman, had died. She had been ill, but her unexpected death was a shock.

Through 1928 Neyman lectured as a docent at the University of Warsaw but continued merely as a special lecturer in statistics at the SGGW. His application to have his habilitation "spread" from the university to the latter institution was not approved until February 1930.

He now also headed another small laboratory, sponsored by the Nencki Institute for Experimental Biology, named for Marceli Nencki (1847–1901), a wide-ranging Polish scientist who had obtained his doctor's degree in Berlin and had headed his own large laboratory in Petrograd at the time of his death.

"But he had always worked for 'Polish science,'" Neyman explains, "so there was a society, scholarly society concerning research and so forth, named in his honor, and there were rich people who financed it by donations.... Now when they noticed my existence, then they thought that I should be in this Nencki Institute, but somehow my specialty must be reflected. And so a small unit was created; and they asked me, 'What do you want to call it?' I told them, 'All right. Let's call it biometric—because it's statistics relating to biological phenomena.'"

Initially the Nencki laboratory was located in the building of the scientific society, where Przeborski had lived with his family, since coming to Warsaw, in a one-room apartment. Neyman recalls how he used to stop by for a cup of tea and an enlightening discussion of contemporary poetry with his old teacher. Later the laboratory was moved to the SGGW and for all practical purposes combined with the laboratory there.

The following summer of 1929, the International Statistical Institute was scheduled to meet in Poland. The ISI had been founded in London in 1885 at the fiftieth anniversary meeting of what later became the Royal Statistical Society. Its purpose had been to meet the need (already recognized by a number of international statistical congresses) to establish uniform practices of collection and compilation of national statistics so that these could be meaningfully compared. Although the institute was originally planned as a private, scientific and unofficial organization, most of its members were government statisticians—a breed quite different from statisticians of today—collectors rather than interpreters of data.

Neyman, as a Polish academic specializing in mathematical statistics, although not a member of the ISI, was a member of the planning committee and Lola was a member of the social committee. He was eager that he and Pearson present a joint paper at the meeting. Pearson, also not a member, held back—mainly because he was extremely uncomfortable about the "joint" paper which Neyman was preparing.

"You had introduced an a priori law of probability...," Pearson wrote him many years later, "and I was not willing to start from this basis. True we had given inverse probability as an alternative approach in our 1928 Part I paper, but I must in 1927–28 still have been ready to concede to your line of thought. However, by 1929 I had come down firmly to agree with Fisher that prior distributions should not be used, except in cases where they were based on real knowledge, e.g. in some Mendelian problems."

He had drafted their first two joint papers, and he agreed that it was Neyman's turn to draft a paper.

"But I could not bear to write a paper involving prior probability, [a feeling] which I thought he had agreed to respect," he explains to me.

On his side, Neyman felt that the question of the use of Bayesian methods was one which he and Pearson should treat in their theory. He did not believe in the Bayesian approach either—except in cases where the probabilities a priori were given by the practical problem under consideration. But, he argued, since there were statisticians who did believe in it (for instance, the English statistician Harold Jeffreys), it seemed important to show that even if a statistician started from the point of view of inverse probabilities he would be led to the same integral which he and Pearson had used in some cases to estimate the probability of rejecting a true hypothesis.

The preparation of the ISI paper took Neyman quite a long time because, as he explained to Pearson, "the point of view on results changed very frequently." When he finally finished it, he felt that perhaps he had omitted things which he should have included; but he had been typing all day, and he was very tired. Twice in his cover letter to Pearson he reminded him, "...it is not the last paper we publish." He begged him to read the manuscript as quickly as he could "and add some 'talking.' Don't be offended for that expression—I understand that it is important, but I am at present very tiered and cannot write any more. Besides—these difficulties in English. You will see on the last page that I tried to write some 'talking,' but I couldn't."

In March 1929, as in February, one letter from Neyman followed another: March 6, March 8, March 11, March 17.

Once, when an answer from Pearson was a little slow in coming, he pleaded, "My dear Egon, don't be so silant as a sphinks," and added, "It is horrible to be so far away from a man who's opinion it is so interesting to know.... Will it be a letter from you to-morrow?"

In London, Pearson mulled over his feelings about the paper Neyman had drafted. It was on the subject of Bayes's theorem—as he was acutely conscious—that he "had first recognized K.P. could be wrong." At last he wrote firmly to Neyman that he simply could not put his name to the proposed paper.

This letter came only a few days after the "don't be silant as a sphinks" letter. Although disappointed, Neyman did not entirely give up hope that Egon would ultimately agree to join him.

"Of course if you do not like to have your name signed under the paper—I shall publish it only under my own. But it is always possible to change in proofs. At present you have time to think it over, and probably to alter a good deal, if necessary. I hope however that finally you will agree."

He tried once more to convince Pearson of the usefulness of treating inverse probability.

"If one believes in our Part II, the circumstance that this a posteriori method gives the same result will not stop him. If somebody believes in the a posteriori things and does not believe in our Part II and [in] Fisher's work— the Part III (if it will be called so) will show him that all the same he must use the same formula."

He reminded him of the words of Student: "that the most important thing in our Part I [where both approaches were treated on a par] is that starting from different points of view we get the results only slightly differing one from another—you quoted these words in a letter to me and I think you have been quite right. Now when the same agreement is shown in a more general case—you are not happy—I suppose you did not noticed it. I must confess that at the beginning I did not noticed the thing myself."

Pearson remained adamant. He also pointed out to Neyman that if they published the proposed paper, with its admission of inverse probability, they would find themselves involved in a disagreement with Fisher, who had come out decisively against it. It is not clear how much Pearson had told Neyman, at that time, of the touchy relationship between himself and Fisher. Many years later he explained, "The conflict between K.P. and R.A.F. left me with a curious emotional antagonism and also fear of the latter, so that it upset me a bit to see him or hear him talk." In 1929, however, he was a very reserved young Englishman. He recalls himself as "a silly ass" who could not communicate directly with others. Neyman nevertheless realized that Egon's reluctance to put his name to the paper was due primarily to his concern about Fisher's reaction.

"I think that your doubts come from a misunderstandniss," he told him, mixing English and German as he more frequently mixed English and French, "and propose you to send Fisher the enclosed letter—in case you think it to be useful."

The letter was two and a half double-spaced typewritten pages.

"My dear Dr. Fisher," Neyman had written, "you are probably informed that a good deal of time Dr. Pearson and myself we are studying your work. I write to say how it is exciting for me. Very often I am not able to follow your arguments and even the theorems seem at the beginning to be false. Then Dr. Pearson or myself we are looking for a proof and finally the theorem appears to be perfectly correct."

(Neyman, reading this letter aloud to me half a century later, stops at this point and says drily, "Sometimes.")

"At the present moment... we are considering ourselves as being in perfect agreement with your ideas about the Likelihood etc. It would be very interesting to know if you are in agreement with us."

He then went on to explain the modification that he and Pearson had made of Fisher's notion of the likelihood of a hypothesis.

"Do you see that the difference from your definition is not a very important one but perhaps you will believe me that it is very convenient."

He then repeated essentially what he had written to Egon about mathematicians who believed in the Bayesian approach.

"The question I would like to ask consists in the following: do you think that publishing the mentioned result I shall make a wrong tactical step. The question arose from a letter from Dr. Pearson in which he says that publishing such result I may 'at once loose any agreement' with you.

"I think that it is rather doubtful. First of all I do not know if you are in agreement with some of our ideas. Further, asuming that you are, I do not see why would you stop to be in agreement if it is found a new reason for believing in your formula./Assuming that the result is correct./"

There is no record of Egon Pearson's reaction to Neyman's ingenuous letter to Fisher except that it remained, unsent, with Neyman's other letters.

Questioned about it almost fifty years later, he says, "I imagine that in my judgment I had not seen that it would do any good. R.A.F. did not think much of the upstarts Neyman and Pearson at this stage and would hardly have minded whether we were agreeing with him or not. J.N. was too sanguine."

1929
—
1930

It was a great disappointment to Neyman that Pearson would not put his name to the paper on hypothesis testing to be presented at the ISI meeting.

"The only thing—," he wrote, "I hope that the circumstance that I wrote a paper, which I think is not in any sort of contradiction to our preceiding ones, will not stop our friendship and our interest in work of ours (it is: your interest in my work and vice versa) even in the case when you will think that some contradiction does exist.... I think that the cooperation is an extremely important thing."

He was excited at the prospect of a big international meeting but even more excited that Egon would at last be a guest in his own country.

"Write please what you would like to see in Poland," he urged, "mountaignes, forests, old towns, big eastates."

From some of his "bad work for money" he thought he might soon have saved about fifty English pounds. He proposed that Egon buy him an automobile in London and bring it to Warsaw.

Pearson still marvels

"...he suggested that E.S.P., who had never driven a car, should buy one & start off in it across Europe! It was characteristic of him to make optimistic suggestions, & not to realise how he was tilting against the impossible. When [rereading] this letter..., I noted down: Here we see his intrepidity hoping to surmount all barriers in me, to carry through an impossible transaction—it reminded me of that optimistic spirit, so characteristic of the Polish race, which led them 10 years later to think that they might use their cavalry against the German tanks!"

Nothing came of the proposal.

In London before leaving, Pearson published a paper with N.K. Adyanthaya answering in certain cases the question of a hypothetical "working statistician":—How sensitive were the 'normal theory' tests to changes in population form? Might he use some with less hesitation than others?

In Warsaw, Neyman turned again toward Paris. That May he sent two short notices in French to Borel for presentation to the Paris Academy. He began to prepare (in French) a paper to be presented at the first Congrès des Mathématiciens des Pays Slaves, which would be held in Warsaw following the ISI meeting. He also wrote a report on the work to date on hypothesis testing in Polish, with a French summary, for publication in Poland.

Neither Neyman nor Pearson now remembers much about the ISI meeting in the summer of 1929 other than the fact that Corrado Gini, the editor of *Metron* and "a dedicated Bayesian," vigorously attacked parts of the Neyman paper to which Pearson had refused to put his name.

("The question about testing hypotheses is not only a mathematical one. It depends highly on some rather philosophical considerations," Neyman had pointed out. "The mathematics will deduce formulae for testing hypotheses provided that there are given sufficient principles to start with. The foundation of these principles does not belong to mathematics—they must be found as a result of analysis of conditions in which a common mind would or would not believe in a given hypothesis. And if a correctly proved theorem of mathematics cannot be rejected by any mathematician who understands the proof, the principles on which the testing of hypotheses is based may be rejected by any person, simply because they may seem [to him] to be wrong.")

Pearson does remember that on this trip to Warsaw he was introduced by Neyman to a "traditional" Polish toast: "To all the ladies present and some of those absent!" He also remembers, after the ISI meeting, a memorable trip to the Tatra Mountains. Then he and Neyman went to Madralin, the house for quiet university research in the woods east of Warsaw, and began to work on a new paper, which again they thought of calling "Part III." In spite of the pleasures of the trip, Pearson remembers also that he was dismayed at the general conditions in which Neyman and other academics lived and worked in Poland.

Professionally—their collaboration back on track again—the visit was highly satisfactory to both friends. For Pearson it was also personally significant. While he was with the Neymans, he had come to an important decision in regard to his personal life. Two years before, he had fallen in love "at first sight" with a young woman who was at the time engaged to his cousin George Sharpe. Although the next day she had returned Sharpe's ring, Pearson had felt that it would not be good form for him to declare himself until his cousin had had an opportunity to try to win her back. Before leaving Poland, however, he had made up his mind that he had given George more than enough time to reestablish the relationship. When he got back to England, he would approach the young woman and try to win her for his own.

Neyman accompanied Pearson to Gdynia, where Pearson was to catch the boat which would take him back to London. He then returned to Warsaw for the Congrès des Mathématiciens des Pays Slaves. At this meeting, in addition to his own paper, "Méthodes nouvelles de vérification des hypothèses," he introduced two of his students—Janina Hosiasson and Stanisław Kołodziejczyk—both of whom delivered papers on hypothesis testing.

At the beginning of November he spent a weekend at Mądralin, working again on the joint paper begun in August. This dealt with the problem of determining whether two samples come from the same normal population.

"The life is really beautiful," he wrote to Pearson, whose old room at Mądralin he was using. "Outside the window a nice little rain is making a nice little noise, besides there is quietness and smell of fur trees....and I have had a very good breakfast (though without porrige) and in general everything is very nice....

"Now—ad rem."

The letter is technical, but it conveys much more than the earlier letters a sense of two against the world—a new view of statistics. There are other people—"people who do not believe or do not understand our principles"— and then there is "our point of view."

That November, letter after letter went from Warsaw to London. Neyman spent as much time at Mądralin as he could: "Only in Mądralin one can work quietly."

Pearson had suggested that they try to publish another paper as soon as possible; and Neyman wrote, as he had written before, "It seems that (as always!) you are right....If I am not wrong, it will be much more work and time until we finish the question...."

He suggested that they could have the paper on two samples out within a month or so if they submitted it to the Polish Academy in Krakow.

"I have also another idea, but I do not want [to] express it in a definit form unless I shall find a firm support from you. In Poland every scientific laboratory is trying to publish its own if not journal then some thing of that sort—memoirs etc. I think that it is not a good idea because there are so many different journals treating the same questions and appearing [rarely] that very many even valuable papers disappear—you cannot have or even know about all these journals. With mathematical statistics of course it is a quite different thing as there are only very few journals consacrated to this science. Should I try?"

The kind of journal he had in mind would not have to appear regularly.

"[But] I think that perhaps some time—not at present—we want a propaganda—thus a journal having readers."

Then, apparently embarrassed at the presumption of his proposal, he concluded abruptly: "—all I have written about it is rather nonsense. Shall we accept the Academy?"

He was eager to see and talk to Pearson again. What about Christmas in Paris? Lola would like to go there. She was stimulated by the exhibitions and always painted "more and better" when she got back to Warsaw. He himself would like to see Paris again, and to see Pearson.

"If I do not see you at present, I shall probably not do so until Christmas 1930—of course I was able to live without seeing you many years from 1894 to 1925, but still."

In London, Pearson was struggling unhappily with his heart. His family felt that he had "stolen" his cousin's fiancée, even though the engagement had been broken off long before he had approached the young woman. Doggedly, he continued to work on the two-sample paper, modifying his draft as new results poured in from Neyman during November and December 1929. Before Christmas he sent the finished manuscript to Warsaw.

Neyman was delighted at "the pleasure of having your last letter and the paper, which appears to be good and, what is still more important, to be the beginning of new series..."

He would "love" to have the title the same as that on their two 1928 papers with "Part III" appended. When he saw Czesław Białobrzeski, a theoretical physicist who had agreed to present the paper, he would ask if it would be possible to present a "Part III" when Parts I and II had not been presented to the Academy.

"Write quickly if you agree. ('Quick—quick!')"

The new paper was of a quite different character from the one for the ISI to which Pearson had declined to put his name. In it the authors applied the likelihood principle to the question whether two independent samples had been drawn from a common normal population: "certainly [as Pearson was later to say]...a textbook example of how to follow through with the likelihood ratio procedure." Their starting point had been a paper by V. Romanovsky; and although initially Neyman had not been much impressed with that work, he had ultimately concluded, "I think we owe Romanovsky quite a bit."

By the time Neyman delivered the new paper to Professor Białobrzeski, the year 1929 was drawing to a close. Professionally, it had been a productive year. In addition to the second part of the long paper, "The theoretical basis of different methods of testing cereals," in which he had laid out his own method of parabolic curves, he had written five papers connected with hypothesis testing—all under his name alone. His everyday personal life, which is rarely glimpsed in the letters to Pearson, had not been nearly so satisfactory.

Letters pose a problem which the historian of Penguin Island neglects to mention. Because they are there—numerous, fresh, in many instances quite detailed, full of personality, it is easy to place more emphasis upon them than they deserve. They are not by any means so complete a record as they seem. Most of Neyman's life in Poland during the decade of the twenties and the thirties is missing from his letters to Pearson: the politics of the time (which interested him greatly), his Polish relatives and friends, his students and colleagues, his concern about his failure to advance to a professional "chair," the ups and downs of his relationship with Lola, the strains and struggles of "existence" in a country with a growing feeling of international insecurity. From time to time, however, one does come across a letter which offers at least a glimpse of this other life. At the end of January 1930 he and Lola had just returned from a holiday at Zakopane, a popular ski resort in the Tatra Mountains. Lola had had a new "ski suit" for the occasion and, as she had said, they had been "perfectly happy" with beautiful weather and magnificent scenery, snowy slopes and deep dark forests.

"Unfortunately we are not at present 'perfectly happy,'" Neyman wrote to Pearson. "...Lola has a small, but constant temperature above normal.

Probably it is the consequence of living always in a single room where one—even two—sleep, eat and one is working all the time."

To make enough money to afford a flat, he would probably have to "commit" a book. The two-sample paper was already in proof, but he was apologetic because he had not yet done anything on the new paper. There were simply too many problems in Warsaw.

In London, too, there were problems—of a different sort—about which Pearson did not write. He was utterly miserable—"disconsolate," he says. When the American statistician H.L. Rietz, whom he had met at the ISI meeting in Warsaw, invited him to come to the United States and lecture for a semester, he could not make up his mind whether to accept.

Neyman's problems were solved more promptly than Pearson's. A doctor who came to examine Lola was so struck by the miserable circumstances in which the couple lived that he offered to lend them his flat for six months—at the same rent they were paying for the room they had—while he was in France. Neyman communicated this news to Pearson with enthusiasm. While he and Lola were living in the doctor's flat, he would write a book— then he might have money enough for a flat of his own! He and Lola would take a vacation. Perhaps they could meet Pearson somewhere.

"What about committing a book together? We could print it both in English and in Polish? I hear all your objections! But no book can ever include all possible knowledge and even the knowledge of the author. Of course the readers of the eventual English book are a little different from the readers of the Polish one! And because [of that] the English book cannot be written so quickly. But some thing from the Polish book can be used in the later English editions. Perhaps it will be two different but similar books?"

He could not arouse any answering enthusiasm from the unhappy Pearson, who had finally turned down the American invitation.

On February 12, 1930, Professor Białobrzeski presented the first Neyman-Pearson paper in over a year to the Polish Academy in Krakow. It was not entitled "Part III," as the two authors had hoped, but simply "On the problem of two samples."

In 1978, when I talk to Pearson, he tells me that he had been a little doubtful about having the paper appear in Poland, but he had decided that perhaps it would be helpful in promoting Neyman's career to have some of their joint work published in his own country.

"Of course nobody saw it. We circulated an enormous number of offprints. My father said, 'Well, why did you publish in Poland?' And I said, 'Because I thought it would help Neyman.' I just had a faint idea, you see, that K.P. wouldn't like what Jerzy and I were doing…"

1978   Today is Wednesday, the day Neyman's weekly seminar meets. For him it is the high point of the work week. Announcements of the speaker and the subject have been placed in the boxes of all members of the department. Anyone who wants to come is welcome. The usual audience consists of students who are registered for credit as well as students and faculty members who, like Neyman, are interested in the applications of statistical methodology to what Neyman likes to call "Societal Problems."

Appropriate visitors to the campus are invariably pressed by Neyman "to give a little talk to my seminar," and on occasion there are as many as three seminars scheduled in one week.

As a rule there is just a single speaker, but today the format is different. Three students from Neyman's winter quarter class ("Probability Models in Biology and Public Health") will be making the first of a two-part presentation of their class work during the quarter just passed: a modification ("a substantial modification," Neyman points out) of a classic 1964 work by Neyman and Scott, "A stochastic model of epidemics."

Before we go upstairs to the room where the seminar meets on the tenth floor of Evans Hall, Neyman briefs me in his office on the background of the presentation the students will make. It is obviously a story he has told several times, but he enjoys telling it again.

"Something a little bit unpleasant happened," he begins.

After he presented the Neyman-Scott theory to the class, one of the students—"a very nice lady, Mrs. Florence Morrison, she works for the State Department of Health"—pointed out that the assumptions upon which he and Professor Scott had based their theory were not in accord with present day knowledge of epidemics and how they spread.

"Things change, you know," Neyman says, "but still it is not nice to find out that something you have done— Well. All right. Too bad. But then— something pleasant happened!"

He shuffles among the papers on his desk and locates a copy of the "Take Home Final Exam" which he gave to the class. He reads aloud from it:

"'Criticize the basic assumptions of the theory'—that's Betty's and my theory—'with reference to communicable diseases spread through personal contacts between infectious and susceptibles....How should the basic assumptions be modified to make the theory more realistic?'"

He lays down the paper.

"So then something pleasant." His face reflects his delight. "The students begin to nibble at a new theory!"

Tremendously pleased with their work, Neyman generously dispenses "A's" to everyone.

At today's meeting of the seminar two of the students—Dr. Odoom, the visitor from Ghana, and Paul Wang, a teaching assistant from Taiwan—will give an overview of the Neyman-Scott theory and describe the assumptions on which it is based. Then Mrs. Morrison will present the current thinking on epidemics and show how the assumptions must be modified to be brought into line with it.

("Mrs. Morrison will get up and say, 'NO—that is all wrong!'" is the way Neyman puts it.)

At a later meeting of the seminar other students from the class will present results of Monte Carlo simulations based on the new assumptions. Neyman has invited the university's Director of Teaching Innovation and Evaluation Services to attend both of these meetings "and see how we teach in this department."

The new theory of epidemics, when it is finally developed, will be quite a bit more complicated than the Neyman-Scott theory. Some of the results of the original theory, however, will still be valid. Neyman is especially pleased that even under the new assumptions what he likes to call "the democracy theorems" will hold. According to these, the best strategy for the upper classes to protect themselves from epidemics is to improve the conditions of life, "sanitation and like that," among the lower classes.

"Now I call that an appealing result!"

As we make our way to the elevator, I ask if his original interest in the theory of epidemics stemmed from his own experience of plagues in the Ukraine during his youth. I am thinking of the two sisters who died before he was born.

"No no no no," he says. "Epidemics are very interesting." He obviously means from a statistical point of view. "You see, they are another manifestation of the process of *clustering*, which you find in many many physical phenomena."

When we arrive at the room where the seminar is to be held, twenty or so people are already enjoying the refreshments which Neyman personally takes care to provide. He leads me over to the table. Several cakes and pies are laid out. There is one which is made with fruit—the students who purchase the refreshments for him know his preference—and he selects that one, explaining, "My mother used to make."

He always sits beside the speaker so that he can hear, and he takes his usual place, to the right, pulling an ashtray over in front of him and laying out his Marlboros and his Zippo lighter. Waiting for it to be time for the presentation to begin, he continues to talk with enjoyment about his students' work in updating his own. It seems that not only is he pleased with them but he is also pleased with himself. As he puts it, he has not tried to keep them down.

"You see," he confesses, "I am always a little bit afraid that I will become like Karl Pearson was with Fisher."

1930
—
1931

From the beginning of the collaboration, even in the midst of his initial misunderstanding of some of the ideas that Egon Pearson was communicating to him, Neyman had found the principle of likelihood not nearly so compelling as Pearson had. This was, as Erich Lehmann had pointed out to me, fortunate. For it was to be Neyman's efforts to find a logical justification for the principle—something more than its simple intuitive appeal and demonstrated effectiveness—that would lead him eventually to the development of the point of view which would constitute the core of the Neyman-Pearson theory and would become the starting point and foundation for the subject of mathematical statistics as it is known today. The steps along this path, which were by no means firm and unhesitating, are reflected in the letters which Neyman wrote to Pearson in February and March 1930.

Lehmann finds these letters much easier to read than Neyman's earlier letters.

"First of all, the situation is exactly reversed from that in the first part of the collaboration when all the ideas came from Pearson. Now all the new ideas are coming from Neyman, so the absence of Pearson's side of the correspondence is less important. Secondly, he is developing a point of view which is very clear and with which we are familiar."

In a letter at the beginning of February 1930, Neyman told Pearson, "It seems that we can have an experimental *proof* that the principle of likelihood 'est fait pour quelque chose,'" by which he seems to have meant that there really was something to it. This remark foreshadowed his statement at the end of the same letter: "If we show that the frequency of accepting a false hypothesis is minimum when we use [likelihood] tests, I think it will be quite a thing."

At this point he apparently had the problem almost formulated. He was still, however, convinced that the likelihood test was the right answer ("the best test"). His efforts were directed toward trying to prove that this was so rather than toward trying to find the test that had the desired property of being best.

The first real step in the solution of the problem of what today is called "the most powerful test" of a simple statistical hypothesis against a fixed simple alternative came suddenly and unexpectedly in a moment which Neyman has never forgoten. Late one evening in the winter of 1930 he was pondering the difficulty in his little office at the SGGW. Everyone else had gone home, the building was locked. He was supposed to go to a movie with Lola and some of their friends, and about eight o'clock he heard them outside calling for him to come. It was at that moment that he suddenly understood:

For any given critical region and for any given alternative hypothesis, it is possible to calculate the probability of the error of the second kind [the acceptance of a false hypothesis]. This is represented by a particular integral, subject to a side condition specifying the permitted probability of an error of the first kind.

He recognized immediately that he was faced with a problem in the calculus of variations, "probably a simple problem." The whole sequence had passed through his mind before he reached the window and opened it to signal to Lola and their friends that he had heard them and was coming down.

("But you see," Lehmann points out, "it is still the likelihood principle he's concerned with. He is not asking yet what is the best test but rather how to establish that the likelihood test is best. He is convinced that he already has the right answer, and he's trying to justify it.")

On February 20, 1930, Neyman wrote to Pearson, "At present I am working on a variation calculus problem connected with the likelihood method." It is the first mention in his letters of what was to become the celebrated Neyman-Pearson Lemma.

"I considerably forget the calculus of variations," he complained in a later letter.

(Several of Neyman's colleagues have expressed to me their surprise at the disparaging manner in which Neyman on occasion refers to the Neyman-Pearson Lemma, the fundamental tool of hypothesis testing. They think that he does so because the mathematics required for its proof is not on the mathematical level of the "conceptual" researches which he has always so admired.)

The letters of this period indicate that Neyman was still sensitive to Pearson's somewhat possessive feeling about the principle which he had originally proposed. In this particular letter he prefaced what he had to say about his new ideas with the statement that his results were "a vigorous argument" in favor of the likelihood method. Two paragraphs later, having stated the new formulation, he again tried to mollify Pearson.

"I feel, you start to be angry as you think I am attacking the likelihood method! Be quiet!"

In all cases he had considered, he wrote, he had found that the contours obtained by the new method were the likelihood contours!

("Now so far he is considering only special cases," Lehmann says. "He looks at lots of cases. It's a simple hypothesis against a simple alternative, and he always finds that you reject when the likelihood ratio criterion is too large. Then he says, 'In all cases I have considered I have found the following—.' And at this point he states what will ultimately become the Neyman-Pearson Lemma. But he still doesn't have the lemma.")

This letter of February 20, 1930, is also noteworthy for containing for the first time in the correspondence, a reference by Neyman to "confidence intervals"—a subject which was later to absorb him as thoroughly as the theory of hypothesis testing.

One of the students in his laboratory, Wacław Pytkowski, was currently occupied with the question of "The Dependence of Income of Small Farms Upon Their Area, Outlay and Capital Investment in Cows." In the course of this work he had asked Neyman how he could formulate, "non-dogmatically," the precision of an estimated value. Neyman had promptly diverged from his planned lectures to take up Pytkowski's question. His first solution (which he was later to dismiss as "quasi-Bayesian") was to define a random variable, which he called a "confidence interval," so that, whatever the a priori distribution of the parameter to be estimated, the probability of its being covered by the defined interval would have a preassigned value, less than one but as close to one as desired.

While—in Warsaw—Neyman was thus coming to grips in a fundamental way with the problem of the lemma and beginning to explore interval estimation, Pearson—in London—was trying desperately to solve his personal problems.

"[I decided] that I could not go against my family's opinion that I had stolen my cousin's fiancée. This was my father especially. He had had· a brother who had had many unfortunate love affairs and, believing in heredity as my father did—at any rate my courage failed...."

In March, without giving any details, he wrote to Neyman that something very disagreeable had happened to him.

"I am almost angry," Neyman scolded—"two days ago you wrote a letter in which you mentioned that it happened some thing very disagreeable with you, and now I have seven pages with not a single word about it. Well, I am sorry—you may not like to tell what it is and I am silly enough asking. Excuse!"

He then plunged into a discussion of the future of their collaboration.

"I think we have to introduce a little more order in our work. 1) Do you want to publish some more joint papers? 2) If so, we must fix a certain plan, as we have lot of problems already started and then left 'in the wood.'"

There were at least six problems that should be attacked—"...we could certainly make this list as long as one wants. But we can not do anything at onse." The first problem was the hypothesis that an arbitrary number of samples greater than two come from the same normal population—a natural extension of the two-sample paper just presented to the Polish Academy. The last problem was to finish what he himself had already started to do with the variational calculus.

"You will understand it in a moment," he assured Pearson.

"To reduce to a given level the errors of rejecting a true hypothesis, we may use any test. Now we want to find a test which would 1) reduce the probability of rejecting a true hypothesis to [a given] level...and 2) such that the probability of accepting a false hypothesis should be minimum.—We find that if such a test exists, then it is the likelihood test. I am now shure that in a few days I shall be ready. This will show that the likelihood principle is not only a principle, but there are arguments to prove that it is really 'the best test.'"

("This sentence now shows the philosophy very clearly," Lehmann points out. "The essence of the new thing.")

Neyman concluded his letter by saying that possibly there was no need for him and Pearson to work out everything together "and even to publish things in joint papers. Only what I would like it is to have a sort of companionship in the work. I would like to know that this and this you are doing and this I have to do. Just as you like."

He was still concerned about the personal problem that was bothering Pearson: "it is very—very disagreeable to know that it is some thing which makes you unhappy and not to know what." Pearson, however, was unable to confide his troubles. He continued to work, taking on the first problem which Neyman had listed: the problem of k samples. The data represented the breaking-strength under tension of small bricks of cement mortar mixed on each of ten different days. Practically speaking, the problem was to determine whether the variation from sample to sample was no more than might be expected by chance or whether there appeared to be significant differences from day to day, suggesting that the manufacturing process was not yet under satisfactory control. The theoretical problem was to show that the test developed in the two-sample paper could be extended to the general case.

It is clear from Neyman's letters that the two friends were working quite independently, for at one point Neyman wrote of "the purpose of *my* efforts" with "my" firmly underlined. But it is also clear that Pearson was following Neyman's work attentively.

"The question in your last letter is exactly the same on which I am working, only from a little other point of view—," Neyman replied on March 12. "Does not matter the order! Let us find the answer."

His next letter indicates that Pearson had sent him a counterexample where, in some sense, the likelihood ratio test was not the best. He responded that he had said only that if there existed a best test, then it was the likelihood test.

("Now by this time the issue must have come up," Lehmann points out. "You see, you test a simple hypothesis and in general you have a large class of alternatives, not just one. Now for each of these alternatives there is a best test, but the best test for the different alternatives in general cannot be expected to be the same test. So that's the problem. If a common (uniformly) most powerful test does exist, then it is the likelihood ratio test. But in many cases such a test does not exist. In Pearson's counterexample it obviously didn't. That foreshadows what is going to come later.")

This same letter, dated March 24, 1930, and running to eleven pages written across the length of the sheets, brings at last the proof of the fundamental lemma. For Neyman the proof concluded the first stage of the development of "a new point of view" toward the problem of hypothesis testing. He was eager to see Pearson and talk with him.

Although Pearson had been working regularly, he was still miserably unhappy over the breaking up of his romance. He refused an invitation to come to Warsaw at Easter and, in a mood for solace rather than statistics, decided to go again to gaze on the great paintings and sculptures of Italy.

Before Pearson left, Neyman suggested another possibility for a meeting. The First All-Union Congress of Soviet Mathematicians was to be held in Kharkov in June.

"Do you like to go. I do not know if bolsheviks will give us the visa, but I already rote the application. I am sure it will be most interesting—and for you also—to see Nu Russia alive as it is."

Again, Pearson could not be tempted.

In the last sentence of a later letter, just before telling Egon to "have a good time" in Italy, Neyman referred to the problem of composite hypotheses, which was to be the next complication in the joint work.

(In Lehmann's opinion, Neyman's letters make it clear that Pearson had no major part in the development of February and March 1930. "Pearson himself is quite frank about this. In 'The Neyman-Pearson Story' he writes of the fundamental 1933 paper in which these ideas were to be presented: 'My part was to help in testing the material; sharpening the arguments; standardizing the notations and terminology; working out the 'examples'; and deciding on the best form for the diagrams.' From what I have now read in Neyman's letters, I don't believe that there was much need for sharpening arguments—maybe for expressing them more effectively in English—but the arguments are all there in the letters.")

By June, before the Neymans left for Kharkov, Pearson had essentially completed the *k*-sample paper and sent a draft of it to Warsaw.

"I agree to have my name under the paper you constructed," Neyman responded but added the condition "that we start immediately another with a bigger share of mine." As a title for the paper he had in mind, he suggested "New Point of View on the Likelihood Method in Testing Hypotheses."

("Still he is attached to the likelihood method," Lehmann notes.)

"Now if you like this and if you wish to add things—illustrations—what you like—let us publish it—where you like."

No decision was made by Pearson. Neyman and Lola set off for Kharkov, still urging: "Come with us. It *will* be interesting." And if Egon could or would not come, what about a rendezvous later in the summer? "Do you like France, Italy, Switzerland? Perhaps Yougoslavia?"

After reading these letters, I am of course eager to hear how Neyman found "Nu Russia" thirteen years after the revolution; but he has few memories of his return to the city in which he grew up. He and Lola found her mother and her aunt—ladies who had no skills that were useful in the new working world—doing menial clerical tasks and living under miserable conditions. They saw a number of old friends and acquaintances, among these Leo Hirschvald, to whom Neyman had enjoyed explaining Zermelo's axiom of choice fifteen years before. Hirschvald had become a professor at the university; and during the period of "Ukrainization," when all lectures were to be delivered in Ukrainian, he had begun to translate into that language the Bernstein lectures on probability theory which Neyman, as a member of the Presidium of the Mathematics Club, had helped to compile during his student days. Neyman brings Hirschvald's translation from his study and looks with bemusement at the title.

"I cannot read it," he says, "and I can speak Ukrainian. I very much doubt if there is a word in Ukrainian for 'theory of probability.'"

Bernstein, he tells me, steadfastly refused to lecture in Ukrainian; and by the time of the congress in Kharkov in 1930, the policy of Ukrainization had been abandoned. Bernstein, as the head of the local mathematical center, third only to Moscow and Leningrad, was president of the congress. He asked Neyman to chair one session.

"And so there was a paper given on the Marxist definition of an average. Well, obviously, this was a paper to be ignored. 'Thank you. Any questions?' And then suddenly, to my great surprise, Bernstein stood up. He said that, yes, in studying various problems it is occasionally relevant to compute an arithmetic average and then in some other problems the average is geometric or harmonic. And so forth. But he is wondering what relation to Marxism does this have. Well! I was surprised that he spoke. But a further surprise was that there was a tremendous explosion of discussion, accusing Bernstein of being capitalist this and that, enemy of humanity, and what not. It lasted long. Eventually ended. Then soon thereafter Bernstein caught me in the corridor and said, 'I thought it would be safe for you to come, but I am not sure. You may wish to leave.' Well. I decided not to leave."

I comment that Bernstein seems to have been extremely independent, given the political situation in the Ukraine during those years.

"Well, Bernstein was a gentleman. As a gentleman—it was obvious nonsense, a Marxist definition of an average—and if he would keep silent— he was responsible for the whole meeting—then he would not have been a gentleman. It is one of the qualities which comes into my mind as the definition of gentleman. A man who would not lie in important things. Polite lie, that is different. But who would stand up and be counted."

Neyman has no published report of his talk on hypothesis testing at the congress, nor does he remember any reaction on Bernstein's part.

Pearson did not join the Neymans later that summer. Neyman pressed him to come to Warsaw at Christmas, but again he would not. He was absorbed in his personal problems, the painful process of self-discovery through psychoanalysis. In 1978 he says: "All that winter I must have been working on 'k-samples' and sent him drafts. A good deal of mathematics was required from him. I sampled and calculated tables."

What was Neyman doing in this period when there are no letters from him to Pearson?

"Well. All right. It must be in *Statistica*."

Statistica is the "journal" which Neyman created for his two little laboratories—a consequence of his feeling that some day he and Egon might want a place for "propaganda." It was made up of reprints of all the papers published during a given period by Neyman and the members of his laboratories. These were simply bound together with a cover which listed the contents on the front, and on the back requested—in three "international" languages—"Please exchange."

He goes to his study and returns with the first two volumes. There were to be only half a dozen, the first dated 1929–30 and the last, 1938. He handles each with the same affection with which he takes up the notebooks containing his thesis on the Lebesgue integral.

"The idea of *Statistica* came from the Latin quotation *Cogito ergo sum*."

You mean that when they published as a group they were a group?
"Yes."

The 1929–30 volume consists largely of Neyman's own work. There are only three papers by students; but since the papers are arranged in alphabetical order according to the last names of the authors, all the students' papers come before his. By the second volume, dated 1932, the number of student papers has doubled. Neyman runs his finger down the titles. Although sometimes, as I have noted, he seems to have erased from his mind much of his early career in Poland, he remembers in detail each of the papers in *Statistica*. During the academic year 1930–31 the workers in the Warsaw laboratories dealt with a variety of subjects. Bacteriology, eugenics, anthropology, agriculture, economics. All are represented in the yellowed pages in addition to the theory of hypothesis testing.

I try to find out how Neyman felt about his scientific future in 1931, ten years after he had come to Poland.

He is silent for a moment and then says:

"*Statistica* shows that at that time there was a big problem. You understand? I can't explain exactly when—"

But, finally, he felt that he had found a big problem in statistics?

"Yes. I was a member of a community, and also I was head of this community. Yes, small. I was the head of a small group of somewhat younger people also to whom had caught the bug, you know. And," he emphatically taps the volumes of *Statistica* on the desk beside him, "they were published!"

Was he reasonably satisfied with his life?

He is silent again.

"Slightly different," he says at length. "I was emotionally involved, all right, in the big problem. That was emotion. Yes, it is convenient to have a little more money and I tried to get it, but that was secondary. The big problem was really the non-dogmatic foundations of mathematical statistics. That was the big problem."

Did he have any desire to leave Poland and join Egon in London?

He smiles.

"That looked as if it were, at the time, ambition to go to the moon."

1931
—
1932

In London, at the beginning of 1931, Egon, unhappy anyway, was finding it increasingly difficult to work with his father, who was approaching his seventy-fifth birthday. He accepted a second invitation from Rietz and set sail for America. He thought he might even take a job there.

In Warsaw, Neyman was also having difficulties with the Professor. He had submitted a paper (written with a student) to the elder Pearson for publication in *Biometrika* and had received a long, but negative, response.

The rejection of the Iwaszkiewicz-Neyman paper on counting virulent bacteria put Neyman in an awkward position. He especially wanted to publish it "because of the pup." Karolina Iwaszkiewicz—"Karolinka"—was a young person, his principal assistant in the lab, and also someone of whom he was especially fond.

"And then if to publish it, then where?" he wrote helplessly to Egon in America. "I think it is necessary to get a man to present the paper to a journal. The only man I can ask is Fisher, or might be Student? Probably Fisher will not refuse, but then I shall not be among the persons which are considered as friends by Professor Pearson. The Biometrika [will be] shut for me—that's all right, but there are you." If the Professor would see the paper in print without the changes he had suggested, "he will not be very glade."

Desiring to avoid trouble and especially trouble for Egon, he decided to publish the work in Poland, since then the Professor would never see it. Whatever Egon wrote in response, it delighted Neyman and considerably bolstered his spirits.

"At present I am ready to kiss everybody, even 'K.P.' as you write...but before I read your letter I was rather redy to kik—almost everybody. It is so good to feel that you are not alone and that not all people around are homini lupi."

Life in Poland in general and at the SGGW in particular was frustrating.

"You may have heard that we have in Poland a terrific crisis in everything. Accordingly the money from the government given usually to the Nencki Institute will be diminished considerably and I shall have difficulties in feeding the pups."

He saw "wolf men" all around him.

"...one very kind gentleman wrote that in his laboratory has been done a paper, which he describes as a very important one, staying above all papers written by students, oppening some new horizons etc. Well, this paper has been done by a girl in my laboratory, the girl has been paid for doing it and I spent an owful lot of time teaching her.... The old man did not even understand what and why has been done.... Moreover he is one of members of a committee appointed as it were to inspect my teaching statistics, and pretends to be able to give me advices."

He was not sure, but he suspected that still another professor was trying to take credit for the paper which he had written with Karolinka.

"A pretty society! Well, I'll show them."

He longed to go to England again—there was a possibility of a trip in connection with a study of the government's health insurance program. In the meantime he had to be satisfied with devouring volume after volume of John Galsworthy, whom he and Lola had recently discovered. He especially enjoyed the trilogy *Modern Times*, which dealt with the saga of the Forsytes during the early nineteen twenties. ("You see, I had tasted a little bit when I was in England," he explains, "and I wanted more.") He urged Egon to read the books, too, "at least *The White Monkey*."

During most of the year 1931, while Neyman struggled to support his little labs in the midst of increasingly widespread and severe economic depression, Pearson was in America. He still had not finished the k-sample paper. It was more than a year since the publication of the last joint work, and Neyman was eager for another.

"I did some thing which, I am afraid, you will not like," he confessed in a letter to America, "and perhaps 'will get my goat'...."

He had mentioned to Professor Białobrzeski, who had presented the two-sample paper, that it would be good if Pearson could have offprints of the new k-sample paper while he was still in America. To his dismay he had learned that this would be possible only if he himself summarized the paper for presentation to the academy.

"I thought that one month of time is some thing and promised to see him tomorrow and to give him a short abstract. At present I feel uncomfy and am afraid of you and about my goat. (What does the expression mean exactly?) It is no good to decide anything for somebody else—well but you did not suggest anything definit! And besides—one month!"

After the middle of March 1931 there are no letters from Neyman to Pearson until the middle of August; however, Professor Białobrzeski did present the k-sample paper at the beginning of May to the Polish Academy.

In America, Pearson while at Iowa City found himself in the company of Fisher. He remembers with wry amusement that Rietz, eager to amuse his English visitors, invited the two of them to go with him to the circus.

In Warsaw, the practical work of the little laboratories continued: Józef Przyborowski, a professor of agriculture from Krakow, came to discuss statistical problems which occurred in his work on the removal of dodder from red clover.

"And so he used to come, discuss with me, this and that. And then one day he said, 'You know—this is not enough. I cannot come every week to here. I need an assistant. Could you recommend an assistant, who would be working with me and I would pay him?'"

Neyman thought immediately, "Ah, Wilenski."

Henry Wilenski was a poverty-stricken Jewish student, the son of an orthodox cobbler who had beaten him so thoroughly when he had violated the prohibition against travel on the Sabbath that the boy had run away from home.

"Now Krakow, the old capital, was the seat of, all right, nobility and, all right, anti-Semitism. And so I said yes to this Professor Przyborowski of Krakow, that I had someone I could recommend for an assistant to him. 'What is his name?' 'Wilenski.' 'Jewish?' 'Yes. [Pause.] Well. You wanted an intelligent man—' 'This—is he the best?' 'Yes, at this time, he is the best.' [Sigh.] 'Well, all right.'"

Neyman arranged for another Jewish student, the son of a banker, to help the cobbler's son to obtain some respectable clothes; and Przyborowski found Wilenski "not unreasonable." In a few years he and Wilenski published a paper together in *Biometrika*.

In spite of the necessity for practical assignments to provide small sums for his "pups," most of whom had no sources of income other than their own efforts, Neyman believed that success in the application of statistical methods came only with the simultaneous development of statistical theory. He and several of his students, notably Stanisław Kołodziejczyk, continued to attack problems of hypothesis testing. During the spring and early summer of 1931, however—what with his lectures at the university and at the SGGW as well as the work of the two little laboratories—there was less time than he would have liked for the theoretical work that interested him. He went when he could to Mądralin, but he missed what seemed in retrospect the easy communication that had existed when Pearson was in England.

"My most intense desire is to see you," he wrote.

He was now considering the problem where there was no uniformly best critical region with regard to a given class of admissible hypotheses. "What region should we [then] choose?" Studying a simple example in which a "best" test does not exist, he suddenly exclaimed, "I think I have got the point!" He imposed an additional condition, later called "unbiasedness" and playing an important role in the theory of hypothesis testing. ("I remember I suggested 'unbiasednicity,'" Neyman recalls, "and I remember Egon smiling.") Subject to the condition, a "best" test existed in the example, but in a closely related problem the likelihood test did not satisfy the new condition.

"I should like to be wrong somewhere. If it is all correct, then the principle of maximum likelihood seems to loose its generality. Exceedingly interesting to know what are the conditions of its *applicability*."

The next surviving document is a "summary as to critical regions," written in September by Pearson, who had by then returned to England. Referring to the disappointing discovery in regard to likelihood, he noted that he had been prepared for the contour not to be exact.

"It looks as if I will have to cross the continent of Europe to talk to you."

Neyman responded delightedly. There was so much for them to discuss. The work on the best critical regions was being pushed "very much forward"; but he squeezed in, writing sidewise at the end of the letter, "The position where there is no B.C.R is still obscure!"

It was Christmas 1931 before Pearson was able to talk to Neyman. Outside Warsaw, in the quiet surroundings of Mądralin, the two men got down to work and got a great deal done. Once classes took up, however, life became again "more complicated." There were lectures, beginning early in the day and lasting until late in the afternoon, students to be helped with papers, people to be contacted about money for the labs. Neyman stayed at the college at night and tried to work "on good work on theory," but without much result. The December issue of *Biometrika* arrived with a short joint paper entitled "Further notes on the chi-square distribution." It seemed to him that he and Egon had published it simply to keep the work going.

When he wrote to Fisher for permission to reproduce a portion of the latter's *t*-table in the little book he was writing "for agriculturists" (and to extend his personal apology for a reference in *Biometrika* which attributed a result of Fisher's to himself and Egon), he added a postscript:

"The crisis in Poland is terrible. Would it be any hope for me to get any occupation in England for a year or two?"

He urged Egon, "Hurry up with the 'Big Paper,'" adding, "...I see that our philosophical considerations about statistical tests are *not* common places and, as we use to say, mere 'demogogical talking.'"

Fisher's response to Neyman's letter was so friendly—"I had not heard of the incident you refer to, but in any case I should know that it was not with you that I had any cause to be offended."—and so encouraging about the possibility of his being able to find some sort of temporary statistical work in England—that Neyman wrote again:

"I do not know whether you remember what I said, when being in Harpenden in January 1931, about our efforts to built the theory of 'the best tests.' Now it is done more or less. It follows from the theory that the 'Students' test, the 't-test' for two samples and some tests arising from the analysis of variance are 'the best tests,' that is to say, that they guarantee the minimum frequency of errors both in rejecting a true hypothesis and in accepting a false one. Certainly they must be properly applied." He added, "You will see that these results are in splendid disagreement with the last articles of Professor Pearson in Biometrika."

He and Egon were eager to have their big paper published in the *Philosophical Transactions* of the Royal Society—Neyman told Fisher—but it was going to be of considerable size, "[and] we do not know whether anybody will be willing to examine a large paper and eventually present it for being printed." It also contained much mathematics, "and not all statisticians will like it just because of this circumstance." In his opinion and that of Egon, "the most proper critic are you."

Although Fisher did not offer to present the Neyman-Pearson paper to the Royal Society, he did write that he would be very interested in seeing it. The question of tests of significance seemed to him of immense philosophical importance.

"It is quite probable that if the work is submitted to the Royal Society, I might be asked to act as referee, and in that case I certainly shall not refuse."

From February 9, 1932, when this letter was written, until October 16, there seems to have been no further correspondence between Neyman and Fisher.

Neyman met Pearson at Easter in Paris and worked at polishing the "Big Paper." Again, back in Warsaw, he found himself unable to do any work or even to answer Egon's letters.

"Don't be angry with me," he begged. "....I had so much to do and lost so much energy that I could not force myself to write....During the beginning of the vacations I shall be incapable to do anything but fishing. In August I shall try to work."

In a later letter he wrote, "You seem to be a little angry with me: in fact you have some reasons as I do not answer properly your letters. This however is really not the result of the carelessness or of anything which could be offensive. I simply cannot work. The crisis and the struggle for existence takes all my time and energy."

In the meantime Egon had approached his father about presenting his and Neyman's paper to the Royal Society.

In 1978, I write and ask Egon Pearson how it happened that Karl Pearson—whom Neyman has described as "skeptical and hostile" about what he and Egon were doing—had agreed.

"He would naturally present his son's long joint 1933 paper to the Royal Society," Pearson replies. "This did not involve his giving an opinion on it. Fellows who 'present' papers do not necessarily agree with all their contents.—But I suppose he was a bit proud of E.S.P. He, at this stage, knew that I was coming to the top, and was too sensitive to discuss his views on the paper with me—we were a bit estranged."

The Royal Society received the Neyman-Pearson paper on August 31, 1932; and the editor of the society's *Philosophical Transactions* arranged for it to be refereed. On October 16, Neyman wrote again to Fisher:

"E.S.P. writes that you have recomended our paper for publication. Although it may be considered ridiculous to thank a judge, I have intense feelings of gratefulness, which I hope you will kindly accept."

"I was a good deal interested in your paper," Fisher replied, "but I ought not to have passed your statement that a distribution is completely determined by its moments. Do you not know Borel's distribution with all its moments zero?...You can always add a bit of Borel to my distribution without altering the moments...."

"As usual," Neyman complained, apparently quoting a phrase that had often passed between him and Pearson in regard to Fisher's writing, "I do not quite follow—especially with these integrals.

"Unfortunately: (1) I am a little sick and (2) have to finish a paper for which I shall have some money to go to Paris [where Lola was again]. So I cannot think of it."

He was sure that if the paper was rejected, they would have plenty of time to think about Fisher's comment; but by the end of the first week of November he learned from Pearson that the "big paper" had been "taken as read."

What—Neyman immediately demanded—was the date of the next meeting of the Royal Society?

"Write quick what is going on."

What was "going on" was that, just two days later, Karl Pearson was presenting the paper "On the problem of the most efficient tests of statistical hypotheses," by his son and his former Polish student, to the Royal Society at its meeting of November 11, 1932.

To Neyman it has always been a source of satisfaction and amusement that his and Egon's fundamental paper was presented to the Royal Society by Karl Pearson, who was hostile and skeptical of its contents, and favorably refereed by the formidable Fisher, who was later to be highly critical of much of the Neyman-Pearson theory. Ironically, when I write to the librarian of the

Royal Society to find out the name of the other referee—customarily there were two—I am informed that Fisher did *not* referee the paper—that there was only one referee—and that he was A.C. Aitken of Edinburgh.

(I am at a loss to explain the conflict between the records of the librarian of the Royal Society and the statement of Neyman in his "thank you" letter to Fisher.)

The paper—published in the Society's *Philosophical Transactions* the following year—represented the second decisive step in the Neyman-Pearson theory of hypothesis testing. In it the authors, no longer satisfied with an intuitive solution to their problem, took on the task of determining the test which at a given significance level would maximize the power of the test against a given alternative. For the case of a simple hypothesis the problem was completely solved by the now celebrated Neyman-Pearson Lemma, the proof of which established that for testing a simple hypothesis against a simple alternative the solution was indeed given by the likelihood ratio test with the specified significance level.

In the case of composite hypotheses—the next problem—they introduced the restriction to similar tests—essentially tests for which the probability of rejection was the same for all distributions in the hypothesis—and solved this problem for families of distributions satisfying certain differential equations.

The work quite literally transformed mathematical statistics.

"The impact of this paper and those which followed in the next few years has been enormous," Le Cam and Lehmann wrote on the occasion of Neyman's eightieth birthday in the *Annals of Mathematical Statistics*. (On another occasion, Le Cam, more exuberantly, compared the effect which the Neyman-Pearson work has had on mathematical statistics to the effect of the theory of relativity upon physics.) "It is, for example, hard to imagine hypothesis testing without the concept of power....And the optimum properties of the classical normal-theory tests are not only aesthetically pleasing but serve as benchmarks against which the performance of simpler or more robust tests can be gauged. However, the influence of the work goes far beyond its implication for hypothesis testing. By deriving tests as solutions of clearly defined optimum problems, Neyman and Pearson establish a pattern for [Abraham] Wald's general decision theory and for the whole field of mathematical statistics as it has developed since then."

Karl Pearson's presentation of the "big paper" took place exactly half a dozen years to the month after Egon Pearson had sent his notes on his doubts about statistical inference to "a foreigner I liked." A new point of view had been established in mathematical statistics. For the first time, as Le Cam puts it—"There was some logic in the subject!"

1932
—
1933

By the fall of 1932 there appeared to be several reasons why Neyman might never become a professor in Poland. One was his subject matter, which was not generally recognized as an academic specialty. Another was the fact that he was married to a Russian—and an independent, outspoken Russian who lived on occasion apart from her husband, worked and painted in Paris, traveled on a freighter as a nurse for the adventure of it, and sometimes led tourist excursions into the Soviet Union. Still another, and probably the most important reason, was that he was known as a nonbeliever who attributed to its Roman Catholic faith many of his country's troubles.

He continued to have to eke out his academic salary by taking on other work. For a while he was employed by the widow of a wealthy landowner, who wanted to learn something about statistics and agricultural experimentation. She was an intelligent and eager student—"also she did not cheat on her experiments"—but she insisted, much to Lola's disgust, that Dr. Neyman's wife be present at all lessons. Most often Neyman obtained the extra work he required from the Institute for Social Problems, a quasi-governmental organization—in his opinion, "the only place in Poland where people are interested in questions in mathematical statistics." During 1932 one of the institute's projects was the sampling of Poland's first postwar census data in such a way that data on types of workers eligible for social insurance could be obtained quickly and economically. Neyman was employed to develop the sampling scheme. In the course of this activity (as he later commented to Pearson) he "pushed a little the theory"—something of an understatement in regard to a work which was later to be designated in the history of sampling as "the Neyman Revolution."

Strange as it may now seem, up until the beginning of the twentieth century there had been heated debate in the International Statistical Institute about the scientific validity of sampling—in any form. By the time the century reached the quarter mark, however, the members of the institute had generally agreed on "the representative method" for situations in which a census of the entire population was out of the question. Two approaches were recognized. The first—purposive sampling—consisted in dividing a population into a comparatively few large groups and selecting some of these as the sample on the basis of their similarity, in regard to certain characteristics, to the population as a whole. For the second approach—random sampling—the population was usually divided into a much larger number of smaller groups from which a sample of many more groups was selected, either entirely at random or at random with some restrictions.

Upon investigating these two different approaches, Neyman concluded that the method of purposive sampling was "hopeless." Coincidentally, he had at hand an impressive illustration of its "hopelessness" in the recent work of two Italian statisticians, one of whom was Corrado Gini, who had so vigorously attacked the paper on hypothesis testing which Neyman had presented at the 1929 Warsaw meeting of the ISI. Gini and another Italian statistician, Luigi Galvani, had been faced with the problem of selecting a representative sample of census data which could be retained for further use after the rest of the data had been destroyed. For this job they had chosen to apply the method of purposive selection, dividing all Italy into a couple of hundred administrative districts and then selecting approximately thirty for a sample on the basis of the fact that their means in respect to certain controls were practically identical with those for the whole population. When, however, they tried to verify the goodness of their sample by comparing the estimates of other characteristics with their values in the population as a whole, they discovered that there was very poor agreement. They concluded that it was impossible to obtain a sample reproducing the properties of the general population.

Neyman disagreed.

"...both theory and experience indicate that, whenever we have in mind a truly statistical problem of estimating means of any size, or regressions, etc., a properly drawn sample is, for all practical purposes, sufficient."

The system which he set up for the Institute for Social Problems utilized a stratified random sample. To estimate the accuracy of the tables he obtained, he applied his method of "confidence intervals," upon which he had been lecturing since 1930 although he had not yet published his ideas. By this method, he was able to show that the stratified random sample was definitely superior in almost all situations to the purposive, or balanced, sample.

Many years later, in a paper on "The Foundations of Survey Sampling," T.M.H. Smith wrote of this work by Neyman, "...since that day random sampling has reigned supreme with any advocate of balanced sampling having to take a defensive position." At the time Neyman, telling Pearson about the work, which he was just completing ("rather with pleasure"), said only, "Of course not everything is new: wanting to have a booklet presenting the whole of the question, it is necessary to present some results of other people....[He was referring here to the pioneering work of the Norwegian A.N. Kiaer and the Englishman A.L. Bowley.] Nevertheless there are many things, which seem to be new. Such is the point of view on the accuricy of the results....Well—it would be too long to tell everything....I suppose that some paragraphs of the booklet are worth publishing somewhere in an international language."

The work appeared in Polish in 1933 with a short summary in English.

In addition to statistical projects, extra money, and congenial co-workers, the Institute for Social Problems provided travel funds on occasion. In December 1932 it sent Neyman abroad to conduct a survey of the use of statistics in health insurance programs in Germany, England, and France. Also included were several weeks for study and "to get inspiration for further work."

He had last been in Berlin in the midst of the general disruption and despair which had followed defeat in the war. Now he arrived during the demonstrations which preceded the appointment, six weeks later, of Adolf Hitler as chancellor of Germany. He found the officials at the Health Insurance Office pleasant and cooperative, but he concluded "that the main purpose of the statistics in German health insurance institutions consists rather in illustrating the activities of the same than in solving problems of practical importance." Much more to his taste was his visit with Richard von Mises, the head of the Institute of Applied Mathematics at the University of Berlin. Von Mises asked him to give a talk on his and Egon's work. Since, as a Pole, Neyman was not especially welcome in a city where National Socialist orators were daily urging a push to the east, von Mises arranged that the talk be given, not as customary in the lecture hall, but in a small seminar room. He also took care to "translate" Neyman's German into his own German as a way of reminding the students of his personal and professorial sponsorship of the speaker. Years later, in a letter to Hilda Geiringer von Mises after her husband's death, Neyman recalled: "...I have before my eyes the faces of something like a dozen students which to begin with appeared not very friendly. Also I remember the pleasure I felt when, as my talk proceeded, some of this unfriendliness began to disappear." Someone asked him what, in his opinion, was the relation between the mathematization of frequencies— the goal of von Mises's efforts—and the efforts of others, such as Bayesians, to build a mathematical theory based upon intensity of belief. He replied promptly that there was no relation—they were totally independent—"and everybody smiled."

In London he visited the National Health Insurance Office and concluded that the British system was "cheaper and more liable to rational control" than the Polish system. (When this comment was later published, the British objected to his use of the word *cheaper,* which to them implied *stingy* rather than *inexpensive* as it did to him.) The brevity of his time in London did not permit much new work with Egon. The proofs of the paper for the Royal Society had to be finished off; and the exact form of the footnote acknowledging Fisher's comment, decided upon. Before leaving Poland, Neyman had received a letter from the latter to the effect that "you and Dr. Pearson are not indebted to me for anything more than, as it were, the correction of a verbal slip" which it was not worthwhile to mention in the published paper.

"However, I think it *is* worthwhile," Neyman insisted, and Pearson agreed.

In Paris, Neyman investigated the French system of health insurance and determined that it was much like the Polish. He concluded that the "should be" position in relation to the use of statistics in the health insurance systems of the various countries he had visited was "in a sense contrary to the present one."

Not long after his return to Poland, the joint paper on the most efficient tests of statistical hypotheses appeared in the *Philosophical Transactions.* Concluding it, the authors had remarked that "owing to the considerable size the paper has already reached, the solution of the same problem for other important types of hypotheses must be left for separate publication." That spring they met in Paris to carry on the work.

"[The new paper] was put together very speedily...," Pearson recalls. "My impression is that we met without any clear plan as to what we should work at as there were so many possibilities; on my arrival he suggested that it would be good to write a paper discussing 'what statements of value to the statistician in reaching his final judgment can be made from the analysis of observed data, which would not be modified by any change in the probabilities a priori.' So we sat down to this and had almost finished when we started."

Back in Warsaw, with Lola remaining again in Paris until summer, Neyman found it harder than ever to work on theory—it seemed he could do that only when he was abroad. Perhaps, he suggested, the paper he and Egon had just completed could be dated Paris, January 1933—"This would in some way increase the probability of my getting money in the future, since the ministry will see that I do not spend it sollely to enjoy myself."

Life in Poland was more and more difficult and unpleasant. The SGGW, which had originally been only a fifteen or twenty minute walk from the university, had been moved some two hours out of town. Accommodations for faculty members were provided nearby; but when I refer to these as "housing," Neyman objects: "Not housing. Just shacks. But were closer."

Anti-Semitism—which had become government policy in Germany—was also on the increase in Poland.

One morning students "occupied" the college and, climbing on top of a high wooden fence which surrounded the building, shouted, "Out with the Jews! Out with the Jews!" When the police were summoned by the rector, the students tried to hide by dropping out of sight; but the fence was so high that many of them had to continue to hang by their hands "and the police beat these hands with rubber sticks." Some professors, as anti-Semitic as their students, spoke of "police brutality." Neyman smiles as he uses the phrase which, like "occupy," brings up memories of much later events he experienced at Berkeley.

He complained to Egon: "Our idiots of students are behaving with Jews as Americans do with Negroes...."

Then suddenly, in the middle of June 1933, a possibility of getting out of the chaotic situation in Poland opened up. University College had selected Fisher to succeed Karl Pearson as Galton Professor and Director of the Galton Laboratory. This dramatic piece of news was transmitted to Neyman by Egon. (It is not clear whether he also mentioned that he himself had been selected to succeed his father as the head of the Department of Applied Statistics, which consisted of four persons.) Neyman immediately wrote to Fisher.

"I know there are many statisticians in England and that many of them would be willing to work under you. But improbable things do happen sometimes and you may have a vacant position in your Laboratory. In that case please consider whether I can be of any use."

Fisher replied by informing Neyman of the division of Karl Pearson's old duties between himself and Egon Pearson.

"This arrangement will be much laughed at, but it will be rather a poor joke, I fancy, for both Pearson and myself. I think, however, we will make the best of it."

He wished that he had "a fine place" for Neyman; but it would be a long time before he would be able to unify his new department, "and you will be Head of a Faculty before I shall be able to get much done."

The new plan of organization at University College was one with which nobody was happy. K.P. felt, according to Egon, that the scheme was almost a breach of trust "since nearly the whole equipment and funds [of both laboratories] had been supplied in the past for use in...a single institution." Fisher felt that as a statistician as well as a geneticist he was uniquely qualified to succeed K.P. both as Galton Professor and as head of the Department of Applied Statistics. It was well known that he looked upon the younger Pearson as a nonentity hanging onto Neyman's coattails. As for Egon, he was uncomfortable even in the mere presence of Fisher. In his new position he would be at a particular disadvantage because of his youth and his lower academic rank.

Pearson seems to have written little or nothing about the situation at University College to Neyman during the summer of 1933. Neyman's own letters treat only statistical subjects. During the first part of the summer he finished the little book on the statistics of health insurance institutions, a task which he found "rather long...and tedious." He had become familiar with Fisher's paper on "fiducial distributions" and "fiducial argument," which had been published in the same year—1930—in which he himself had started to lecture on confidence intervals. He cheerfully recognized Fisher's priority for a theory of interval estimation independent of probabilities a priori, but it seemed to him that Fisher's approach involved "a minor misunderstanding" and could be improved upon. After a little fishing in the summer, he thought he would be refreshed enough to do some work on fiducial things, "or—better to say—on confidence intervals and regions." He planned to publish what he wrote "with compliments to Fisher."

He and Egon had submitted their new paper, "The Testing of Statistical Hypotheses in Relation to Probabilities *a Priori*," to the Cambridge Philosophical Society. The comments they received from the referees led them to think one of these had been Harold Jeffreys—the leading English proponent of the Bayesian point of view and a man planted firmly on the other side of the fence from Fisher.

"The question of a priori probabilities is of course a complex one," Neyman remarked in a letter to Pearson upon receiving the referees' comments. "...I have no doubt that...Jeffreys is rather wrong confusing two different things: the intuitive feeling about the probability of an event (or of a proposition, which is really equivalent) and the mathamatical probability. The [latter]—though an abstraction—may be measured and is an attribute of certain things as f.i. their weight. The former is much less the attribute of a thing outside us."

He thought it was unfortunate that Fisher had introduced the notion of fiducial probability in his controversy with Jeffreys. "If Fisher would not introduce this word, I suppose everybody would understand what he means. I think I do." While he and Egon certainly didn't want to become involved in a quarrel between fellows of the Royal Society, it might be worthwhile for them to discuss the problem in order to make clear that Fisher and Jeffreys were disagreeing because they were using the same words in different senses. At the same time maybe he and Egon could also make clear what they meant when they used certain words. "Otherways we must be prepared that people will attack us as they do with regard to Fisher." He thought "it would be extremely good to put in order the whole question of the problem of estimation."

That fall, the beginning of the academic year 1933–34, Neyman's professional status and financial situation continued unchanged. Pearson seems to have held out some hope in regard to a position at University College; and Neyman, to have communicated such a possibility to the SGGW; for in October the rector sent an "urgent" memorandum to the minister informing him that Poland might lose Dr. Docent Neyman to England. There was, however, no action on the part of the minister to improve Neyman's position. With his fortieth birthday less than half a year away, he was increasingly pessimistic about his future in Poland. He dreamed of going to England as he tried, without much success, to finish his paper on confidence intervals.

"...I do not know whether [or] when it will be finished. I am living for the money from the Institute for Social Problems, which unfortunately requires some work to be done for the money. The amount of this work, together with the lectures etc. take practically all my time. The only means to get time for theoretical work is to declare that I am sick and to stay at home and work. I suppose that before long I shall do so."

He begged Egon to come to Warsaw for Christmas.

"Without this, I feel, there is no hope for my working theoretically this year. We should probably do a lot.... I think there is full evidence that viribus unitis we are able to produce useful things."

In a postscript he added, "You may say to Fisher that he was very lucky in inventing fiducial limits, but his method of approach [to] the problem was less lucky. The question with regard to any number of unknown parameters is equally easy when properly stated. Of course the above sentence is [then he crossed out *is* and replaced it with *may seem*]—may seem rather rude but it is so by accident and in no way on purpose."

At this point, for the first time in the collaboration, there is a surviving letter from Egon Pearson, who in his new position as a department head had a secretary to keep carbon copies of his correspondence.

"Christmas...is an impossibility," he wrote; "the trouble is that while it is a very pleasant change coming to Warsaw, and I want to see you and to get on with interesting work, it provides no holiday. And the work in term is at present so strenuous, that some kind of break from work is necessary, otherwise it would be difficult to carry on." ("I had this horrible feeling that Fisher was above me," he explains to me. "Really terrible.") He did not think he would say anything to Fisher about Neyman's "fiducial" work, "as it is all too vague. But I do hope some time you can tell me your ideas about the method of approach to the problem. I am frequently talking about it in my lectures, though I do not use the idea of fiducial *probability,* only of the limits."

Pearson's decision not to come to Warsaw at Christmas was a great disappointment to Neyman. He says to me of his situation at the end of 1933:

"I was almost crazy."

1978   Neyman has been invited to spend the month of July 1978 in Brazil, lecturing at the Instituto Mathematico Pura et Applicione (IMPA) in Rio de Janeiro and attending the Second Brazilian Symposium on Probability and Mathematical Statistics in São Paulo. He has not been in Brazil since 1961, when he was invited to come and give advice on the establishment of a statistical institute there.

He plans to take Galsworthy's *The White Monkey* with him to read on the airplane. It is still his favorite among all the novels about the Forsyte family, the one in which Fleur has her baby; but he says that he is most attracted by the interaction between "Old Soames," her father, and "Old Mont," her husband's father.

I remind him that fifty years ago he was recommending the book to Egon. Why does he like it so much?

"Because," he says, "rightly or wrongly, it seems to me to be real."

He and Betty Scott leave Berkeley on July 2. The flight is long with a six-hour layover in Lima; but finally they are in Rio, installed by IMPA in a luxurious hotel on the Copacabaña. Barry and Kang James, former graduate students at Berkeley, now at IMPA, come to see them at the hotel that evening. The Jameses are worried about the logistics of the Neyman-Scott visit. IMPA is in an old area of the city which is hard to get to. They don't have a car, but even if they did they would find it impossible to park anywhere near the institute.

"We usually take buses," Barry James explains to me, "but we didn't think it would be a very good idea for Mr. Neyman and Miss Scott to have to ride a bus to our institute. We were originally planning that our group—we have only five statisticians—would meet at the hotel and talk over our program with Mr. Neyman. We wanted his advice on how to set up a statistical laboratory, his advice on courses we should give and the structure of the program, how we should go about working on applied problems, things like that. Then we thought he could go to IMPA just a couple of times and give lectures on his recent research interests.

"Well, he wouldn't have any of that. He didn't want to stay at the hotel. He wanted to go to our institute. He wanted to become a member of the institute, to act as one, become part of our daily work routine, come to the institute when we came, eat lunch at the places where we ate lunch—that kind of thing.

"So what we did is we went to his hotel in the morning, and then we caught a cab—we all just piled in, Kang and I, Miss Scott and Mr. Neyman, and the taxi driver. We arranged an office for him at our institute, and he gave a series of five or six lectures.

"At the beginning anyway he wanted to go to lunch with us. So there's a little restaurant, Gibi—it's named after a comic strip character. It's about two blocks from our institute, but in Rio that can be a long way with the traffic and the crowds. But he went with us, walking, and we were so scared when we crossed this very busy street! We had to wait a long time for a break in the traffic. Then we went out in the middle of the street and motioned to the cars to stop. After that day—he got safely back—we went out and bought food and brought it back to the institute.

"After he had been there for a while, the students wanted to take him out. They wanted to get to know him better. So they invited him, and Miss Scott, too. They went to this old German restaurant—it's kind of a famous hangout in Rio—noisy, but it has some of the best food in the area. So they went there in the afternoon, and they had some beers.

"Then Mr. Neyman wanted to repay the kindness so he invited all the students to a 'hoopla'—that's what he called it—a party at the end of his stay—in the tearoom of our institute. This was about thirty, forty students. It was a very nice party. The only unfortunate thing was that the city had decided to tear up the street that day, and the tearoom is right at the front of the building with just metal bars separating it from the street.

"After he left our institute, he went to São Paulo for the symposium. There were five hundred people. They dedicated the symposium to him. That's a little story, too, because when he got the symposium volume he insisted that everybody present come up and sign it."

All five hundred people?

"Yes!" Barry says.

"It was quite a long line," Kang volunteers.

Neyman had planned to stay in Brazil for the entire month of July, but toward the end of the month he begins to be impatient to get back to his lab. He and Betty Scott leave a little earlier than they had planned. The flight home is broken by an overnight stay in Caracas. One of Neyman's students, the Venezuelan, Luis Perrichi, comes to the airport to meet them and take them to their hotel. He returns again at five in the morning to take them to their plane.

On the flight home Neyman finishes reading *The White Monkey*.

How does he like it this time?

He is thoughtful.

"Somehow—I don't know why—it didn't seem so real this time."

There was yet another reason in December 1933 why Egon Pearson did not feel so urgently as Neyman the necessity for a meeting at Christmas. Gingerly dividing his father's old preserve with Fisher, he was in the process of easing out one member of the little department which he had inherited; and he hoped to be able to replace him with Neyman. Before the end of the year the provost gave permission for the proposed change. Pearson invited Neyman to come to University College for three months that spring as an assistant in the Department of Applied Statistics.

The response from Warsaw was by cable.

"OK HIP HIP HURRA THANKS = JUREK."

The same day that Neyman sent the above cable he also wrote to Pearson, beginning his letter "My dear chap." He had just been "in a wave of pessimism and depression" when Egon's offer had arrived. He understood that his duties at University College would be somewhat similar to Egon's "in the good old days," but it would be easier for him to get permission to turn his lectures over to Karolinka in the middle of the year if the official offer did not mention that he was to be merely an assistant. He was sure that Egon could "formulate the letter in a splendid way as you always do—even when you have to write desagreable things.... You understand that there are troubles with the Ministry etc. As for myself, I should jump at once for as long as I could have money to live on."

Visions of another joint paper in the *Philosophical Transactions* of the Royal Society danced before his eyes!

"Oh, Egon," he wrote, "you do not imagine what your proposal means to me."

Neyman obtained a three-month leave of absence fom the SGGW. He and Lola agreed that she would go with him to England. There may have been a promise on both sides "to try again." Going ahead of her, he took a cheap boat from Gdynia to Hull and arrived in London in time to attend the meeting of the International Statistical Institute, which was celebrating its fiftieth year at the same time that the Royal Statistical Society was celebrating its centenary. The sessions were being held at University College, and it was there that Neyman met Ronald Aylmer Fisher again.

Fisher, forty-four years old, had by that time been elected a member of the ISI. Not a tall man—in fact quite small, shorter than Neyman—he stood out in the crowd, first by virtue of his moustache and beard—a neatly trimmed goatee but a disturbing adornment to Egon Pearson and some of the other cleanshaven English statisticians—second because of the extremely thick glasses which he had worn since childhood—and third because of a vital and somewhat arrogant quality in his manner and bearing. Although Neyman passed his fortieth birthday during the ISI meeting, he definitely thought of himself as a young person in relation to the well established and well known Fisher.

When the term began, in the old building where he had come in 1925 to study with Karl Pearson, he found a tense situation. Fisher and his department were ensconced on the top floor—Egon Pearson and his department on the floor below. The Common Room was carefully shared. Pearson's group had afternoon tea at 4; and at 4:30, when they were safely out of the way, Fisher and his group trooped in. Karl Pearson had withdrawn across the college quadrangle with his young assistant, Florence David. He continued to edit *Biometrika;* but, as far as Miss David remembers, he never again entered his old building.

Americans studying at University College during the 1930's found that famed center of statistics sadly lacking in comfort. "To put it bluntly," one öf them wrote, "I have never labored under such disadvantages." The chairs in the library—which looked harmless—could give him a backache in fifteen minutes. Many a time he wore his overcoat inside "when study without it would have been impossible," and often in class he could not read what was written on the blackboard because the room was so dark. He wondered what a person would do if he ever got really thirsty in the statistics building, since there was no provision for drinking water. Although the calculating machines on Fisher's floor were electric, K.P.'s old handcranked Brunsviga was all that was available downstairs. And so on. But to Neyman, fresh from the discomforts and frustrations of life in Poland, his college accommodations and his and Lola's new living quarters in Bloomsbury were eminently satisfactory.

"Jerzy—to start with—got on quite well with Fisher," Egon Pearson tells me. "It's a very curious thing that he, Jerzy, was trying to bring me and Fisher together. And he took us out—Jerzy took us out to a nearby cinema together. It was most extraordinary, and Fisher thought, 'Oh, Jerzy's here for only a term. All right.'"

The film, Neyman remembers, was *Congress Dances*—its subject, the Congress of Vienna, which had resulted in the central part of Poland's being placed under the Russian czar as king of Poland. After the film the three statisticians shared "a not unreasonable dinner" with Neyman as host. At one point in the course of the meal, Fisher absented himself for a few moments and Pearson said, musingly, "I wonder if now I will be invited to Rothamsted...." (Fisher still lived near the Experimental Station there.) But when Fisher came back to the table, he turned to Neyman and, pointedly ignoring Pearson, said, "Now, next time, you must come to me at Rothamsted."

Neyman had been looking forward to another paper with Egon, but joint work did not begin immediately. Pearson was burdened by his new administrative duties, the tension of coexisting with Fisher, and the "bit of estrangement" that existed between himself and his father. In addition, almost immediately after Neyman's arrival, he became engaged to a young woman named Eileen Jolly.

Neyman was greatly disappointed at the lack of expected companionship. Later Pearson was even to say that he thought Jerzy might not have come if he had known that "I was going to do something silly like get engaged and get married"; but, seriously, he adds, "...it was rather a blow to J.N., but he obviously would have come to England even if he had known what was impending."

The elder Pearson joined the younger in welcoming Neyman and his wife.

"…Karl Pearson and Egon Pearson, and what was the name of Egon's sister [Sigrid]—oh, they all were wonderful people!" Olga Neyman recalls with enthusiasm. "This Karl Pearson invited us for several days to his country place. That was very shortly after we came. They were extremely nice to us, these Pearsons."

(In this case it is not difficult to reconcile Mrs. Neyman's memories of gracious hospitality on the part of Pearson with Neyman's description of him as "hostile and skeptical"; for, as George Udny Yule wrote of K.P., the temper in controversy was the more remarkable because there was no such temper in relation to anything but matters intellectual.)

What was Karl Pearson like? I ask Mrs. Neyman.

"Oh, he was wonderful! Like some old English duke, you know. Oh, a wonderful face! Just pure blood English face, you know. He was very tall, extremely beautiful old man. Just gentleman, very polite and nice. Oh, wonderful, wonderful!"

Was Egon as impressive looking as his father?

"No no no no! But he was awfully nice."

In Neyman's lectures at University College in the spring of 1934, he did not treat hypothesis testing, as Pearson had originally suggested, but instead concentrated on estimation, which increasingly attracted him as a consequence of his work on the representative method for the Institute for Social Problems. The problem of the representative method was, as he saw it, "the problem of statistical estimation par excellence." The statistician was concerned with a situation where it was impractical to look at every member of a population. As a result he had to try to estimate the characteristics which interested him on the basis of a sample. Up until recently it had been generally assumed that the accurate solution of such a problem required the knowledge of certain probabilities a priori; but Fisher had shown in his work how, under certain conditions, what could be described as "rules of behavior" (Neyman's term) could be employed which would lead to results independent of such probabilities. But most unfortunately, in Neyman's opinion, Fisher's work was not really understood by many statisticians, some of whom even questioned the validity of some of his statements. The difficulty seemed to Neyman mainly due to Fisher's very condensed form of explaining his ideas—and perhaps also to his method of attacking the problem. Neyman himself had developed a different approach—one which he thought would get around these obstacles. He was especially pleased that his work on the problem of estimation—like his and Egon's work on the problem of hypothesis testing—seemed to consist mainly in a rigorous justification of what in the past had been considered correct on more or less intuitive grounds.

As Laplace had said, the theory of probability was just good common sense reduced to formulas.

A student who heard Neyman's early English lectures on confidence intervals was P.V. Sukhatme, later a statistical director with the Food and Agriculture Organization (FAO) of the United Nations.

In 1978 Sukhatme is visiting Berkeley, lecturing in the School of Public Health, and I take the opportunity to question him. He speaks especially of Neyman's concern for his students and, as an example, describes the dinner which regularly preceded Neyman's seminar.

"And he called ahead to the restaurant to make sure the dinner would not cost more than 2s. 6p. He had asked the students—I remember he asked me— about how much they had to live on, and so he had decided that 2s. 6p. was all they could afford to pay for dinner. He usually brought a cake. Sometimes the meal would go on for an hour and a half. Then we would all go back to the college for the seminar."

Sukhatme attended lectures by both Neyman and Fisher and found their approaches, from a mathematical point of view, quite different. Neyman emphasized rigor. "It is not so obvious as it looks," he would say in his heavily accented English. He would then take the students through the whole line of reasoning, step by step—a process which sometimes seemed to them unnecessary. Fisher, on the other hand, trusted intuition. On one occasion when Sukhatme was struggling to prove some convergence "with rigor," Fisher came by and inquired about the difficulty. "There is nothing wrong," he assured the young Indian. "N = 1, N = 2, N = 3. Write it up and I'll publish it."

Throughout his term in London, Neyman continued to be on good terms with Fisher; and he was invited, as he recalls, several times to Rothamsted. Fisher also took the lead in proposing Neyman for membership in the International Statistical Institute. A "Proposition de Candidature" furnished me by Dr. E. Lunenberg, the present director of the Permanent Office of the ISI, lists the names of the "Membres proposants" in the following order: 1) R.A. Fisher, 2) Arthur L. Bowley, 3) E.B. Wilson, 4) Maurice Fréchet, and 5) E. Szturm de Sztrem (the director of the Central Office of Statistics in Warsaw). The nomination is dated May 8, 1934.

A little more than a month later, Neyman presented his first paper before the Royal Statistical Society. This was an English version of the pamphlet, already published in Polish, "On the two different aspects of the representative method." It was Neyman's custom, Sukhatme says, to bring his papers to class before a meeting and discuss their contents with his students.

"He prepared us for the future—gave us goals—encouraged us to go to the meetings and participate in the discussions. Some day we would present papers ourselves. Now I have presented three papers to the Royal Statistical Society, and I have always harkened back to Neyman in my mind. He gave us great confidence."

For the English presentation of the paper on representative sampling, Neyman made some changes, adding a little more of the history of the problem and lauding more especially the role of Bowley, whose fundamental work had been the starting point for his own work. As a member of the Royal

Statistical Society, Bowley would be present at the meeting. Neyman also went into a little more detail about his method of confidence intervals—an approach which he considered essentially equivalent to the Fisher approach of fiducial limits but simpler and more direct and hence more practical. Fisher would of course also be at the meeting.

In connection with the papers which it accepted for presentation, the Royal Statistical Society followed a procedure which Neyman very much approved. The secretary circulated the accepted manuscript in advance among members known to have a special interest in the subject so that they could have an opportunity to prepare their comments before the discussion period that would follow the formal reading of the paper. Remarks and the author's reply were then carefully recorded by the secretary—sometimes in the first person and sometimes in the third—and published with the paper itself in the journal of the society. It is therefore possible to reconstruct the meeting of June 19, 1934, in considerable detail.*

Neyman began his talk with a generous tribute to Bowley; and, during the course of the short section which he devoted to an explanation of the confidence interval approach, he also spoke in highly complimentary terms of the work of Fisher. He regretted only that the problems of fiducial limits were considered by Professor Fisher as something like an additional and perhaps minor chapter in the theory of estimation.

"However, I do not agree in this respect with Professor Fisher. I am inclined to think that the importance of his achievements in the two fields [of the theory of estimation and fiducial limits] is in a relation which is inverse to what he thinks himself."

After he had explained his own approach as an alternative to Fisher's, he went on to his main topic: the two different aspects of the representative method.

"By modern standards any one of Neyman's ideas would have been worthy of publication," T.M.H. Smith has commented in his history of survey sampling, "but in one paper he develops a new theory of inference, introduces the ideas of efficiency and of optimal allocation, provides a framework for inferences from cluster samples, and presents a powerful case for rejecting purposive sampling. The only major features of current survey design that he failed to introduce were multi-stage sampling and variable probability sampling, but these followed logically from his work."

The lively discussion which followed Neyman's presentation, however, concentrated almost entirely on the part devoted to confidence intervals rather than on the part devoted to sampling.

Bowley, who had been asked to make the motion of a vote of thanks to the speaker, opened a little apologetically.

---

*In the following passage and those in later chapters which deal with papers presented by Neyman to the Royal Statistical Society, I am indebted to the society for permission to quote from the papers themselves and from the secretary's records of the discussions which followed their presentation.

"After Dr Neyman's very courteous references to my work..., it is somewhat ungrateful that I feel it is my duty to criticize..., and I am very glad Professor Fisher is present, as it is his work that Dr Neyman has accepted and incorporated.... I am not at all sure that the 'confidence' is not a confidence trick."

Egon Pearson, who rose to second the motion of thanks, tried to defend his friend. In the last few years, he pointed out, there had been a determined effort to clear away some of the uncertainty that existed in statistical reasoning. In this process many, among whom he included himself, owed a great deal to R.A. Fisher for the stimulus they had received from "wrestling" with ideas he had put forward—"If I purposely use the word 'wrestling,' Professor Fisher will, I think, take no exception when I add that the stimulus is all the greater because it has been necessary to wrestle." Dr. Neyman had also been stimulated by Professor Fisher's ideas, as he had frankly admitted; but, in Pearson's opinion, he had brought a very real contribution of his own into the field of statistical inference. "This conception of the problem of estimation is not exactly Professor Fisher's conception, but it seems to me that some of the interest lies in just those points where there are differences."

Still another speaker, Isserlis, expressed doubts about the method of confidence intervals. Then Fisher himself rose.

He began by referring to "the luminous account" which Dr. Neyman had given of sampling methods. After some discussion of sampling as it applied to agricultural experimentation in Great Britain, he also went into the subject of confidence intervals. The quoted material which follows is taken directly from the secretary's report of his remarks:

"It would be expected [Dr Fisher said] that he should comment on those applications of inductive logic which constituted so illuminating and refreshing an aspect of the evening's paper....Dr Neyman, as he had explained, differed from Dr Fisher in the relative importance he attached to the two stages in which he had attempted to develop a theory of estimation, independently of all assumptions as to probability *a priori*... This difference was not entirely one of perspective. Dr Fisher's own applications of fiducial probability had been severely and deliberately limited.... Dr Neyman [on the other hand] claimed to have generalized the argument of fiducial probability, and he had every reason to be proud of the line of argument he had developed for its perfect clarity. The generalization was a wide and very handsome one, but it had been erected at considerable expense, and it was perhaps as well to count the cost."

As to Dr. Neyman's thinking that the term fiducial probability had led to misunderstanding—Dr. Neyman must be mistaken: he himself had not come upon any signs of misunderstanding in the literature. He concluded by saying that his criticism "did not affect the value of Dr Neyman's advice on the sampling problem."

The report of the discussion of Neyman's work contrasts with the enthusiasm of future statisticians for the sampling portion of the paper. To Morris Hansen and William G. Madow, writing on "The Historical

Development of Sample Surveys," it is a "pioneering contribution."
William Kruskal and Frederick Mosteller, concentrating on the history of
representative sampling, refer to it always as "the Neyman watershed." It is
in their view "chockablock full of so many important ideas and insights that
we must restrain ourselves to deal only with representativeness." As has been
mentioned, T.M.F. Smith, reviewing "The Foundations of Survey Sampling," discusses the paper at length under the heading of "The Neyman
Revolution."

At the time, Neyman himself was surprised that the sections of his paper
dealing with the new form of the problem of estimation had played such a
large part in the discussion. It was at least gratifying, he said, "that the
criticisms were so divergent that one of the speakers would say that everything
in it is doubtful, and another that it is nothing new." Obviously a separate
publication was needed to clear up the matter.

Egon was quite satisfied. Fisher had on the whole approved of what
Neyman had said. If the impetuous Pole had not been able to make peace
between the second and third floors of University College, he had managed at
least to maintain a friendly foot on each!

1934
—
1935

In the summer of 1934 Pearson was able to offer Neyman an
opportunity to remain at University College the following year
as a regular lecturer. Neyman consulted Antoni Przeborski, his
old friend and teacher. Przeborski, as he was later to recall,
asked only one question about the English offer. Would the salary be enough
for him to live on? Informed that it would be, he said—with emphasis, "All
right! *Accept the job and stay there!*"

Still, Neyman hated the thought of abandoning his Polish "pups."
Perhaps, he suggested to Egon, some of his young collaborators from the
Warsaw laboratories might also come to London—Karolinka could be his
assistant again and study, not only with him, but also with Fisher and
Pearson. This proposal was agreeable to Pearson and was included by him in
the letter formally offering Neyman the position.

Pearson's letter—translated by Neyman into Polish and the translation
certified by a notary as "accurate"—is still in the archives of the SGGW. It
concludes with the statement—here Pearson's English is translated from
Neyman's Polish into Neyman's English—"I consider that it will be
completely appropriate for your Biometric Laboratory, the #2, to be visiting
our Biometric Laboratory, which has been for a long time the #1."

The dean of the School of Horticulture immediately wrote again to the
Minister of Religion and Public Instruction. He reminded the minister of his
earlier uneasiness that Dr. Docent Neyman might go permanently to
England and informed him that the situation was now serious. Dr. Neyman

119

had just been offered a permanent position as well as accommodations for his students in London. Under such circumstances would not the minister "be inclined to consider favorably" a proposal to establish a chair of mathematical statistics for Dr. Neyman in the School of Horticulture?

Again, there was no action on the part of the minister.

Neyman accepted the English offer—"not reluctantly," as he says—but he continued to consider that he was merely on an extended leave of absence from the SGGW. He spent most of the summer of 1934 in Warsaw, working with his former students and giving lectures on mathematical statistics for anyone interested. *Statistica* for 1934–35—and for as long as it continued to be published—carried reprints of all his papers as if he were still a regular worker in the Warsaw laboratories.

That fall—at University College—the new academic year began pleasantly. Neyman continued to receive friendly invitations to Rothamsted, and Egon had happier things to occupy him than his relationship with Fisher:—on the last day of August he and Eileen Jolly had been married in a simple ceremony at the Old School House.

P.C. Tang, who came to University College from China in the fall of 1934, has sent me an account of his first memories of Neyman at that time.

It was a time when "modern statistical theories, methods and applications were in deep, broad development," he writes. "A great part of the work was originated by the professors; and up-to-date, clear written textbooks were almost unavailable. Dr. Neyman, with his strong Polish accent, lectured in the class, walking to and fro, without writing a word on the blackboard. The subjects he talked about were mostly his own ideas. I found it difficult to grasp all that he lectured—even the English-speaking students had more or less the same difficulty. Dr. Neyman often called me to his office and explained to me more in detail."

During the term Neyman moved ahead on a paper, "On the problem of confidence intervals," in which he took up Fisher's remark that his generalization of fiducial probability "was a wide and very handsome one, but it had been erected at considerable expense...." Before he published the paper, Fisher presented a paper of his own on "The Logic of Inductive Inference" at the December 18 meeting of the Royal Statistical Society. Neyman had already left London for Poland, where he would spend Christmas; but, having received a copy of the talk in advance, he commented upon it in writing before he left.

The discussion which followed Fisher's presentation turned out to be somewhat acrimonious, according to the report of the secretary. Bowley opened by expressing his bewilderment at being chosen to speak first and commenting that it was "not the custom, when the Council invites a member to propose a vote of thanks on a paper, to instruct him to bless it"; and bless Fisher's paper, he did not. Isserlis, who followed Bowley, remarked that "Professor Fisher, like other fond parents, may perhaps see in his offspring qualities which to his mind no other children possess; others, however, may consider that the offspring are not unique."

In contrast to these and other comments, Neyman's written remarks were highly complimentary. He opened by sketching the history of mathematical statistics, "born as an independent discipline" with the first papers of Karl Pearson.

"Now the period of 'solving problems' is over. The next period of criticism and laying foundations has been started by R.A. Fisher in his Phil. Trans. paper of 1921. A series of other revolutionary papers followed and today we are discussing one of them."

Often Professor Fisher's papers were especially interesting because of the many hints and questions which the author did not have either the time (or interest perhaps) to follow up. Some readers were probably stimulated to think, "What an interesting problem is raised! How could I develop it further?" Others, like Neyman, couldn't help thinking, "What an interesting way of asking and answering questions, but can't I do it differently?"

Neyman's comments, which were read at the end of the discussion, drew grateful words from the beleaguered Fisher.

"The choice of order in speaking, which puzzles Professor Bowley, seems to me admirably suited to give a cumulative impression of diminishing animosity which I should be glad to see extrapolated," he began. Later he added, "...however true it may be that Professor Bowley is left very much where he was, the quotations show at least that Dr Neyman and myself have not been left in his company."

He devoted only a brief portion of his response to Neyman's suggestion that still another theory of statistical inference might be possible—"a system of mathematical statistics alternative to that of Professor Fisher, and entirely based on the classical theory of probability." It had been, he said, of great interest to him to follow the attempts of Neyman and Pearson to develop a theory independently of some of the concepts he himself had used.

"That wherever unequivocal results have been obtained by both methods they have been identical is, of course, a gratifying confirmation of the hope that we are working along sound lines."

Neyman did not deliver another paper to the Royal Statistical Society until after the beginning of the new year. In his position in London he hoped to reach a wider audience for the collaborative efforts of his Warsaw laboratories, and he brought several such papers back from Poland after Christmas. In February 1935 he presented to the society's newly created Section on Industrial and Agricultural Research a paper which he had prepared with Jadwiga Supińska and Tadeusz Matuszewski. It was designated by its authors as the first part of a work on statistical studies in bacteriology and dealt with the accuracy of the "dilution method"—one of the oldest methods of estimating the concentration of bacteria still alive and developing under given conditions. The paper, an early example of the application of the method of confidence intervals, passed without discussion.

Neyman's next paper, "prepared with cooperation of K. Iwaszkiewicz and St. Kołodziejczyk," was not to be so fortunate. From the report in the *Journal of the Royal Statistical Society* it appears that F.L. Engledow, who chaired the meeting, suspected the tone of the discussion which would follow, for he prefaced his introduction of Neyman's paper with a careful tribute to Fisher, recorded by the secretary essentially as follows:

The title of the paper—"Statistical problems in agricultural experimentation"—had impelled him, he said, to reflect upon the setting such a paper would have had even ten years ago. It had been recognized then, of course, that precision was important in agricultural experiment, but investigators had been content if they could achieve merely some sort of estimate of the reliability of their results. Their attitude towards the kind of estimate had been uncritical. All this had changed with the publication, almost exactly ten years ago, of Professor R.A. Fisher's book *Statistical Methods for Research Workers....* The intervening ten years had not only made clear the scope and power of these methods of Fisher, but had engendered a much bolder spirit of investigation in agricultural circles.

Clearly, the Pole was going to take a stand on ground which the Englishman considered his own.

Neyman opened by paying his compliments to Fisher and Student for their pioneering work in agricultural field trials. He then got down to the subject to which he was to devote the larger part of his paper—what he described as "one of the most important achievements of the English school"—the development of certain randomization designs for testing the effects of agricultural treatments.

Both the method of Randomized Blocks and that of Latin Squares were quite effective in reducing the inevitable random error; and certainly during the early stages of their application it had not been wise to raise questions as to the absolute accuracy of such arrangements, the statistical soundness of which surpassed that of all previous work. By now, however, the new methods were sufficiently well established that it would be useful to discuss more fully the nature of the errors involved in each—and also to compare their efficiency. The method of Randomized Blocks was usually considered as a first step toward an ideal and the method of Latin Squares as a step further in that direction. It was his purpose to show by means of models and numerical examples that such a conclusion was not always justified.

To this end, he set up an explicit mathematical model for the response— something which hadn't been done in the past. Utilizing this model, he showed that, when the method of Randomized Blocks was used, Fisher's test for the hypothesis that the treatments under consideration had the same "true yields" would give reasonably accurate results. When the method of Latin Squares was used, however, Fisher's test could be off in certain ways which would lead one to suspect that it wouldn't always give accurate results. In short, things were a little more complicated than had been thought.

To make his point intuitively clear, Neyman also displayed physical models which had been reproduced for him at University College from others originally built at the SGGW. By means of an arrangement of colored pegs and blocks, the models represented two different situations in regard to soil heterogeneity: one where the soil would be entirely unsuitable for the use of Latin Squares and the other where it would not be a factor that had to be considered.

The discussion of Neyman's paper was opened by Fisher. Entirely ignoring his role as proposer of the vote of thanks to the speaker, Fisher said bluntly—it was his first sentence—that he had hoped Dr. Neyman's paper would be on a subject with which he was fully acquainted and on which he could speak with authority, as in the case of his address to the society the previous summer. He concluded his comments by remarking, "were it not for the persistent efforts which Dr Neyman and Dr Pearson had made to treat what they speak of as problems of estimation, by means merely of tests of significance, he had no doubt that Dr Neyman would not have been in any danger of falling into the series of misunderstandings which his paper revealed."

Frank Yates, who had succeeded Fisher as the director of the Experimental Station at Rothamsted, also contributed critically to the discussion; and when Egon Pearson rose, the mild tone in which he had commented on Fisher's paper on inference in December was gone. While he knew—he began—that there was a widespread belief in Professor Fisher's infallibility, he must beg leave to question the wisdom of accusing a fellow worker of incompetence without, at the same time, showing that one had succeeded in mastering his argument...."

The hour was late, and Neyman promised only a few remarks and a written response later. Fisher interrupted him twice and, according to the published report of the discussion, managed to have the last word.

Some thirty years after the meeting described above, Oscar Kempthorne, an American statistician who has been particularly active in the continuing development of explicit randomization models, commented:

"The allusion to agriculture is quite unnecessary and the discussion is relevant to experimentation in any field of human enquiry. The discussion section...is interesting because of the remarks of R.A. Fisher which are informative in some respects but in other respects exhibit Fisher at his very worst...The judgment of the future will be, I believe, that Neyman's views were in the correct direction."

In the days following the meeting, Neyman worked hard on his written response, which was to run to four large pages of fine print in the published discussion. He began by saying that he was grateful to all those who had honored his paper by discussing it. "Whatever the tone and the intentions of the discussion, it always helps to clarify the position." He then took up Fisher's criticism. He wanted to thank Fisher for a sentence in the third part of his contribution to the discussion—and he quoted him: "'I suggest that before criticizing previous work it is always wise to give enough study to the subject to understand its purpose.'" He did not think that this sentence of Professor Fisher's could apply to him. He had looked carefully through his paper for any criticism of relevant previous work. No. The sentence he had quoted applied rather to its author, Professor Fisher himself.

"[He] not only criticized my paper, but blamed me for a variety of sins of which I am not guilty—all this before apparently taking the trouble to discover what [my] paper is about and what are the results. According to him: I was unwise in the choice of my topics, I have been speaking of things with which I am not fully acquainted, I deceived myself on so simple a question, I forgot the meaning of facts, I confuse the questions of estimation and the tests of significance...."

This last criticism was particularly hard to understand, since only last year, discussing the paper on the representative method, Fisher had blamed him and Dr. Pearson "for thinking and writing (which we do) that the problems of estimation are *different* from those of testing hypotheses!"

It seemed to Neyman that Fisher's remarks had been based on two mistaken assumptions: "(1) that I tried to understand his intentions when he was creating the *z* test, etc., and (2) that I considered the hypothesis that 'differences of treatment make no difference to the yields.' Instead of guessing the desires of Professor Fisher, I was interested in the problems of agricultural experimentation as I understand them and in the adequacy of the methods of their solution in frequent use."

It took a long time for Neyman to write his response, not only to Fisher but also to the others who had spoken. By the time he finished, he felt that some revision of his opening statement was in order.

"I started my reply with a welcome to any kind of discussion. Yet after this long journey I feel that it is a pity that so much valuable time should be given to unfriendly criticism, based on misstatements and errors, and, subsequently, to the necessary corrections."

After the Royal Statistical Society's meeting of March 28, relations between workers on the two floors of K.P.'s old preserve became openly hostile. One evening, late that spring, Neyman and Pearson returned to their department after dinner to do some work. Entering, they were startled to find strewn on the floor the wooden models which Neyman had used to illustrate his talk on the relative advantages of Randomized Blocks and Latin Squares. These were regularly kept in a cupboard in the laboratory. Both Neyman and Pearson have always believed that the models were removed by Fisher in a fit of anger.

1935
—
1936

The *London Times* of May 13, 1935, carried two items of special interest to Neyman.

On the first page it reported the death of Marshal Joseph Pilsudski. Just nine years before, Pilsudski had entered Warsaw at the head of his troops and had taken over the government. During the years since then, he had ruled Poland behind a somewhat disguised dictatorship and, recognizing that no other European nation was prepared to oppose the new Nazi government in Germany, had come to terms with his neighbor to the west. What would happen in Poland now that Pilsudski was gone?

On a back page, in accordance with the regulations of the University of London, the *Times* carried an advertisement: "UNIVERSITY OF LONDON.—The Senate invites applications for the University Readership in Statistics tenable at University College. Salary £500 a year...."

A readership was a tenured position, the salary one hundred and fifty or maybe even two hundred pounds more than Neyman was receiving as a lecturer. His family responsibilities would soon be increasing with a baby expected at the end of January. His prospects in Poland were dim. Only the year before, he had provided data showing the extent to which the Roman Catholic church controlled the wealth of the country; and Bassalik—the former director of the National Agricultural Institute at Bydgoszcz—had been very angry and had told him that as a result he would never become a professor in Poland! Yet he hated to give up even the fiction that he was only on leave from Warsaw and his little labs. He did not immediately submit his application for the position of reader.

Between May 13—the date of the advertisement in the *Times*— and May 31—the deadline for applications—he was again involved in a skirmish with Fisher. The controversy, as before, was over the application of statistical theory to agricultural experimentation.

On May 23 Frank Yates presented a paper on "Complex Experiments" (known today as factorial designs) to the Section for Industrial and Agricultural Research of the Royal Statistical Society. Such experiments, a creation of Fisher, provide an economical and convenient method for investigating the main effects and the interactions of a number of variables.

Neyman opened his remarks on Yates's paper in a complimentary vein, pointing out that he had expressed on several previous occasions his "high appreciation" of the work done at Rothamsted by Fisher and his colleagues. Indeed it was his opinion that Fisher's concept of the complex experiment was likely to become a powerful tool of agricultural experimentation. He was nevertheless inclined to think "that before trusting it so entirely as Mr Yates and many of the other speakers have done, it is desirable to produce some further evidence as to its validity." He then proceeded to take up an instance where the interactions under consideration, although substantial, were not likely to be found significant because of insufficient replication of the experiment.

"It may be argued," he conceded, "that my example is too artificial and that in reality we shall never find a similar system of interactions. This, of course, is a possible point of view and one commonly held. But even if the experimenter is inclined to believe in the absence of [such] troublesome interactions and thus in the soundness of the method of complex experiments, I think it will be useful for him to realize that the method is based on a belief. Further, he should recognize what curious answers may be given if the number of replicates is small and if Nature chooses to be frivolous...."

(In the intervening years, this point of Neyman's has been generally ignored by practicing statisticians who make use of factorial designs. Robert Traxler wrote in 1976, "...the present writer was not able to find a single book or research paper offering an interaction oriented methodology beyond the traditional one, found to be inefficient 40 years ago." Yet during this period, in pharmacology, medicine and biology, increasing attention had been paid to a particular type of interaction—the "synergism" which occurs when two or more agents combine to produce an effect greater than the sum of their separate effects. "Adaptation of the design and novel methodology, in some intelligible sense 'optimal,' remains a challenging problem of mathematical statistics.")

One week after Neyman criticized Fisher's concept of the complex experiment and at the last permissible moment, he submitted his application for the advertised readership in statistics at University College. He gave the names of Karl Pearson and George Udny Yule as references. Many years later he is not sure of the details of the administrative process involved in the appointment; but he does remember vividly that late in June, on his way to the meeting of the committee which would pass on the appointment, Fisher stopped by his room at the college.

"And he said to me that he and I are in the same building and he is going to this meeting. And so he said that, as I know, he had published a book—and that's *Statistical Methods for Research Workers*—and he is upstairs from me so he knows something about my lectures—that from time to time I mention his ideas, this and that—and that this would be quite appropriate if I were not here in the college but, say, in California—(Neyman gives a little chuckle: "I must say that was kind of prophetic.")—but if I am going to be at University College, then this is not acceptable to him. And then I said, 'Do you mean that if I am here, I should just lecture using your book?' And then he gave an affirmative answer. Yes, that's what he expected. And I said, 'Sorry, no. I cannot promise that.' And then he said, 'Well, if so, then from now on I shall oppose you in all my capacities.' And then he enumerated—member of the Royal Society and so forth. There were quite a few. Then he left. Banged the door."

At the meeting which followed (as Neyman later gathered from Egon), Fisher objected to the proposed appointment but was overruled by the others present. At the same meeting, Egon was promoted from reader to professor. He was now at last of equal academic rank with Fisher.

On July 17, 1935, a couple of weeks after Neyman was made a reader, Fisher wrote to H.W. Methorist of the International Statistical Institute:

"...perhaps you will tell me whether Dr Neyman's name will automatically re-appear as a candidate [for membership in the ISI] at future elections, or whether he would need to be again proposed and seconded...."

He explained to Methorist that he no longer desired to propose Neyman: "...since Dr Neyman has now 'become domiciled' in England it is important that Poland should be represented by a mathematical statistician resident in that Country rather than by a non-resident."

Methorist replied that Neyman's name would be dropped unless he were renominated.*

*Neyman was not elected to membership in the ISI until 1948.

126

Although the readership at University College was in fact a permanent position, Neyman did not feel that he had committed himself permanently to England.

"I would like to work in Poland in the department organized as a result of my twelve years of effort," he reiterated to the rector of the SGGW. His new English post was equivalent to "second to best in the Polish system," but—according to his contract—he could resign at any time if he informed the college authorities before the end of April. "If an appropriate post at the SGGW or any other institution in Poland is offered me, I'll accept...."

That summer, however, he spent most of his time in England rather than in Poland. The daughter of the novelist Rider Haggard had lent Egon and Eileen a little house in the country, and the newlyweds had invited the Neymans to come and stay with them. The women sketched, and the men started to write a book together on their new approach to statistics. Pearson has a number of manuscripts, some in his handwriting and some in Neyman's; and on what he calls "a containing sheet" he has written— "Papers as left 29 August 1935."

Neyman was "very keen" on the idea of a book or journal publicizing their ideas. Already in England he had had difficulties, or had thought that he would have difficulties, in getting papers published. He had sent his paper "On the problem of confidence intervals" to the *Annals of Mathematical Statistics* in America because he had felt certain that K.P. would not accept it for *Biometrika* and that R.A.F. would be able to prevent its publication anywhere else. Asked whether Fisher really had that much power (for he had absolute discretion only over the *Annals of Eugenics*, published by the Galton Laboratory), Neyman replies, "Maybe not. But, rightly or wrongly, that's the way I felt." He saw English mathematical statistics as controlled by two "publishing empires"—an "establishment" that was trying to keep him down as (it was said) Pearson had tried to keep down Fisher.

He tells me that K.P. had even been hesitant about publishing the work of young Kołodziejczyk on the linear hypothesis (later recognized as a landmark paper) and had agreed finally only because, as he said to Neyman, "you and Mr. Pearson think it is important." During the summer of 1935 he had rejected a paper on which Neyman had collaborated with Palmer O. Johnson, a professor of education from the University of Minnesota, although he had offered to publish Johnson's non-theoretical part.

"We shall not, I know, agree about the theoretical method of approach to the solution of such problems, but I am sure we may agree to differ," Neyman had responded. "Johnson's problem and my solution are, however, closely connected together and, as I do not think he would wish his work published separately, the best course will be to arrange for publication elsewhere."

But where?

By the time the new academic year began at University College, the unpleasant relations between "upstairs" and "downstairs" had become a scandal, knowledge of which had spread even to the United States. On the rare occasions when the two groups happened to be having their tea in the Common Room at the same time, Fisher contrived to be surrounded by his

127

disciples and Neyman and Pearson by theirs. In lectures there was a certain amount of sniping back and forth between Fisher and Neyman. According to Churchill Eisenhart, an American who came to the college as a doctoral student in the fall of 1935, through some informer whom nobody was ever able to identify, each man seemed to know what the other had said in his lectures and would then try to discredit it in a way which seemed to Eisenhart unproductive and very confusing to students who were not scientifically sophisticated enough to understand what was going on—"cross criticism, at cross purposes," as Eisenhart was to write of the earlier controversy between R.A.F. and K.P. over the relative merits of maximum likelihood and the method of moments.

At this point in Neyman's career he usually lectured in the standard fashion although, more often than most lecturers, he encouraged students to ask questions. These often resulted in lengthy digressions. A question by Pytkowski in Warsaw had been the stimulus for his lecturing on confidence intervals, and a suggestion by Eisenhart in London also made a contribution to the theory.

What happened, Eisenhart explains, was the following:

"He gave the original development with all the a priori distributions in it, and it washed out. So I figured that if it washed out, you must be able to do without it. So I asked him whether he would come after tea to a little seminar at which I was going to try to give an alternative development. So I tried, but it was wrong. And Neyman sat there looking at it and looking at it and looking at it. And finally he said, 'It's wrong and it's wrong here, but you're on the right track.' Then about a day or two later he came into class, very enthusiastic, and said, 'I've done it!'"

(Although Neyman did not mention Eisenhart's contribution in his published paper, he has, as Eisenhart reminds me, mentioned it in print and in talks on a number of occasions. "He will even bring it up in a case of statistics and the environment or some such thing—I'm just inventing, you know—but he will point out how difficult it would be to calculate the standard error, but you can calculate the confidence intervals easily and then transform them—and, by the way—and then he brings in me. Now when I see him, I tell him, 'Jerzy, I appreciate your saying this about me, but I think you're overdoing it.' But he loves to give credit.")

Eisenhart also recalls how he ran afoul of Neyman on occasion. Once he was asked by him to coach a student for her presentation to the seminar. He rehearsed her a number of times and was convinced that she was well prepared. On the appointed day, when she was about a third of the way through her talk, Neyman began to pepper her with questions.

"Please sit down, Professor Neyman," said the young American, who was the son of a distinguished Princeton dean and not so intimidated as most students by academic rank. "Let her finish her presentation. Then you can ask questions."

"Well, I'm telling you, he just blew his top. He just got enraged, you know. It wasn't the custom on the continent for a student to tell a professor to sit down.

"Afterwards, Florence David [K.P.'s assistant] buttonholed him in his office and told him that he was wrong and I was right—the lady should have been permitted to make her presentation before he began to ask questions. Florence David wasn't afraid of him at all!"

Another student in the fall of 1935 was Yoong Tang, who had come to London a year after her husband. "When her father died in China, she was so sad that she wanted to give up her studies and hurry back home," he tells me. "Dr. Neyman consoled her and encouraged her to stay on. Later, when my thesis on Power Function was nearly in shape, Dr. Neyman advised her to investigate the possibility of applying my resulting tables to plant breeding problems. When she got the materials and results, Dr. Neyman helped her to organize and put the paper in order.... Aside from the study, Dr. Neyman talked to [both of] us about the important points in dealing with western culture and customs, which are greatly different from those of the Oriental. Sometimes, we went out together for social gatherings. Frequently, we felt that we were taking up too much time out of his precious, highly productive period during our stay in London."

Although the Tangs, struggling to cope with a new language, had considerable difficulty in understanding Neyman's English, Sukhatme found it easier to understand than the English of many of the English professors. Eisenhart, reared in multilingual Princeton, scarcely noticed the accent—he was more concerned, as notetaker in several courses, that Neyman changed his notation from week to week. Less cosmopolitan Americans, however, found the accent "the most interesting thing" when they first heard Neyman lecture, writes Norman L. Johnson of the University of North Carolina, who came to University College the year after Eisenhart. "[But] we soon...forgot [the accent] in the excitement of becoming acquainted with new and still developing fundamental ideas on the very bases of statistical inference, only moderated by the ever-present threat of a sudden invitation to demonstrate some result to the class in person."

One day, to Neyman's surprise, he received an invitation to take tea with the provost.

"We talked and the tea was not unreasonable, but then suddenly I see that he feels uncomfortable. So then I said, 'Provost, I see that you have something unpleasant to tell me. What is it?' Then he said that some people had complained to him about my English. 'This conversation has been most pleasant,' he says to me, 'but I must say that I think they have reason!'"

The result was a short course in phonetics during which, as Neyman recalls with satisfaction, he learned the difference in pronunciation between *bitch* and *beach*.

During the fall of 1935, he continued to work on the project he had first publicly proposed in his discussion of Fisher's talk on inference in December 1934—the possibility of formulating the problem of estimation in such a way that its solution would stand entirely on the ground of the classical theory of probability. He was also working on a further contribution to the theory of hypothesis testing, this one dealing with unbiased tests. Although he was doing by far the larger share of the work, he intended to publish the paper with Pearson.

In spite of his improved professional standing and the increase in his salary, he was not completely happy with his circumstances. When, a month before the baby was due, he saw another advertisement by the University of London in the *Times*—this one for the University Chair in Statistics, tenable at the London School of Economics—he immediately applied, giving Gosset, Egon, and Georges Darmois of the Institut Henri Poincaré as references. The position was most alluring, paying double that which he was currently receiving. Egon understood his feelings; but, as he wrote in his New Year's note, "I can't really wish that you should go to the London School of Economics, though I can heartily wish you had £1000 a year.... I don't think you will get the same chance of developing the work you are interested in, and from what I know of being head of a department, it involves ties and problems which will absorb a good deal of your time. So that the development of 'our' school of thought will suffer."

At the time Pearson wrote this note he had just received the manuscript of Neyman's paper on unbiased critical regions of Type A and Type A_1, and it disturbed him somewhat. He and Neyman seemed now to be up against one of those difficulties which occur when a large portion of "joint" work is not really joint.

"Your attitude, I think, is that in order to get out Student's test rigourously we must establish all these general conditions... however long. My attitude is: if all this is required to show that one little test is Type A, is it really worth hurrying into print, when what seems really important is not whether the test is unbiased (Type A), but 'best unbiased'....

"I am sorry to be troublesome, but if joint work is to mean free play of individual views, it is quite certain that you must have my opinion frankly; otherwise it only means that I am suppressing my thoughts, and underneath getting less & less satisfaction out of co-operative work."

Neyman had just learned of the rejection of another paper, "Sufficient statistics and uniformly most powerful tests of hypotheses," which he and Pearson had submitted to the Royal Society through a friendly professor in the Department of Pure Mathematics. Although he was in Paris, delivering a series of lectures at the Institut Henri Poincaré, he took the time to whip off ten foolscap pages replying to the referees' criticisms. It was no good, he insisted to Pearson, "to wind the matter in cotton."

"I cannot remember what happened next," Pearson says, "but very probably it was agreed... that it was best to accept the Referees' Report. This must have been the critical moment when Jerzy and I decided to look into the possibility of printing a Department Journal."

The framework for such a journal was already at University College; and the name which Neyman and Pearson selected—*Statistical Research Memoirs*—continued a tradition established by K.P. in 1904 with the first *Drapers' Company Research Memoir.** Subsequent *Memoirs*, edited by K.P., had continued to appear for some thirty years and had covered a wide variety of subjects, ranging from theoretical topics to applications in biology, medicine, anthropology, and eugenics. There was, however, to be a difference in character between the new series and the old one. This would be partly due to the breaking up of the unit which K.P. had headed, but also to other factors.

"It is widely felt," the new editors wrote in the foreword to what was to be the first volume, "that in spite of the existence of a large number of special problems for which perfect solutions exist, statistical theory in general in its present state is far from being completely satisfactory from the point of view of its accuracy." It was their ambition that their new journal would "contribute towards the establishment of a theory of statistics on a level of accuracy which is usual in other branches of mathematics."

They began enthusiastically to gather contributions. Ultimately the first volume included the paper rejected by the referees of the Royal Society and also the paper on unbiased critical regions of Type A and Type A_1, which Neyman had been able to generalize as Pearson had hoped that he would. Also included was the Johnson-Neyman paper on the application of statistical theory to educational problems, which K.P. had rejected. In addition, there were papers by several students in the Department of Applied Statistics: P.P.N. Nayer, P.V. Sukhatme, and Robert W.B. Jackson. These efforts should be more fruitful, the editors explained, if they were published mainly in one place rather than spread over a number of journals. "In this way it will be easier for those interested...to find the work and to judge, as a whole, a number of investigations linked by a common purpose and often following a common approach."

The first volume of *Statistical Research Memoirs* had not yet appeared when, on April 27, 1936, Karl Pearson, whose editorial intransigence had been at least partially responsible for its creation, died at the age of seventy-nine.

Speak only good of the dead.

Karl Pearson had been the "founder of mathematical statistics, originator of its various applications, teacher and inspirer of innumerable research workers of many nations and races," the two editors wrote, dedicating their new journal "To the memory of our Professor...."

At his funeral the lines by Robert Browning beginning, "This man decided not to Live but Know—," which he had at times applied to others, were applied to him:

*From 1903 to 1932 Karl Pearson's department received a series of generous grants for unspecified activities from the Worshipful Company of Drapers, one of the old chartered companies of the City of London.

Lofty designs must close in like effects:
  Loftily lying,
Leave him—still loftier than the world suspects,
  Living and dying.

It was "A Grammarian's Funeral." Sometime afterwards it was suggested to Egon that it would be appropriate to reissue *The Grammar of Science* as a memorial to "K.P."

1978 Neyman loves to play the host, whether at a banquet for the hundreds who come to the Berkeley symposia or sharing lunch with me on Saturday at the Moveable Feast, a combination deli and sidewalk café, just down the street and around the corner from his house.

There are two regular social events in his week. One is the "little drink" for a miscellaneous group, including the speaker and some of the students, at the Faculty Club after his Wednesday seminar. He feels that the students are "a bit shy" and that in a social situation they will relax and ask questions. If the speaker is a visitor to the department or someone from out of town, Neyman also takes everybody to dinner. No one can ever get the check away from him.

The other regular social event is the midday Sunday dinner at his house, which Betty Scott prepares. A frequent guest, in fact almost a regular when she isn't at the cottage she still maintains in the south of England, is F.N. David.

"F.N.—Florence Nightingale," Neyman explains, adding that some parents should be shot. "I call her 'Her Grace.' She is retired now and a little lonely so I telephone every Sunday morning at seven o'clock—she is up, too—and tell her what we are going to have for dinner. Then if she likes it, she comes."

I tell Neyman that I would like to talk to Professor David. I hope that, being English, she will be more of a link to the English statistical past than Neyman himself.

"Do come and have dinner with us then," he urges.

On the following Sunday I am introduced to a woman of seventy with still fresh English skin, keen eyes, and an English voice. Her hair is cropped very short. She wears a jacket and trousers and smokes a slim cigar.

Betty Scott calls her F.N.D. How should I address her? "Professor David" seems too formal in Neyman's house. She says she prefers David—"which is how they address you in English universities."

Over drinks, I am unsuccessful in leading her into conversation about University College; later, however, when I telephone and ask for an appointment, she agrees to talk to me.

The first thing she says when we meet in her office in Evans Hall is that she does not approve of what I am doing. She is writing a book about statistics and statisticians, but she will not publish it until everybody concerned is dead. I explain that I am trying to take advantage of the opportunity to learn from Neyman himself about his life. I simply want to get from her something of the flavor of his London days, not to talk to her about him.

I begin by asking about her own career. Her father was a classicist. Since he thought there shouldn't be two classicists in one family, she "did mathematics" at a women's college. To escape the female's then inevitable teaching career, she tried to get into actuarial work; but a job offer was withdrawn when it was discovered that she was a woman—she had applied as F.N. David. She then went to University College and obtained a position in Karl Pearson's laboratory as a research assistant. That was in 1931. Two years later, when he retired, she moved with him over to Zoology "as far away as he could get from statistics."

What was her youthful reaction to Karl Pearson?

"Difficult to say. Probably I was fond of him. Sometimes he was really grumpy. Rather pigheaded. But O.K. if you spoke up to him. For two years I had his undivided attention. They were the hardest two years I ever spent. *Biometrika* papers would come in, and he would not quite see the point so he would rewrite them. I would do the mathematics. I was a little better at that than he was.

"Finally [this was in 1935] he more or less pushed me out. Didn't tell me to go or anything. I came to the Statistics Department as an instructor. Most of the time I was babysitting for Neyman, explaining to the students what the hell he was up to.

"I did my stuff as far as university regs were concerned and decided I didn't need a doctorate. Then one day Neyman said, 'Miss David, you will have to get your Ph.D.' 'Well, hell, I don't want a Ph.D. It costs twenty pounds.' But I got one anyway. Neyman was my 'internal examiner.' The 'external examiner' was A.C. Aitken.

"I saw the lot of them. Went flyfishing with Gosset. A nice man. Went to Fisher's seminars with Cochran and that gang. Endured K.P. Spent three years with Neyman. Then I was on Egon Pearson's faculty for years.

"Fisher was very vague. Karl Pearson was vague. Egon Pearson vague. Neyman vague. Fisher and Neyman were fiery. Silly! Egon Pearson was on the outside.

"They were all jealous of one another, afraid somebody would get ahead. Gosset didn't have a jealous bone in his body. He asked the question. Egon Pearson to a certain extent phrased the question which Gosset had asked in statistical parlance. Neyman solved the problem mathematically."

I say that many people seem to think that the solution was more logical than mathematical.

"Well, isn't mathematics just logic?"

As I leave, she says, "He won't like what you write, you know. He never does. I've seen it happen too many times."

1936
—
1937

In the summer of 1936 Neyman and Pearson were working completely apart. Egon, taking with him a huge hamper of papers, had withdrawn to Halsway Combe to compose a survey of his father's life and work for *Biometrika*. Neyman, staying in town, was bringing to conclusion an extensive paper developing a theory of estimation which, unlike Fisher's fiducial theory, would be based entirely on the classical theory of probability.

Neyman worked with passionate intensity, eager to get the paper into print. No longer did he feel that he would be hindered in publishing in England. With K.P. gone, the trustees of *Biometrika* had appointed Egon editor of that journal; and he and Egon agreed that in the future things would be different in regard to publication there.

Toward the end of the summer, Pearson, who was being kept informed by Neyman about the progress of his paper, began to be disturbed about its length. The paper also seemed maybe somewhat more mathematical than it should be for the readers of *Biometrika*.

There were—Neyman explained in his opening paragraph—two situations in which a practicing statistician finds himself faced with the problem of estimating some unknown parameter. In one, he is dealing with a population which he can study only by means of a sample. In the other, he is dealing with a number of experiments which, when repeated under apparently identical conditions, give varying results.

"It will be noticed that the problem in its practical aspect is not a mathematical problem," he pointed out, "and before attempting any mathematical solution we must substitute for it another problem having a mathematical sense and such that, for practical purposes, it may be considered equivalent."

This was the problem which he formulated and then solved by using the "confidence intervals" which he had been pondering since 1930.

The idea of an interval which has a given probability of containing a certain unknown value was not, in itself, a new one. Confidence intervals, or limits, had been utilized—by Laplace, for instance—but without exploring the underlying concept. Important work had recently been done by an American, Harold Hotelling. Neyman's approach, however, was entirely original and very characteristic. As in the theory of hypothesis testing, it was "the best"—optimality—which he sought. In this situation he proceeded to formulate as "the best" the confidence interval that has the smallest probability of containing any false values. He was then able to take all the concepts and results from the theory of best tests (unbiasedness, uniformly most powerful, and so on) and translate them into corresponding ones in the theory of estimation. It was a formidable unification.

"[I] think a lot of [your paper] is extraordinarily good," Egon assured him when he received the manuscript in the middle of September; "some of the initial statement I don't find as clear as it is meant to be, but the whole conception of the approach pleases me very much."

134

He had some qualms about publishing it in *Biometrika,* however.

"...it is not quite what I had suggested..., partly because it is too long—when I said 20 pages I meant it—partly because what I had in mind was something simpler which would have appealed to the practical statistician."

Five days later he commented again in another letter.

"I think your development of the theory of estimation belongs to the category of things that are fundamental & lasting, & so if I don't do what your impetuous spirit (with its longing to get things off your mind and into print) desires, you will know that it is not from any lack of appreciation of the work....

"The trouble is that I have in mind a way of explaining your ideas to practical statisticians which will make them say, 'Oh this stuff of Neyman's is much better than Fisher's,' but then I expect you to write in the peculiar way I want, which naturally you don't."

Six days later, writing again, Pearson was still taking the publication of Neyman's paper more or less for granted, although maybe not in the upcoming issue. On Saturday he would be back in town—"...we will see each other and can discuss more." But by Saturday he had reverted to his original opinion: Neyman's lengthy and mathematical paper was simply not appropriate for *Biometrika.* He, Egon, was going to have to start on an editor's "career of inevitable unpopularity"—and with his close friend and longtime collaborator. He packaged the manuscript and took it with him to Neyman's house.

The paper which Pearson was rejecting was one for which Neyman felt a deep emotional attachment. It had cost him much thought and effort, and it seemed more completely his than any of his other, earlier work. He has always placed it even above the joint work on hypothesis testing (although it is doubtful whether most mathematical statisticians would agree with him). A note of bitterness comes into his voice, forty years later, when he describes Egon's arrival on Saturday morning at Brentmead Place with the package containing the rejected manuscript. It was, in his words, the beginning of a "divergence" between them.

It seemed clear that there were still two publishing "empires" in England. Egon had closed off *Biometrika* as arbitrarily as K.P. had been accustomed to do in the past, and Fisher was so powerful that he would be able to prevent publication of the paper in any other important journal.

Neyman felt that he needed advice, and he decided to go to Cambridge and talk to George Udny Yule, one of the first English statisticians with whose work he had become familiar even while still in Poland.

Yule was a curious choice, for—as Neyman later wrote—"Yule's own attitude toward mathematical statistics was distinctly nonmathematical...." In fact, five years earlier, he had resigned his readership at Cambridge on the grounds that more and more mathematicians were coming into statistics and he did not feel competent mathematically to give them what they wanted and needed. Neyman, however, had been much interested by some pioneering work of Yule's on accident proneness. He also had—and still has—a great deal of affectionate admiration for a man "who thought he was too old to teach statistics but young enough to buy an airplane and learn to fly."

Yule encouraged him "to try" with his new theory of estimation. "If not this place, then that place. Eventually I would find." So he decided to *try* the most prestigious scientific journal in England—again the *Philosophical Transactions* of the Royal Society. Expecting that his paper would meet opposition from Fisher, he gave considerable thought to the appropriate fellow to present it. At length he decided upon the leading Bayesian, Harold Jeffreys.

"There was a dispute between him and Fisher," he reminds me, "so I thought he might—and he did."

Unlike Fisher, who dismissed Jeffreys's *Theory of Probability* with the comment that on the first page the author had made a logical mistake (the adoption of the Bayes postulate) which invalidated all the formulas in the book, Neyman wrote tactfully, in a footnote to his paper, that the theory of probability in which one chooses to work is a matter of personal taste.

"It may be useful to point out that although we are frequently witnessing controversies in which authors try to defend one or another system of the theory of probability as the only legitimate [one], I am of the opinion that several such theories may be and actually are legitimate, in spite of their occasionally contradicting one another. Each of these theories is based on some system of postulates, and so long as the postulates forming one particular system do not contradict each other and are sufficient to construct a theory, this is as legitimate as any other."

Many years later, Jeffreys was to tell Fisher that the two of them were closer to each other in their view of estimation than they were to Neyman; but in 1936, approached by Neyman, he agreed to present the paper on the new theory of estimation. The secretary of the Royal Society received the manuscript on November 20 and passed it on, as was customary, to a pair of unnamed referees. Neyman was positive that one of these would be Fisher, who had just returned to University College after an extensive lecture tour in America.

This tour of Fisher's had been arranged by S.S. Wilks of Princeton, who had studied in London a few years before. During the course of it, Fisher had delivered a three-day series of lectures at the Graduate School of the United States Department of Agriculture in Washington, D.C. These in turn had been set up by W.E. Deming, who had visited University College in the spring of 1936. The Fisher lectures at the Graduate School had been such a success that when, shortly afterwards, Deming heard that Wilks was trying to arrange a series of lectures for Neyman at various American colleges, he assured him that if Neyman did come to America the Graduate School would like to have him spend a week in Washington, giving a series of formal lectures similar to Fisher's and also participating in smaller "conferences" for more advanced people.

During the early winter of 1936–37, as the paper on the new theory of estimation was being read by the referees of the Royal Society, Neyman received the invitation to visit America. Wilks had lined up eleven colleges and universities willing to pay $50 a lecture, and Deming had arranged a week in Washington. According to Egon, the provost was getting a little "fed up" with people running off to America; and Egon himself was of the opinion that the Americans "rather overdo themselves with conferences & things—at least, for all the excitement they make, they ought to be the most scientifically trained people in the world." But Neyman was thrilled at the prospect of crossing the ocean and also of making some extra money in America. "Hard up" was becoming a rather permanent state with him since the birth of baby Michael in January 1936.

(Sometimes even Egon bristled at his borrowing: "You are rather a mutt getting caught out in this way…," he scolded on one occasion. "I know it is difficult for you, but I do think that you ought to make out some kind of an estimate of how much your official salary is short of your expenses. I do not mind lending, but it would make it more satisfactory if one knew a little in advance, and so felt confidence that you were not just living from hand to mouth, always hoping for the best!")

Neyman sent Wilks and Deming a list of suggested topics, "containing mainly items which are not yet published." He was eager to learn the needs, tastes, and abilities of the different American audiences he would have.

"When arranging, please have in mind the following principles," he wrote to Deming: "(a) I am very hard up and therefore I should like to earn as much as possible.…(b) Once a lecture is arranged anywhere and the topic is chosen, I should like my audience to understand properly what I have to say and therefore to give a sufficient number of lectures, independently of the fee granted."

He repeated to Wilks, "…even if an institution is prepared to pay for one lecture only, if there is interest, I would prefer to lecture more than once (for the same pay) so as to be able to get the audience to understand and be satisfied."

The six weeks which Neyman spent in America in the spring of 1937 were for him a kaleidoscopic jumble of strange sights and sounds, new impressions, and an exhilarating enthusiasm for what he had to say. Helen Walker showed him around Manhattan, taking him up to Harlem and telling him that he would not know America until he knew the American Negro. Wilks escorted him to Princeton where, strolling one afternoon, they came upon Einstein, who had already left Hitler's Germany for a position at the newly created Institute for Advanced Study. At the University of Chicago there was a reunion with Otto Struve, his old friend of Kharkov days, now a professor of astronomy and the director of that university's observatory. At Ann Arbor he received an offer of an associate professorship at the University of Michigan from T.H. Hildebrandt. When he refused, Hildebrandt countered by saying that he might be able to offer a full professorship at a later time. At

Illinois there was a young Negro in the audience who listened very attentively. This was David Blackwell, the president of the mathematics club at the university.

Professionally speaking, the most significant portion of Neyman's visit to the United States was the time he spent at the Graduate School of the Department of Agriculture. It is a happy circumstance that the Americans' reaction to his lectures is reported in detail in a letter which Deming wrote, immediately after Neyman left Washington, to his friend Raymond T. Birge, the chairman of the physics department at the University of California at Berkeley.

He could talk a week about Dr. Neyman's visit, Deming told Birge. Neyman had given three lectures on three successive days, the audience for each numbering over two hundred. He had also conducted six "conferences." These had been planned to give experts the opportunity to question him on specific topics, the attendance to be limited to only about forty people; but it had been "hard to exclude anyone who really [wanted] to come & interest ran so high that the conferences turned out to be additional public lectures." The attendance was invariably over a hundred.

To Deming his guest (who was also his houseguest) seemed to possess all the marks of a great man—"marks that distinguish him from one who merely *thinks* he is important...the personification of humility & grace... continually on the alert to see if he could be of any assistance to anyone in any way...deeply appreciative."

He was most impressed by the way in which Neyman handled "controversial" matters.

"He never hesitated to say when he was on a battleground, & he presented his own views in such a way that the audience was given credit for the ability to form their own opinion. In some cases, he merely said that he fails to understand Fisher's point of view. I think this is the device he used whenever he thought Fisher is absolutely wrong."

In addition to participating in the lectures and conferences which Deming had arranged, Neyman spent two days as a consultant on sampling methods for the Department of Agriculture.

"He was much pleased to have the money (he is really hard up) but more pleased to think that he had been a consultant paid by the U.S. government...."

Neyman's salary at University College (the equivalent of $2500) did not seem to Deming enough for a man with a family.

"He told me he may go back to Warsaw next year; his salary there will be the same in sterling, but will go much further....I asked him if he would consider a job in this country, & he seemed pleased."

From Deming's personal point of view, Neyman's visit had also been a tremendous success.

"I think most people thought at first I was bringing a second rate statistician to town. Few people have pretended to keep up with his work (his fault, I'd say, because he writes poorly), though of course everyone knows of

Fisher. But before Neyman left, most everyone realized that we had in our midst a man whose ideas are not to be taken lightly. They saw, I think, that Fisher's methods are by no means the last word, & that in some ways he has led them astray."

Neyman returned to London in the middle of May. His long paper on the theory of estimation had been sent back by the secretary of the Royal Society with the statement that one referee's report was unfavorable, the other favorable. He has always assumed that the unfavorable referee was Fisher; but, according to a communication I receive from the librarian of the Royal Society, it was in fact George Udny Yule. He guessed (correctly) that the favorable referee was A.C. Aitken.

"The Secretary had written to me that this man, this favorable referee—he didn't give me the name—had suggested that I should read this paper by Kolmogorov," he explains.

The work recommended was the famous monograph "Grundbegriffe der Wahrscheinlichkeitsrechnung," published in 1933—the slim volume which Neyman has earlier described to me as "a jewel," adding when I comment on its small size, "Jewels are not big." In it Kolmogorov had brought to a successful culmination the movement, in which Bernstein had also been working, to provide a rigorous formulation of the foundations of the theory of probability. He had done this by laying down clearly and "out loud" what a few other mathematicians had merely hinted at in their work—the relationship between the classical theory of probability and the modern theory of measure which had developed from the work of Henri Lebesgue.

"I had not known about it. So I learned about it for the first time in '37 from this referee, and so I looked at it, and then I *grabbed* it!"

But how was he able to *guess* that it was Aitken who had suggested that he look at the Kolmogorov volume?

"So there was a monograph which this Aitken published which attracted my attention. It was reasonable, attracted my attention favorably....His mathematical level was 'way above London, and that was obvious from his monograph. The relevance of this book, for example—the Kolmogorov. Frequency, connection with measure, and so forth. It's obvious that they are the same ideas which were inspiring me; but I didn't know about it and so someone suggested that I should read it. And if I think of the people in the field relating to probability in England at the time, he was the only one that I can think of."

The favorable report by one referee outweighed the unfavorable report of the other, and the Royal Society accepted Neyman's paper, "Outline of a theory of statistical estimation based on the classical theory of probability." It was presented by Jeffreys at the meeting of June 17, 1937.

Although he had not been appointed to the professorship at the London School of Economics, the academic year 1936–37 had been a year of exhilarating recognition for Neyman. A triumphant American lecture tour. An invitation to present a paper at a high-powered international congress on the theory of probability at the University of Geneva. The acceptance of an important paper by the Royal Society for publication in the *Philosophical Transactions*. He felt that it was time for University College to give him the academic position and the money that these successes warranted. Egon agreed that the invitation to the congress in Geneva was certainly a sign of European recognition; but he didn't think that it would have very much influence with the provost, or with the English people, "who on the whole, perhaps in an insular way, are inclined to look on congresses & conferences as bilk and not solid science." There was really no immediate chance of Neyman's being made a Professor of Mathematical Probability ("or some such title") at University College.

Egon himself was beginning to feel overburdened by his duties as department head and editor of both *Biometrika* and *Statistical Research Memoirs*. Sometimes he looked back wistfully on the carefree days when he had been a junior lecturer under K.P. And then, of course, he always carried the psychological burden of having Fisher on the floor above him.

Fisher had been offered a job in the United States that spring—a position at Ames, jointly sponsored by Iowa State College and the United States Department of Agriculture—and Pearson had fervently hoped that he would accept. The idea of moving had also been attractive to Fisher, according to his daughter:—"...in contrast to the halfhearted support at University College, it seemed in America he could expect that the energetic people of that great land would, in their phrase, 'run with the ball,' welcoming, adopting, and supporting his schemes with some alacrity." Having already experienced two Iowa summers, however, he had requested that the contract include an option for the Department of Agriculture to transfer its contribution to another university if that proved desirable. The department had agreed, but the State College, fearing to lose him, would not agree. He had "reluctantly" decided that he had better remain at University College.

Throughout the spring and early summer of 1937, Neyman and Pearson debated the question of Neyman's academic future. There are a number of references in Pearson's letters, and one in Deming's letter to Birge, suggesting that Neyman had an opportunity to go back to Poland as a professor; however, when I specifically ask if he was ever offered a professorship in Poland, he says firmly that he was not.

"I think you have yet to decide for yourself," Egon told him, "between the advantages of (i) staying in England and (ii) having independence & a professorship in Warsaw. I know it is a horribly difficult decision, but I am inclined to think that though you don't really want the routine job of a Department, you really won't be quite happy without the independence that it would give you."

He did not *like* the idea of Neyman's leaving London.

"But I believe that unless you could be quite independent, so that you did not in a way have to get permission from me to do this or that, there will always be a risk of something coming in between us."

Despite this intuitive feeling, Pearson worked diligently to improve his friend's financial position and thus keep him at University College. As a result of the deaths of Karl Pearson and of the widow of F.J. Weldon, some funds had become available to the department. Pearson proposed to the provost that these should be used in part to supplement Neyman's salary from the college by £150 (or $750) a year over the next three years (carefully arranging that the money go to Neyman at the rate of £25 every two months).

The provost agreed to the arrangement. Neyman was satisfied. Egon was delighted.

"With a possible prospect of keeping you in England, a great weight begins to go off my mind. I must draw up my 'statement' for the Provost tomorrow. What a lovely day!"

That same June—perhaps on that same lovely day—a special committee on the teaching of statistics on the Berkeley campus of the University of California was making an urgent recommendation:

"There should be added to the mathematics department a member ranking as assistant professor or higher whose special field is statistics and who in addition to a thorough command of theory is definitely interested in the applications."

1934
—
1937

To tell the story of how the University of California—in its search for a statistician—came to settle on a Russian-born Pole teaching in London, I must resort to a tactic of nineteenth century novelists: leave Neyman at University College and begin this chapter, "Meanwhile, back in the United States…"

The series of events which resulted in Neyman's invitation from Berkeley can be said to have begun in 1934. In July of that year Griffith C. Evans, formerly of Rice Institute, came to the University of California to take over the chairmanship of its mathematics department. His appointment had been at the instigation of influential faculty members in other sciences who were dismayed at the mediocrity of the mathematics department of the university. That same July a paper, "On the Statistical Theory of Errors," appeared in the *Review of Modern Physics*—a joint work by W.E. Deming (of the United States Department of Agriculture) and Raymond T. Birge (the chairman of the physics department of the University of California at Berkeley). Both of these events—the coming of Evans and the collaboration of Birge and Deming—were to play an important role in Berkeley's ultimate invitation to Neyman—a choice which would dramatically affect the direction of statistics in the United States.

Since Evans's early days at Rice Institute, he had had various contacts with statistics, had read a little of Lévy, and had become personally acquainted with Fisher during summer weeks they had spent together at the University of Michigan in 1931. Toward the end of his life, in a series of notes for "A Brief History of the Mathematics Department," he recalled that "as early as 1935 I envisaged California as the place for a really outstanding statistician, if possible of the level of R.A. Fisher himself."

As a consequence of their joint paper, Birge and Deming had also come in contact with Fisher, to whom they had sent a reprint. He had acknowledged it as "most valuable" and had made some comments which they had arranged to publish.

The collaboration of Birge and Deming began a long friendship which was, fortunately for the historian, to be carried on almost entirely by mail. In addition to treating statistical questions of mutual interest, Deming communicated to Birge the gossip of the statistical world. The center of that world was London—University College. Karl Pearson and R.A. Fisher "disagree almost to the point of taking up arms on some questions in statistics"—"K. Pearson has no use for Student, either"—"Student and R.A. Fisher stand together"—"Fisher can say nothing good of Neyman and Pearson"—"I have heard from all sources that Egon Pearson is really a prince of a fellow."

In the spring of 1936, a fortnight after the death of Karl Pearson, Deming visited University College and wrote in detail to Birge about his contacts with the great men there. He met Egon Pearson first and was on the stairway talking to him when Fisher came by.

"Pearson introduced me, and told Fisher that I would like to attend his (Fisher's) lectures. The discourse was entirely amiable, though not prolonged. Later I learned that this was the only time Pearson and Fisher had been caught in conversation."

Deming was greatly disappointed in Fisher as a lecturer. "[He] frequently invited questions.... The only trouble was that no one could ever ask an intelligent question; a person could only ask that he go back and repeat...." It appeared to Deming that Fisher had very few students compared to Pearson and Neyman, although there were quite a number of research men who had come to London to settle down and work under his direction on data they had brought with them. "For something like this, he must be a splendid man.... He has an extremely penetrating mind, grasps the problem quickly, and always seems to have plenty of time. He has the knack of making a person feel perfectly at ease."

In contrast to Fisher, Egon Pearson was "the perfect lecturer," but Deming stressed that he meant *lecturer* as opposed to *teacher:* "In *his* classes you will see the lecture system at its worst. His notes are written up completely, even to tangential remarks and afterthoughts, and he follows them absolutely. There is no opportunity for remarks, and none is invited."

Neyman was quite different from Pearson as well as from Fisher.

"Dr. Neyman...actually teaches..., invites questions, and talks to the students as if he were really trying to do them some good." It was Neyman to whom Deming felt that he owed the success of his trip. "I attended only one of his classes, since he invited me to talk with him privately at considerable length.... We spent, on the average, perhaps an hour and a half together daily. Dr. Walker [Helen Walker of Teachers' College, Columbia University] joined us before long. Dr. Neyman talked and paced the floor while Dr. Walker and I made notes and remarks and asked questions."

That fall Deming wrote extensively to Birge about Fisher's American lecture tour and the four days he had spent in Washington at the Graduate School of the Department of Agriculture.

"I will take back all I ever said about Fisher's lecturing. He gave us three of the best lectures I have ever heard in statistics."

Fisher was "a perfectly charming fellow" except when he was on the subject of Neyman.

"I asked him what he thinks of Neyman's work. He said Neyman had been mining (he did not say *under*mining) and exploiting his work from the beginning, and unfortunately misunderstanding a lot of it. Fisher feels that Neyman has not done anything fundamentally new but admitted that some of Neyman's diagrams and terms were good pedagogy."

Fisher also referred to Neyman and Pearson's *Statistical Research Memoirs* as "a heap of junk."

Curiously, Deming did not mention to Birge that Fisher was going from Washington to Berkeley, where he would spend almost a month on the campus of the University of California as Hitchcock Lecturer. Birge was surprised, and a little miffed, to get the news for the first time in the student newspaper at the beginning of September 1936. He wrote immediately to Dean C.B. Lipman that he would be "rather interested to know who was directly responsible for getting Dr. Fisher to come here, since, with the possible exception of Professor Mowbray [A.H. Mowbray, an actuary and professor of insurance], I have not yet been able to find anyone here at Berkeley who was directly interested in Dr. Fisher's work."

On the contrary—Dean Lipman responded—members of several different departments had put forth Fisher's name, including Professor Evans of the mathematics department: "Professor Evans told me that he regarded Fisher as perhaps the greatest statistician alive today."

Birge pitched in to do what he could to make Fisher's visit to Berkeley a success. Deming sent him various suggestions—a microphone that Fisher could carry around with him—long walks, several miles a day—tea every afternoon at 4 o'clock. "...[and] if you want to make a hit with Fisher, provide him with a calculating machine.... He likes to compute, and it is no use to offer to do it for him or to have it done."

From Birge's point of view, however, Fisher's visit to Berkeley was a great disappointment. Four and a half single-spaced typewritten pages in the archives of the Bancroft Library record his criticisms of the English statistician and his lectures. The latter, he wrote to Deming, were simply lifted from *The Design of Experiments* and were delivered as if everything in them were self-evident. He thought that Fisher "did less than the minimum amount to earn his money." Contrary to the conditions of the generously endowed lectureship (a $2000 honorarium plus a $1000 travel allowance), which provided for a three or four week stay by the lecturer on the campus so that he could be consulted informally by faculty and students, Fisher had spent the first five days of his visit in San Francisco rather than in Berkeley and had left one day before the minimum three weeks, standing up a dinner in his honor. Everything had ended "as badly as possible" with Fisher's taking a manuscript of Birge's to read and then leaving it in the car that took him on an excursion to the great redwoods.

"My present idea of Fisher's actual idea on science,—from the personal side,—is as follows," Birge wrote Deming immediately after Fisher's departure. "He wishes to discover the fundamental scientific principles. He is only too glad to have others use in *practical* problems the general ideas he has developed, and he is only too glad to explain to anyone how such general ideas should be applied to their particular problem. As you have written, he is glad to discuss such things early in the morning or late at night. *But* he is *not* glad or even willing to have others work on the purely theoretical aspects of his work. He expects others to accept his discoveries without even questioning them. He does not admit that anything he ever said or wrote was wrong. But he goes much further than that. He does not admit even that the *way* he said anything or the nomenclature he used could be improved in any way."

Fisher, Birge wrote firmly, was the most conceited man he had ever met—"and that is saying a lot with such competitors as Millikan et al!"

Later he added that he was surprised it had taken him and Deming so long to recognize Fisher's true character: "You know, Oppenheimer [J. Robert Oppenheimer, who was then a member of the physics department at Berkeley] never even met him while he was here. Oppenheimer says, 'I took one look at him and decided I did not want to meet him,' and it generally takes Oppenheimer about one second to size up a person (and get it correct)."

I have not found any description of Evans's reaction to Fisher's lectures at Berkeley. What is recorded is that on several occasions he talked with Fisher about the situation in regard to statistics and probability on the campus and asked him to suggest the names of some statisticians who would be suitable for Berkeley.

I am interested in whether Evans, given his high respect for Fisher, ever went so far as to suggest that the latter might come to Berkeley. According to Joan Fisher Box, after her father had returned to London the previous fall, "the possibility of removing from England either to the Midwest [i.e., Ames] or to California was definitely in the air [and] was much discussed within the family...." But in response to an inquiry from me, Dr. Box writes that Fisher had not received an offer from a California university. In the Fisher archives at the University of Adelaide, although there is a folder marked "Evans," the name has been crossed out and replaced by another.

Before talking to Fisher, Evans had investigated the American situation and had found "no central figure of sufficiently strong personality to create a school of theoretical statistics." He had then turned abroad and considered Karl Menger, the world famous Austrian mathematician who—he was sure—would welcome an invitation to the United States, "partly because of the prospect of a Nazi development in Austria, and partly because he believes that the United States is to be the center of scientific development in the near future." Fisher, first mentioning half a dozen Englishmen (including "older" men like E.S. Pearson, who was a year and a half younger than Neyman), enthusiastically endorsed his successor at Rothamsted—Frank Yates. He said that of all the men he had mentioned, Professor Yates was the one with the most imagination and theoretical power; and even if he did not remain permanently in America, he would find it to his advantage to have experience there and would develop precisely what Berkeley needed. A day or two after Fisher left, Evans recommended the addition of Yates to the Department of Mathematics.

"As far as the general constitution of the department goes, the field which is conspicuously lacking is that of statistics and probability," he explained to the university president, Robert Gordon Sproul. "...There are many on the faculty...who deal with the application of the theory to concrete problems, but no one to whom such persons can go for adequate mathematical assistance and no one who is taking part in the significant development of the theory itself."

In spite of the fact that he had recommended that Yates be offered a position, Evans continued to gather names of other possibilities. He and Monroe Deutsch, the provost of the Berkeley campus, wrote to several mathematicians and scientists whom they knew in England, having settled on that country as the one in which they were most likely to find what they wanted. Two months later, Evans was mentioning in his correspondence with Sproul several names in addition to that of Yates. He now inclined toward A.C. Aitken of Edinburgh. At the same time Fisher wrote that he had sounded out Yates and that Yates was willing to come to the United States— an act which Evans felt was presumptious on Fisher's part.

At this point, at the end of January 1937, Provost Deutsch, in response to a question from President Sproul, suggested that in view of the university's financial situation such an appointment as the mathematics department was contemplating was unjustified. "Of course, I realize the awkwardness of such an answer after Professor Evans has been writing letters all over the world." The idea of obtaining a statistician for the following academic year was dropped. Instead, the Budget Committee appointed a subcommittee, headed by the astronomer C.D. Shane, to investigate the situation in regard to statistics on the campus and see if the number of statistics courses being offered by different departments could be reduced.

Although Birge had not been asked to serve on this committee, he continued to take a lively interest in statistics. That spring, when Neyman came to the United States, Birge heard in detail from Deming (as has been described earlier) about the success of Neyman's lectures in Washington and the charm of his personality. When Deming informed him that Fisher had been offered the position at Ames, Birge commented that Neyman "would probably make a much better addition to the staff at Ames than would Fisher, but it is probable that those at Ames would not view the matter in that light." He added, "I think the University of California also very much needs a first class statistician!"

It was just a couple of months after the Ames offer to Fisher that the

Berkeley subcommittee on statistics submitted its final report. While it was true, it conceded, that many of the statistics courses on campus were covering the same material, far worse was the fact that most of these courses required no mathematical prerequisites other than the high school algebra and plane geometry required for admission to the university. A still more dreadful aspect of the situation was that there was not a single person on the faculty at Berkeley qualified to teach the theory of statistics at an advanced level. The subcommittee strongly recommended that such a person be added as soon as possible to the mathematics department.

This recommendation coincided most satisfactorily with Evans's long-held views. He was an uncommunicative sort of man, inclined to one-line letters and agreeable but ambiguous "mmmmm's" between puffs of a pipe that was rarely out of the corner of his mouth; and although he must have been aware, after the Fisher lectures, of Birge's interest in statistics, he did not consult him. He felt that he knew what was needed at Berkeley, and he quietly began to add more names to those he had already gathered for the proposed position.

It was not until the beginning of the new term, in the early autumn of 1937, that Birge discovered that Evans had amassed quite a list without ever asking him for his suggestions. He grumbled a little to Deming, for he still felt that, even though he was a physicist rather than a mathematician, he was the only faculty member who knew anything about what was going on in statistics. He mentioned Neyman's name to Evans and found, just as he had expected, that Evans had never even heard of Neyman, "who, I suspect, is probably the best available man for the position."

Birge had not changed his negative opinion of Fisher. He felt that no matter how much a person might think he was in agreement with Fisher, if he said or published anything that had any connection with Fisher's work, "[Fisher] is going promptly to find that it is incorrect, even if [in finding it so] he has to directly contradict his own previous statements." He strongly urged Evans that no one suggested by Fisher should be considered for the Berkeley post "because I believe that Fisher will recommend only those who are so completely subservient to him that they have no original ideas of their own whatsoever."

On October 1 Birge reported to Deming that Neyman was definitely on Evans's list. For whatever reason, the name of Yates had been removed by the time that the name of Neyman was added.

Evans, once he learned of Neyman's existence, was very quickly impressed by his qualifications. It happened that there was a graduate student from Berkeley—Francis Dresch—in Cambridge in the fall of 1937 who had gone there hoping to study economics with John Maynard Keynes. Evans decided to write to Dresch and ask him to attend lectures by the English statisticians who were under consideration. In his "Brief History" he was to admit, still with a little embarrassment, that he was concerned about Mr. Neyman's English.

Of his first meeting with Neyman, Francis Dresch, who is now affiliated with SRI International in Palo Alto, says:

"I think I just dropped by, and he was giving a lecture, and then we had a conversation and went to lunch. We took his secretary, Catherine Thompson, along. Later I was told by Neyman that I had made a great impression on her; but in the process of going to lunch he got her to help him select a pen and pencil set as a gift for someone, and then it turned out that it was a gift for her. I thought he had already picked her out for himself."

Dresch was unbothered by Neyman's English.

"The thing that impressed me about Neyman was the following. Statistics had always seemed to me like a collection of problems looking for a central theme. Of course, that may have been partly my provincial view. I can't recall anyone at Berkeley showing any interest in statistics. Other than Mowbray. So this was the first I had heard of some sort of central theme to statistics which on the one hand was based on probability theory and fairly sound mathematics and on the other hand was directed toward, and seemed applicable to, a wide range of problems that had to be attacked in one way or another by statisticians. Well, I don't suppose that the letter I wrote to Evans after I saw Neyman is still in existence."

I agree that there is not much hope of its turning up. The Evans papers in the archives of the Bancroft Library are a frustrating jumble. Formulas are scribbled on Christmas cards, and manuscripts sometimes have Mrs. Evans's knitting instructions on the back of them. But just when I have given up hope, a typed "COPY" of Dresch's letter turns up in the files of the president of the university. Dated Cambridge, November 8, 1937, it begins: "I have just returned from London with a most favorable impression of Neyman."

Dresch had heard one of Neyman's lectures, part of an introduction to modern probability theory, concerned almost entirely with the theory of point sets.

"Although he is not a particularly colorful lecturer, he presented his material in a clear, systematic fashion with considerable patience and, on the whole, seems to be quite an effective teacher. Coming to a convenient stopping place before the end of the lecture period, he spent the remaining few minutes passing about the class with individual questions summarizing the material covered, and obtained a reasonably intelligent response. He has a very pleasant unaffected manner both in front of his class and in conversation. I would expect him to be very well suited to guide the work of research students, judging both from his personality and from his own research attainments."

After seeing Neyman, Dresch went to call on the Polish economist Michał Kalecki. When he mentioned Neyman, along with several others, as having suggested that he look him up, Kalecki—to Dresch's surprise—launched into a eulogy of Neyman.

"...Kalecki claimed that some rated Neyman as the greatest theoretical statistician on the continent at the present time. Granting that this extreme view might be rather hard either to attack or defend, nevertheless, [he] argued Neyman's reputation is particularly impressive when one considers the great personal modesty in spite of which the reputation was acquired."

Evans had not yet received Dresch's letter when he made up his own mind:—Neyman was the man who could develop a statistical center at Berkeley, "superior to anything west of the Mississippi." Sproul agreed. On November 10, 1937, Evans sent off a letter addressed to "Dr. John Neyman" at University College:

"The University of California is in need of someone to coordinate the work in statistics in the University and, especially, to develop the subject on the mathematical side. We have wondered if we could not persuade you to undertake this task?"

1937
—
1938

During the summer of 1937, while Evans was still gathering recommendations for the position at Berkeley, Neyman was absorbed in preparing his Washington "lectures and conferences" for publication in mimeographed form by the United States Department of Agriculture. Egon's sister Sigrid had lent him a small house in Little Hampden—known in the Pearson family as The Cell—and while Lola painted and the baby Michael toddled around the garden, he worked "furiously" at a table under the plum tree.

The six "conferences" were being written up from a stenographer's notes by Deming, who was extremely enthusiastic about the whole project.

"This set of lectures and conferences with Dr. Neyman will be a real book...," he assured Birge, "and the quality will put it into the class with the most important statistical publications in years. I keep telling Neyman it is the *most* important thing yet written...."

But when the English vacation started in July, he was greatly upset to receive a radiogram from Neyman:—"Before publication conferences require rewriting." It was perhaps his first inkling that his author and his obliging houseguest of the previous spring were two quite different people.

"I do not know just what to infer from your radio message," he hurriedly responded, "but I do hope you don't mean that the material you have received so far really needs to be rewritten."

He conceded that he had had no idea when he began that the stenographer's notes would require as much revision as they had; but he had worked very hard and very carefully to edit them, and he flattered himself that he had been able "to bring your brilliant contributions into the open in a style that has pleased all who have seen the results." Among these he cited "Mr. Milton Friedman."

Neyman was not to be put off.

"I am most grateful for the trouble you have taken to put in order the stenograms.... The opinions of other scientists to whom you have shown the drafts are most encouraging. But a few scores of my own papers and about the same number of other people's that I have formulated in the past produced a habit and an attachment to the words I like and the way of presenting things I like, and it is difficult for me now to leave the formulation of my lectures and conferences to somebody else, whoever he may be."

Although rebuked, Deming struggled to maintain the informality of the conferences, which he was afraid would be lost if Neyman rewrote them, and to move the project along so that the book would get into the hands of the American statistical public as quickly as possible. Letter after letter passed between the two men. "Jerzy" and "Ed" of the previous spring were replaced by "Dr. Neyman" and "Dr. Deming." There were a number of misunderstandings, both linguistic and mathematical. On one occasion Neyman misspelled *smooth* when referring to a not yet published paper, "'Smooth' test for goodness of fit." "I don't understand the reference to 'Smouth,'" Deming wrote. "Is that the name of a statistician?" On another occasion he corrected Neyman's use of the term *any real value*. "I think that you underestimate my knowledge of English in general and in mathematical literature in particular," Neyman shot back. "I will go further and accuse you of not using your own English in quite a precise way—this is a good joke, isn't it?"

Deming was finally moved to describe the tribulations and despair of his own collaboration with Birge in 1934.

"I mention this," he wrote to Neyman, "because in this work that we (you and I) are doing, it may occasionally seem hopeless. I do some very dumb things once in a while…, but after it is all done, there will be something to show for our efforts. We must not get discouraged, or let anything pass until we are both well satisfied.…I know that you will be patient with me and overlook some of the blunders that I am continually falling into.…after all, the final authority rests entirely with you on all matters."

When this conciliatory letter arrived in London, Neyman was temporarily away from University College, attending the international congress on probability at the University of Geneva. The list of invited guests was impressive—Fréchet, Pólya, Heisenberg, Steinhaus, S. Bernstein, Hostinský, de Finetti, Wintner, Cramér, Lévy, von Mises, Glivenko, Kolmogorov, Slutzky, E. Hopf, Cantelli, Steffensen, Feller, Onicescu, Dodd, Jordan, Obrechkoff—and Neyman was delighted to be included. The work which he had chosen to present was his new theory of estimation, which would not appear in the *Philosophical Transactions* until later that fall. It met with a somewhat skeptical reaction from some of those present. Neyman remembers most vividly that Paul Lévy "expressed doubts" and that the Hungarian mathematician George Pólya jumped up and—shouting "Non!"—proceeded to explain Neyman's ideas to Lévy.

Neyman likes this story, and he retells it in 1978 when he is asked to recall "encounters with Pólya" at Stanford University's celebration of Pólya's ninetieth birthday. It still puzzles him that a mathematician like Lévy did not understand. He repeats, shaking his head, "Paul Lévy!"

Harald Cramér, who met Neyman for the first time at Geneva, writes to me that he was impressed by Neyman's work but did not feel, as he told Neyman at the time, that it was yet in a completely satisfactory mathematical form.

Among those absent from the congress was Richard von Mises, who—dismissed from his post at the University of Berlin—had accepted an appointment at the new university in Istanbul. His absence was a disappointment to Neyman, who was currently assisting Eugene Rabinowitch—they had met again at University College—and Donald Scholl with the English translation of von Mises's classic volume on the foundations of the theory of probability. A still greater disappointment was the absence of the Russians, especially Kolmogorov.

The combination of scientific discussion and sociability at the Geneva congress impressed Neyman tremendously. Afterwards, in a report which he and E.L. Dodd wrote for *Nature*, he emphasized that the many social events—receptions and excursions into the Alps—offered "a precious opportunity" for personal contacts and private exchange of opinions and criticisms. It was an experience he was not to forget.

On his return to London, exhilarated and stimulated, he responded agreeably to Deming that he was "not discouraged in the least by the fact that we sometimes disagree." The letter, dated October 18, 1937, included the news that Student had died two days earlier, suddenly and completely unexpectedly, at the age of sixty-one.

(Deming had once met Student at a meeting of the Royal Statistical Society: "with a beard, very humble and of pleasing personality. He stated that he is usually a great disappointment to people that have wondered what 'Student' looks like; 'not very interesting after you find him,' he said.")

Although Neyman had admired and liked Student, he had not had a great deal of personal contact with him. Egon, however, was deeply affected by the death of the quiet, unassuming man from the Guinness brewery. He felt that in many ways his own statistical philosophy had been shaped as much by Student as by Neyman.

"I think that there are so very many things that we owe to 'Student' in the present statistical world....I would like to interest people in him, his practical mindedness & his simplicity of approach. It would be so easy for people to miss in the picture that large part he played simply by being in touch, by correspondence or personal meetings, with all the mathematical statisticians of his day."

But the memoir of Student which he wanted to write would have to wait until he finished his long life of his father.

"I have now got up to 1920 with K.P....so I feel fairly content," he reported. "There are difficulties in the recent years because I want to (i) supply an indirect answer to R.A.F., (ii) not represent K.P. as perfect, (iii) not be unpleasantly rude to R.A.F., (iv) recognize the part that R.A.F. played in stimulating modern statistics, (v) mention R.A.F. as little as possible."

It was not long after the death of Student that Francis Dresch dropped in at University College. There were always American visitors attending lectures and discussing statistical problems with the faculty members, and Neyman had no reason to suspect anything more in his encounter with the young American from Berkeley. It was a complete surprise when, scarcely a week after Dresch's visit, he received a letter from a man named Griffith C. Evans offering him "a lectureship" at the University of California (in those days it was never necessary to add "at Berkeley") and a salary of $4000 a year.

He wasted no time in replying. He was honored—he was grateful—he would consider the offer most carefully—it was very attractive—"But, as in all human affairs, there are certain complications...." He had a family, and he was naturally uneasy about exchanging what he had every reason to think was a permanent post—if not a completely satisfactory one—for a post which might last only two or three years. He thought that before he made any decision, it might be well if he came to Berkeley for a summer.

"It may spare you the inconvenience of having for a few years a person on the staff whom you may see at once is unsuitable. It may also give us both sufficient courage for some arrangement of a more permanent character."

Almost immediately, there was also a letter from Birge who, although he did not know Neyman personally, had asked Evans's permission to write informally to him, extolling the advantages of California and its university.

"You are the first one to whom an offer has been made," he assured him, "...really the first choice of all those concerned."

He went on to explain that the lectureship offered by Evans was in essence the equivalent of a professorship since the salary of $4000 was that normally paid to a professor at Berkeley. The only difference was that a lectureship was not a tenured position while a professorship was.

The letter from Birge was followed by a letter from Deming, who pointed out that "Professors Birge and Evans...have seen that my advice is good. My feeling was, when I was consulted, that if they were looking for a man in mathematical statistics, they might as well have the best—at least try for the best."

Neyman was, of course, tempted by the Berkeley offer. He had always dreamed, he told Deming, "of being in a position to start a cell of statistical research and teaching from the start, not being hampered by any existing traditions and routines which were established long ago, for no good reason, and now still are being respected for no good reason either."

Evans had stated his belief that the opportunity at the University of California in the field of statistics was as great as at any place in the world. Neyman, however, knew from his visit to the United States the previous spring that the academic level of students at American universities was well below that of English and European students. In fact, he says, he does not think he would even have considered leaving London for Berkeley if it had not been that he "did not want to be behind barbed wire again."

"I was at the time citizen of Poland, alien in England, and prior to that time, or about the same time, there were meetings in Poland between the people in the Polish government—it was called 'the government of the colonels'—and the representatives of Hitler. It was clear to me—and many people—that war was coming, but it was not clear whether Poland will be on one side or another one. It came to my mind that when war comes I shall be an enemy alien again...."

If he went to Berkeley and if the arrangement turned out to be unsatisfactory to one or both of the parties, what would happen to him as a foreigner? He thought he would probably be told to leave the country immediately. Birge assured him that "if the University of California were so foolish as to let [him] get away" there would be many other American universities that would be eager to have him on their faculties.

What Birge said was probably true "at the present time," Neyman agreed, but that was because the work he had recently done and was doing was "not very unsuccessful."

"You probably agree that the conception of favorable conditions is a very relative one and what is good for one man may not be good for [another]. All depends on whether a man is expected to do the things he can and that is just what I would like to find out. If it happens that what I can do, you wouldn't like me to do, and on the contrary, what you want me to do is very different from what I can, then my work in California will be a failure, and the facilities of getting a new position after a failure, will be considerably diminished."

His reluctance to accept their offer only made the people at Berkeley more eager to obtain his services.

"All three of his letters [to Evans, Deming, and himself] were almost identical in their essential contents," Birge pointed out, "and so it seems evident that Neyman has no hidden reasons for hesitating to accept our offer. It is certainly proof of his natural modesty that he should hesitate to come for fear that he might get kicked out after a couple of years."

Never before in his career had he found someone who, upon being offered a position, expressed doubts about his ability to fill it!

Everything the Berkeley people heard—for the letters from Dresch and others had by then arrived—ratified their choice.

"In fact, there seems to be no doubt in anyone's mind who is in a position to judge," Evans reported to Provost Deutsch, "and I expect that it will be only a year or two before Neyman will have at his disposal the choice of about any university here or abroad."

He wrote emphatically to Neyman, "There has been no doubt in our minds that you are the person who can best undertake the statistics program as a member of the Department of Mathematics." The advantage of offering a lectureship rather than a professorship had been that the offer of a tenured position would require action by a faculty committee, and Evans had wished to avoid that complication. Now, however, such a committee was being appointed by President Sproul, and he hoped that in a few days he would be able to offer a permanent position.

During all this correspondence across the Atlantic, the normal course of life at University College continued. There were lectures to be given and students to be seen, examinations to be conducted, theses to be approved.

The most outstanding of Neyman's students at that time was a Chinese, Hsu Pao Lu, or P.L. Hsu. (Neyman expresses to me his admiration for Hsu with a Polish phrase which he translates—with a little bow and a gracious wave of the hand—as *Please sit down!*) But it was a Canadian student, an entomologist named Geoffrey Beall, who inadvertently led him to a most useful scientific formulation. Concerned with field experiments involving treatments with insecticides, Beall came to Neyman for help after he had unsuccessfully tried to fit the Poisson distribution to his results. Questioned, he explained how moths deposit eggs in masses of different sizes from which the hatched larvae then move out in search of food. It was from this description of Beall's that Neyman got the idea of a new class of "contagious distributions" which—in its connection with the mathematical model of "clustering"—he was to utilize time and again in the future.

Neyman was turning out a number of research papers in 1937; and the following year, with nine published works, was to be one of the most extensive in his bibliography. Among these was the paper "'Smooth' test for goodness of fit," in which he put forward for the first time a technique which eventually spread to a large part of asymptotic work. The paper was dedicated to the memory of Karl Pearson, "who originated the problem of a test for the goodness of fit and was first to advance its solution."

The earth was not resting lightly on the father of mathematical statistics. Fisher had published an unpleasant attack on one of the last papers K.P. had written, describing it as "unfortunately marred by great bitterness, and by vehement attacks on an Indian writer [R.S. Koshal] whose offence appears to be that of deciding, after trial, that a curve of Pearson's type I could be fitted to his data more successfully by the method of maximum likelihood than by the method of moments." Fisher excused publishing such a critical paper less than a year after Pearson's death on the grounds that in the last few years those close to K.P. had intimated that any controversy would have a detrimental effect on his health.

As in so many instances in the lives of these two great statisticians, what pertains to one could easily be applied to the other. Fisher's paper was also "unfortunately marred by great bitterness." In it the recently deceased Pearson—the founder of mathematical statistics—was referred to as "a clumsy mathematician" of "arrogant temper" and "factious mind"—who did not originate "the simple procedure of equating moments"—a process of which "he could never appreciate the limitations, or master the proper use."

Neyman was "disgusted" at the violation of the ancient precept to speak no ill of the dead. He bristled also at Fisher's treatment of Pearson's early work. It seemed that some protest ought to be made "when a critic overlooks whole passages in the main text of a paper and misinterprets footnotes which are clear on reference to the text on the same page." He wanted to publish a note on Pearson's deduction of the moments of the binomial in the *Journal of the Royal Statistical Society* rather than *Biometrika* "[to] make impossible any doubt of its being strictly objective." He wrote to Isserlis, one of the editors, reminding him that ten years earlier he and Greenwood had published a note inspired by the same motives.

"I am still sorry for being found guilty [of having 'unduly depreciated and ignored' the work of Tchouproff], but I agree with the principle that deceased workers must be defended and that collective bodies like the Royal Statistical Society have the right and, perhaps, even the duty to provide facilities for the purpose."

Isserlis and the other editors declined to publish Neyman's note "on the grounds that its publication would give undue emphasis to an attack that is in our view not worthy of notice," and Egon then agreed to publish it in *Biometrika*.

In January 1938, with Neyman still pressing Evans for an opportunity "to get acquainted," the University of Michigan entered the picture with a possible offer of a professorship at $4500 a year. As a statistical center, Michigan had a leading position in the United States. It boasted "the most completely equipped statistical laboratories possessed by any university in the world,...courses in elementary and advanced mathematical statistics of long standing and wide reputation,...an increasing number of graduate students whose primary interest lies in the mathematical theory of statistics,...staff members of outstanding competence and experience,...a very complete mathematical and statistical library." In addition, the *Annals of Mathematical Statistics,* the only journal in the United States devoted wholly to that subject, had been founded at Michigan by Harry Carver and was still being edited there.

Perhaps, Neyman suggested, California and Michigan might be willing to share the expenses of a summer series of lectures at both universities.

"The next thing I heard," Dresch says, "was that Neyman seemed to have some reservations, and Evans had a rumor—maybe even knew—that Michigan was in the market. I don't remember exactly what he wrote, but it was something like what I am paraphrasing—if it would convince Neyman to come to Berkeley, I could tell him that I would be his assistant. I had no reason to expect any kind of a job offer from Berkeley so it was kind of a bribe. If I brought him back alive and in the process didn't alienate myself from him, why then I would have a job. Not a bad thing in '37, '38."

At almost the same time that Michigan entered the picture, Deming mailed Neyman the first copy of *Lectures and Conferences on Mathematical Statistics:* "Revised and supplemented by the author with the editorial assistance of W. Edwards Deming." The volume ran to 168 mimeographed pages ("on high quality paper," Deming pointed out) and was priced at $1.25. Those who had ordered copies at the time of the lectures were given a special price of 35 cents.

In the next couple of years Deming's faithful optimism about the importance of *Lectures and Conferences* and the need for such a book was to be amply justified. The first edition of seven hundred copies was quickly exhausted. The clarity of the presentation (for which Neyman and Deming had worked so unremittingly, each in his own way) is always mentioned to me by present day statisticians who originally came to the book as young practitioners of the science or even, sometimes, as students. Harvard's Frederick Mosteller describes how everything in it seemed so clear and logical that he had no idea he was being introduced to new and controversial ideas: "I thought that must be the way it was done."

Shortly after Neyman received the first copy of *Lectures and Conferences,* he also received a letter from President Sproul of the University of California offering him his choice of a full professorship at $4500 a year or the opportunity to come for a preliminary visit with an allowance to cover his expenses. Neyman continued to delay. He had set as his goal a salary of $5000—the equivalent of a professor's salary at University College. He went to a number of people for advice, among them George Udny Yule, whom he had recently asked to present a new paper to the Royal Society—Jeffreys, he felt, had become too unsympathetic to be asked again.

"[After two brief experiences in North America], I felt I simply *could not* transplant on any terms," Yule told him. On the other hand, he saw no possibility of a comparable position for Neyman in England.

In regard to Neyman's new paper, Yule explained that the Royal Society had recently ruled that a member who presented a paper should be able to answer questions about it. "As you know, I *cannot* really follow your work," he told Neyman straightforwardly. He had hoped that there might be in the society some pure mathematician interested in modern research in probability theory, "but I gather it has been mostly foreigners who have been working on that line.... It would be simply damned unfair if you were barred from sending in papers to the Royal because in effect some people differed from your views and the one man who was only desirous of being fair couldn't follow your work." If there was no one available, however, he *would* present the paper, "and do the best I can for you."

Neyman decided not to press the old man (who had recently suffered a heart attack which had forced him to give up flying). An examination of his bibliography for 1938 and 1939 suggests that the paper in question was the one published in the series, *Actualitées scientifiques et industrielles.* In it he presented his theory of estimation more clearly, put forward a number of new technical results, and—most significantly—introduced the idea of "inductive behavior." Initially it was a pedagogical aid to help the reader understand the application of confidence intervals, but it was to develop into much more.

"One of the most important trends of the past few decades," the American statistician L.J. Savage was to write some twenty years later, "is a shift in formulation of the central problem of statistics from the problem of what to say or believe in the face of uncertainty to the problem of what to do. It would be hard and unrewarding to seek out the very beginning of this economic movement, as I shall call it because of its emphasis on the costs of wrong decision. It goes back at least as far as Gauss, but Neyman brought it forward with particular explicitness in 1938, coining the expression 'inductive behavior' in contrast to 'inductive inference' [a term popularized by Fisher].

Wald took up the theme with energy and enthusiasm, exploring it in great detail and stimulating many others to do so...."

In addition to consulting Yule—and undoubtedly others—Neyman of course asked Egon about going to America. Did Egon think it would be "reasonable"?

"If I hesitated when you asked me...," his friend replied, "it was because I find it difficult to put myself completely in your position. Because I cannot imagine going there myself....England being my country and not yours. Yes, I am sure it is reasonable *if* you also find out that the duties required of you will give you time for thought and research. American life is rather a hustle."

He concluded his letter: "How vile the Nazi are! Perhaps they will make you keener on America!"

And perhaps they did.

President Sproul remained adamant on the question of salary. Evans completely agreed. "...we do not regard ourselves as engaging in a financial race with the University of Michigan. The competition is in the nature of the opportunities of the respective positions." If they were willing to take a risk on Neyman, then Neyman ought to be willing to take a risk on them! It was "one test...of his fitness."

On April 21, 1938, a few days after his forty-fourth birthday, Neyman accepted the Berkeley offer. He tells me that he preferred California to Michigan for one reason: "There was nothing there. As far as statistics was concerned, it was *tabula rasa.*"

The last months in England were crowded. In addition to his regular students and lectures, he was supervising three Ph.D. candidates and two candidates for the M.Sc. The second volume of *Statistical Research Memoirs* was currently being prepared for the press; and Neyman and Pearson took a degree of responsibility which seemed very strange to their American students, rewriting or even actually writing many of their students' papers— a practice which Neyman has continued to a certain extent over the years.

The last paper which he personally wrote while still in England was a review for *Nature* of the new (fourth) edition of *The Grammar of Science.* With the publication of his own *Lectures and Conferences,* he had begun to think again of the book which he and Egon had so often discussed. As Karl Pearson had said in the *Grammar,* there were periods in the growth of science when it was well to turn attention from its superstructure to its foundations. The book he and Egon had talked of, *Whither Mathematical Statistics?,* would do just that. Although he did not mention the proposed book in his review, he did lay out in his concluding paragraph his thoughts about the need for a critical evaluation of modern statistics, repeating Pearson's epigraph: "La critique est la vie de la science."

His time in England was almost over. It was four years since his "HIP HIP HURRA!" acceptance of Egon's invitation. Although there had not been as much joint work as he had expected, his own achievement had been considerable.

"During the years 1934–38 Neyman made four fundamental contributions to the science of Statistics," George Roussas wrote in 1970 when nominating him for an honorary degree. "Each of them would have been sufficient to establish an international reputation, both for their immediate effect and for the impetus which the new ideas and methods had on the thinking of young and old alike. He put forward the theory of confidence intervals, the importance of which in statistical theory and analysis of data cannot be overemphasized. His contribution to the theory of contagious distributions is still of great utility in the interpretation of biological data. His paper on sampling stratified populations paved the way for a statistical theory which, among other things, gave us the Gallup poll. [His] work, and that of Fisher, each with a different model for randomized experiments, led to the whole new field of experimentation so much used in agriculture, biology, medicine, and physical sciences."

Neyman was scheduled to leave England around the beginning of August, taking a ship to New York and then a train across the continent. Lola and the baby would follow from Poland in a more leisurely fashion, sailing through the Panama Canal, along the western coast of North America, and into the Golden Gate of San Francisco Bay. Almost up until the last minute Neyman and Pearson worked frantically at The Cell on the material for the second volume of *Statistical Research Memoirs*. When they had finished, they sat for a long time on the field gate at the bottom of the lane to Little Hampden, "facing up within ourselves," as Egon was later to write, "to the separation that was inevitably to come."

On August 12, 1938—a date which he was always to remember—Neyman got off the train at the Southern Pacific Station in Berkeley and found five people waiting to welcome him to the University of California.

1978   In the office of the Berkeley statistics department varicolored sheets have suddenly appeared in the boxes of staff members: *Come to the 40-year Hoopla—Friday, August 11, 1978.*

Everyone recognizes the address as Betty Scott's and the date as the fortieth anniversary of Neyman's arrival in Berkeley, moved ahead one day because on the day itself many members of the department, including Neyman, will be flying to San Diego for a combined meeting of several statistical organizations.

A bare ten minutes after the appointed time, Betty Scott's house on Tunnel Road is packed. Betty is in the kitchen carving a leg of lamb. Everybody else is eating, drinking, talking. Neyman gets up to greet me. He is rosy and immaculate. As always, I am surprised at the firmness of his handshake. He takes me over and introduces me to Professor and Mrs. Birge, beside whom he has been sitting. Birge is ninety-one and in a wheelchair. It was he and his wife and daughter who, with Professor and Mrs. Evans, met Neyman when he got off the transcontinental *Challenger* at the Southern Pacific Station on the evening of August 12, 1938. The Evanses are not present. Evans died in 1973, and Mrs. Evans now lives some distance from the Berkeley area.

It seems that every guest must tell me when he or she came to Berkeley. Professor Birge came as an assistant professor in 1918: "That was twenty years before Professor Neyman came!" Betty Scott is distinguished for having studied some statistics at Berkeley "even before Neyman." Joe Hodges says that he arrived at the same time as Neyman—he was an entering freshman in August 1938. Erich Lehmann came two years later as a graduate student, a refugee from Hitler's Germany by way of Switzerland and England. Miriam Scheffé, whose husband Henry was until his death in 1977 one of the leading lights of the Berkeley department, remembers visiting Neyman in his little statistical laboratory at the beginning of the Second World War. The Scheffés were on their way to Princeton, "but Henry always knew that he wanted to end up here in Berkeley."

Peter Bickel, the current chairman of the statistics department, looking still like a graduate student, comes up. He explains that since he got his Ph.D. under Erich Lehmann, who got his Ph.D. under Neyman, he is a Neyman "grandson." The number of Neyman's statistical descendants is scarcely computable. Lehmann, who is only one of his thirty-nine Ph.D.'s at Berkeley, has had thirty-five Ph.D.'s; and Bickel, twenty-one. I ask if there is a Neyman "great grandson" on the faculty. The answer is no. It is now the policy of the department not to keep its own students. "In the beginning it was necessary because nobody else was training people."

Bickel introduces me to David Freedman, the principal author of the text, entitled simply *Statistics,* which is most often used in the course designed to give students in other majors an introduction to the subject. The book—like Hodges and Lehmann's *Basic Concepts of Probability and Statistics*—is dedicated to Neyman and presents his view of statistics as a study of frequentist and stochastic phenomena. This is the view of the majority of the statistics faculty at Berkeley, although other views are also represented.

Guests at the hoopla are not limited to members of the statistics department nor to people from the Berkeley campus. Among those present is Charles Stein from Stanford. He and Herbert Robbins of Columbia have been described to me as the contemporary statisticians whose work Neyman most admires. In 1952, reissuing his *Lectures and Conferences* in a hardback edition, he devoted an entirely new section to "the brilliant recent results of Charles M. Stein." Ten years later he trumpeted Herbert Robbins's work with similar enthusiasm as "Two breakthroughs in the theory of statistical decision making." Stein recalls how, as a young army meteorologist, he became interested in statistics when he read the writings of Neyman and Pearson and their students in *Statistical Research Memoirs.*

At the abundant buffet one leg of lamb follows another. After a while an enormous sheet cake is brought out. It is decorated in birthday-cake fashion with the dates 1938–1978.

Birge is asked to say a few words. From his wheelchair he recalls the pressures exerted by the other sciences in the early 1930's to build up mathematics at Berkeley—the bringing of Evans from Rice Institute to head the department.

"They decided they needed somebody in statistics, and they appointed a committee to get an assistant professor to come; and when I heard about that, I persuaded the committee to change it to a full professor and particularly to get Mr. Neyman. And then I persuaded Mr. Neyman to come. And he's here, you see!" Birge says this last triumphantly to general laughter. "I'm ninety-one years old, and my memory seems to be failing me all the time so I can't remember anything more that happened. It was all a long time ago. But, anyway, that's the way statistics started at the University of California. I know that!"

Mrs. Birge chimes in with details. They took Neyman sightseeing, but he was not really interested in the Golden Gate Bridge or the Bay Bridge, recently completed, or Treasure Island, the artificial setting for the International Exposition which was scheduled to open the following year. He wanted to know about the administration of the university, the courses offered, the students.

Then Neyman, obviously enjoying himself, stands up to "supplement" what the Birges have said. He recalls a whole series of incidents which are connected with Birge. What seems most impressive to him is a physics department picnic to which he was invited shortly after his arrival.

"And there I saw something which was not usual in Europe, in many countries with which I was familiar; namely, what you would call—it might be called a *fraternization* between students, advanced students, and faculty members. This was in a place just off Euclid Avenue."

Somebody fills in "Codornices Park," and Mrs. Birge says, "Yes, that was an annual affair."

"That was something unfamiliar to me," Neyman says, "and I must say that I liked it very much."

Martini in hand, he concludes his recollections of his relationship over the years with Birge by proposing a toast "To Professor and Mrs. Birge!" After the toast to the Birges is drunk, there is a general shout of "Polish toast!"

"Polish toast," Neyman says—"before the war, traditional—at present, no longer. Too bad!"

The guests raise their glasses with him.

"To all the ladies present—

"And some of those absent!"

The University of California was not the biggest university in the world in August 1938; but it was even then a very big university and, for Neyman, a long way from Bloomsbury and University College.

Established by the Birges at International House, he strolled through the wooded campus, foggy on August mornings, and into Benjamin Ide Wheeler Hall. The headquarters of the mathematics department were on the top floor. They consisted of two rooms: one for Mr. Evans and the other for Sarah Hallam, a graduate student who worked as the half-time departmental secretary. There was only one telephone, which was on Miss Hallam's desk. When it rang and she was not there, Mr. Evans stepped around the corner and answered it himself—behavior which seemed undignified to the newcomer from University College.

The week after Neyman's arrival, some twelve thousand students materialized on the campus, and almost immediately Evans asked him to assist with the registration of freshmen who wished to take mathematics courses.

"So I was sitting at one desk on the right, Evans was sitting there, and so there was a crowd of students in front. And so he invited six, I remember—took their names, this and that, and then, 'Please come to the blackboard.' And that was a girl. All right. So she came. She didn't anticipate this. 'Please write down one-half plus one-third.' She looked at him and then wrote. 'Now write "equals" and then see what you get.' She looked at him indignantly and wrote one plus one, two plus three. Two fifths! Well. Evans shook his head. 'You go to the left. You will be in that section. Next please.' That was the way he sorted the students."

"I can't believe it," I say.

"True," Neyman insists. "True."

He did not, however, report the incident to Egon Pearson when he wrote to him a few days later, but commented merely: "My situation here is rather pleasant and the people here are really quite nice."

Evidence of Berkeley's contemporary reaction to Neyman is also scant. Birge was busy with administrative responsibilities in the physics department and wrote to Deming only that Neyman had arrived and was being consulted by people in various departments, "although," he added, "in the only case where I asked him to comment on certain manuscripts which...had been referred to me, he refused to do so on the obvious grounds that they were so poor that he did not want to get in bad with the authors by making frank comments."

("Of course, the situation is quite different if somebody asks me for advice in his research," Neyman explained, "—I would never refuse that.")

Although, in 1938, the University of California consisted of several different campuses, President Sproul resided in the President's House at Berkeley and was a very personal presence there. A forceful man with a booming voice—the university comptroller before becoming president—he was immensely popular with faculty and students, who were to turn out in force later in the year to plead that he refuse the presidency of the Bank of America and remain at Berkeley—as he did. Neyman found himself more comfortable with Provost Deutsch, a classics scholar, considerably older than Sproul, and preferred to consult him on personal matters. On his side Deutsch, at least in the beginning, found the new professor from London "a rather shy person."

Evans's active participation in the creation of a statistical center at Berkeley did not cease with Neyman's arrival. He petitioned Sproul for a half-time technical assistant for Neyman and a new calculating machine. "...a new man on our faculty...would seem to be entitled to the use of a new machine." He pressed for courses in point set theory and the modern theories of integration and abstract spaces "which the best modern work in the theory of probability demands." He was also concerned that Neyman have the necessary data and assistance to carry on his own research.

Lola and Micky had fallen ill just before they were to leave Gdynia; and Neyman did not know whether they were still in Poland, in England, or at sea when he arrived in Berkeley. He went ahead with preparations for their arrival. "Let's hope for the best." It was a phrase he was to repeat often during the next half dozen years. A house with a view was rented in the hills where most of the other professors had their homes. A car was purchased, driving lessons taken.

In Europe the crisis precipitated by Hitler's demands upon Czechoslovakia moved inexorably toward what was to be known as "Munich."

"At any rate I am very glad you have got away first, if it is going to be war...," Egon wrote. "We wait on the wireless."

On the last day of September, as Neville Chamberlain returned to London with an agreement he hoped would guarantee peace, the ship carrying Lola and Micky sailed into San Francisco Bay.

"In spite of a couple months spent in the States, I...accepted all the news [of Munich] as if I still was in London," Neyman told Egon in the letter announcing his family's safe arrival. "It was a ghastly time. The worst of it is that I do not believe in agreements with Hitlers and think that the tragedy is only postponed...But even if the war is only postponed, we must thank the gods for that. I begin to think of some work, and probably you do too."

The university had already purchased a building, a former divinity school, to provide housing for the statistical laboratory which Neyman was to organize. The space which had been assigned to him had served as the chapel of the school, and he was "rather pleased" that his laboratory was going to replace "a centre of superstition." The fact that the building would not be ready for occupancy until Christmas did not disturb him. He was overwhelmed by the events taking place across the ocean. Teaching two courses seemed enough of an effort for the time being.

He did not forget the incident in Evans's office. From time to time, and more frequently as time went on, he interrupted his lectures to call students to the blackboard and ask them questions which forced them to recapitulate the material. From the beginning his custom was "Ladies First." One of the first ladies was Sarah Hallam, the department's part-time secretary. She was able to reconcile herself to the "high school approach" of the new Polish professor, but another student—himself now a professor—stopped attending class and relied upon Miss Hallam's notes to pass the course.

Neyman taught two classes that first year. One, on probability and statistics, was an upper division course which had been regularly offered in the past. After Neyman's class had been meeting for several weeks, a tall attractive young woman named Elizabeth Scott, an astronomy major who had taken the course the year before, dropped in.

"It was known to us as students that Neyman was going to come and that he was a great statistics person. So I arranged with my friends that when he got up to something I didn't know, I would start going to the class. Well, he asked a question of somebody and this person—'I don't know.' Then 'I don't know,' 'I don't know,' 'I don't know.' Pretty soon he came to me, and I said, 'I don't know.' He said, 'Why don't you know?' and I said, 'This is the first time I have come to the class.' Then he *really* jumped on me!"

Her explanation that she had already taken the course was not an adequate excuse: she must, he told her, attend every meeting if she was to benefit at all. Thus instructed she returned the next time the class met and continued regularly thereafter.

Neyman's other class dealt with new developments in probability and statistics, such as the axiomatics of von Mises and Kolmogorov, as well as his own and Egon Pearson's work. He quickly discovered that if students at the lower level were weak in simple arithmetic, graduate students knew little of modern mathematics.

Shortly after his arrival he became the chairman of the Committee on Statistics, the members of which ranged over a dozen different departments having some interest in statistical education for their majors. He found these contacts with other subject matters "very precious," he told Warren Weaver of the Rockefeller Foundation. "Through [such contacts] one is faced by all kinds of applicational problems, which almost invariably throw a new light on various sections of the mathematical theory of probability and frequently suggest new mathematical problems."

His first communications in regard to statistics made clear his intentions for Berkeley—and for America. "This country needs statisticians *sensu stricto*, not mathematicians with some knowledge of the theory of probability and not, say, biologists who have studied a little of mathematics." In the past, Americans who had wanted to become statisticians had gone to London to study. Now it seemed "reasonable" to organize a center "more easily accessible."

His own ties to London remained strong. Late at night he wrote regularly to Egon. There was hardly anything about the American situation, academic or political, in his letters. Instead familiar English names ran through them. He did not see himself abandoning his long collaboration with Egon.

"It was really a good part of my life in London. Let's try to build another good one and fruitful for us both—in spite of our being so far away from each other."

All during 1938–39 he tried to move forward the two projects which he and Egon had been discussing before he left England: the joint publication by University College and the University of California of *Statistical Research Memoirs* and the writing of the joint book, *Whither Mathematical Statistics?*

"We *must* write it!"

According to plan, the book would consist of a mix of Pearson's lectures at University College and Neyman's Washington lectures and conferences, the authors undifferentiated. A yet unwritten chapter on "bad habits" would, in Neyman's opinion, be the really new and important thing. In a draft of the introduction, he explained the purpose of this chapter.

"The desire of the authors to have the level of mathematical accuracy in statistical literature a little raised [will make] it necessary to introduce some criticisms....

"Of course, having in view to exhibit the unsatisfactory state of the general level of statistical literature, the examples [will have to be] selected from the writings of leading scientists, having a considerable and fully deserved authority, of those, in fact, who could not be easily suspected of not being able to do any better."

He wrote to Egon: "But wait! Wait till we publish the 'bad habits.' He will burst!" He was, of course, Fisher.

While letters concerning these two projects were going back and forth across the Atlantic, the American statistical community was eagerly drawing Neyman in. He was invited to present a paper at the joint meeting of the American Statistical Association and the Institute of Mathematical Statistics in Detroit. He was appointed a member of the committee planning the Tenth International Mathematical Congress, which was to take place in Cambridge, Massachusetts, in 1940. He was asked to serve on the editorial board of the *Annals of Mathematical Statistics*.

By the middle of November, Evans was reporting to President Sproul that Professor Neyman had lived up to expectations and had already been helpful to other departments of the university as far away as Riverside. He reminded Sproul that in his letter offering Neyman an appointment he had expressed the hope that after a year his salary would be raised by $500: "I recommend that this be done."

In December, before leaving for the Detroit meeting, Neyman sent Evans four and a half single-spaced typewritten pages outlining his ideas about the organization of teaching and research in statistics at Berkeley. Already provisions for machines, statistical tables, and journals had been made in the budget. He had also been given Miss Hallam as a part-time research assistant, but this arrangement did not seem satisfactory: "She, or some other person, must spend *the whole of her time* in the Lab and *confine her whole attention* to the work in the Lab."

As to teaching, he felt that all the systems currently being followed should be revised. He liked to cite as an example of the chasm which existed between the "mathematical" and the "empirical" schools of statistics, two quite different reviews of Cramér's book, *Random Variables and Frequency Distributions.* ("This is an excellent book!") M.S. Bartlett, "a distinguished representative of the younger generation of English statisticians," found it good but much too short and difficult to follow. A.Y. Khinchin, with Kolmogorov the leader of the Russian school, found it good—especially for beginners! Yet the scientific basis for the treatment of the problems of applications which concerned the English school (read "Fisher") was exactly the recent work of the Russian school. "And to master those results one needs to be well on the level of the book by Cramér."

In addition to the program of research and teaching which he outlined in his letter to Evans, Neyman pressed for the Statistical Laboratory to have its identity officially recognized. The people in charge of statistical offices, he explained, were frequently unfamiliar with mathematics and afraid of it. "Consequently a request for data signed by a 'Professor of Mathematics' or even by the 'Chairman of the Department of Mathematics' is likely to produce more of an unpleasant surprise than enthusiasm."

In Detroit, at his first scientific meeting as an "American" statistician, Neyman delivered, not one, but three papers: "Contributions to the theory of sampling human populations" (stimulated by a question from Milton Friedman during the Washington conferences), "On a new class of 'contagious' distributions" (the outcome of the London work with Beall), and "On the hypotheses underlying the applications of statistical methods to routine laboratory analysis" (work which in its published form was to be extended to sampling inspections of mass productions).

The last named paper was presented at a session presided over by Dr. Joseph Berkson of the Mayo Clinic, a combative figure in American medical statistics. Sometime afterwards Berkson wrote to Neyman with a statistical question. Neyman responded carefully and at length. From that time on, the two men corresponded regularly, the thickness of the folders labeled "Berkson" in Neyman's files easily exceeding those marked with the name of any other individual.

Neyman was unimpressed, however, by the general level of the native American statisticians at the Detroit meeting. "Bits of talk" further convinced him that it was urgent for him and Egon to continue the publication of *Statistical Research Memoirs.*

"These bits of talk are difficult to report, and it is difficult to explain how they contribute to my general conviction that the *Annals* will hardly ever attain what one could call a good level.... This is a small thing but it is a characteristic one: there was a question of having papers by Europeans written in French or German, and the [other] editors decided in the negative because one to two such papers may result in losing scores of subscribers. What is the level of those subscribers?"

On the nonscientific side, he was enthusiastic about his first American meeting: "a big hotel...some evening meetings [and] at the same time and next door there were various parties and dances just as you see in some American films with beautifully undressed glorious girls, intoxicated guys, etc. Great fun!"

By the beginning of his second semester at Berkeley, Neyman was established in the former divinity school and was writing to Egon on a new letterhead: *University of California, Department of Mathematics, Statistical Laboratory*. He had begun to teach an extra fifteen hours a week, utilizing for lectures the time designated in the catalogue for laboratory work in order to give his students some knowledge of what had been happening in mathematics since the World War. They could learn to compute on their own! He badly needed help. The minimum statistics faculty at Berkeley—he told Evans— should be three: himself, an instructor, and a teaching assistant.

It seemed, that spring, that the ideal applicant for the post of instructor had indeed presented himself. Abraham Wald, dismissed from his position in Vienna because he was a Jew, had been spending a semester at Columbia with Hotelling under a fellowship partially supported by the Carnegie Foundation. Although he had left Europe reluctantly, he was now eager to remain in America as a result of recent events abroad. He had first written to Neyman about a job in August 1938; but Neyman, having come himself so recently to Berkeley, had been unable to do anything for him then. In March 1939, however, when Wald wrote again—"I would be happy to be at the same university as you are and to have the opportunity to discuss different statistical problems with you"—Neyman pressed Evans to add the younger man to the mathematics faculty, but without success. Ultimately Wald wrote that, although he would still have preferred to be at the same institution as Neyman, he was accepting the position which Hotelling had been able to arrange for him at Columbia.

Discussions with Egon about the book dragged on. Neyman began to get impatient. He suspected a lack of enthusiasm and suggested abandoning both the joint book and the *Memoirs*.

"...what I wonder did I say that made you think I wanted to write a separate book?" Egon demanded. He could not give much help in starting up the journal again, and he had not wanted to give the impression that he could; but for a long time there had been in the back of his mind "the idea that it woud be a good thing to have [my] version of our philosophical approach down in typescript and passed on to you because of the quite considerable chance that there would be at any moment a sudden suspicion of civilized lecturing & scientific thinking over here."

If only they could talk, they could settle all the troublesome questions between them! Both had received invitations to deliver papers at a second congress in Geneva sponsored by the International Institute of Intellectual Cooperation of the League of Nations—the subject this time to be the applications of the theory of probability. Neyman wanted very much to go, but he was concerned about the risk to his family if war would break out while he was abroad. He consulted Provost Deutsch, who referred him to the dean of the School of Jurisprudence, a specialist in international law, "if there is still such a thing." It was the dean's opinion that if Neyman were to be caught in Europe at the beginning of hostilities, with Germany on one side and France and Poland on the other, he would be drafted into the French army. After much debate, Neyman decided to send his paper for the congress ("Basic ideas and some recent results of the theory of testing statistical hypotheses") to Egon to present for him.

He tried to imagine how Egon and Eileen, in England, must feel under the threat of war.

"…is it that you get insensitive in time or the contrary?"

"To some extent one seems to get used to tension & stop worrying very much about what Hitler, Mussolini & Co. say," Egon replied. "Now that it is so clear that nothing that Hitler says need be believed, it simplifies the situation a bit.…It is difficult what to make of Russia…[and] curious how the Poles, after appearing as villains over Czechoslovakia [Poland had joined Hitler with its own demands for territory]…are now regarded as perfect high minded allies."

In case of war undergraduate teaching at University College would be continued at the University College of Wales, but Egon did not think there were many people—certainly not the whole department—who would go there.

"As a matter of fact, if there were a war and no really suitable statistical work cropped up, I would quite like to do something quite different: i.e. I would rather look after an anti-aircraft gun than have to compute range tables."

On July 1, 1939, a new fiscal year began at the University of California. Although the amount budgeted for statistics was not so generous as Neyman had hoped, he had been given a full-time secretary and research assistant in Miss Betty Scott—"the best of my last year's students, an astronomer," he reminded Egon. "I do not know however whether she will stay. Her attachment to astronomy seems to be considerable."

All during July, Neyman worked daily in the Statistical Laboratory. He pictured Egon presenting both of their papers in Geneva with Fisher in the audience, and waited impatiently for a report.

Fisher had made the final summation, Egon wrote as soon as the congress was over, and had as usual "[squirted] a lot of black fluid over the point like an irritated squid…actually I don't think this 'scandalousness' was obvious to the audience, partly because they would not understand the references in a full 4 days programme.…So in a way I was glad you were not there, as you might have got very excited, & if there had been a scene, we should have suffered as well as R.A.F."

Of the Germans and the Italians, only Gini had been present, "always with a mixture of sourness and a kind of sardonic humour, [trying] to pull people's ideas to pieces & substitute his own."

On August 22 Neyman, responding, wrote to Egon, "The international situation is apalling—will this letter reach you or not?"

It was obvious that Poland was going to be the next victim of Hitler. Great Britain, shocked by Germany's complete about face in regard to Czechoslovakia, had made its position clear. An attack on Polish independence, resisted by the Poles, would mean active support of Poland.

On September 1 the German army marched across the Polish border; and Neyman's first year in America, which had begun with Munich, ended with Warsaw.

1939
—
1940

During the first half of September 1939 radio broadcasts every few hours reported to the American people the beginning of the Second World War.

"I have the sickly feeling forgotten since the days of the Russian Revolution and, for the time being, cannot work very efficiently," Neyman wrote Pearson six days after the Germans had marched into Poland. "...As I write now (11 a.m.) night is approaching in England and you may be being bombed...."

At almost the same time Pearson himself was writing, "I am afraid you will be feeling terribly upset, in fact far more so than we...."

Although Neyman confessed that in such times "one even feels uneasy writing about books started and testing hypotheses, etc.," he agreed with Egon that until something else was demanded of them, they should go on doing what they were doing. The joint publication of *Statistical Research Memoirs* might have to be abandoned, but the possibility of the joint book would remain. On September 11, having just heard that a German battleship had been sunk and that Warsaw was still holding out, Neyman felt "sufficiently cheered up" to think again of editing Egon's lectures, which he had received earlier, "but then I could not think about them."

Throughout the first weeks of the war he tried to avoid dwelling upon what might be happening to his brother and to colleagues and friends in Poland and to concentrate on work. He felt that he partly succeeded in the first but failed utterly in the second.

"...I have made a feeble attempt to interest one American publisher with our book....I had to tell him that the book will be nothing like a textbook....Of course I could perhaps try to persuade him that in spite of being no textbook the book will sell alright, but somehow I didn't and probably shall not have the energy to make another trial."

The next day the Polish capital, which had defended itself heroically, fell.

Neyman still had not heard what had happened to his brother and his little family, who had been living in Gdynia before the war; nor had he heard anything of his many friends, colleagues, and students still in Poland.

There was in existence in the United States a foundation—established as a memorial to General Tadeusz Kosciuszko, the Polish hero of the American Revolution—the purpose of which was to promote the exchange of students and professors between the United States and Poland. Now, with the cessation of academic life in the latter country, the exchange had become of necessity one way. There began to be a continuing correspondence between Neyman and the Kosciuszko Foundation as well as other institutions and individuals who might be able to help his colleagues still in Poland and, later, newly arrived as refugees in the United States. In this, he frequently asked advice of young Dresch, who was now working with him as an instructor.

"I would kind of advise him on strategies, or at least give him my opinion, and try to tone down what I thought were undiplomatic letters. He seemed to rely on my judgment enough to let me see them and make some changes in some of them."

Neyman was also concerned about friends in England. He suggested to Florence David that she come to Berkeley and work in the Lab, but she preferred to try to find useful statistical work in England. In the meantime she was planning to help with the fall harvest.

Throughout the uncertainties of autumn 1939 work in the Statistical Laboratory became ever more intensive.

"We were all kind of learning statistics together—Neyman's kind of statistics at any rate, which is the kind I think of as being real statistics," Dresch says. "All of us, I guess, had some kind of smattering, just as I had, of somebody else's notion of statistics, on a very elementary level, so we all attended Neyman's lectures. He also established a kind of workshop. It went on the university books as a two-hour lab section, but it really lasted more like four. Essentially it was a kind of bonehead session on miscellaneous bits of mathematics that were required for statistics and that he thought were kind of skimped by the mathematics department. I taught that workshop the first year. The next year I gave some of his lectures, sort of repeating what he had done the previous year, while he introduced some new courses. During the time I was giving some of Neyman's courses, George Dantzig [now a professor at Stanford] was running that so-called lab section. At one point, I remember, Neyman was also conducting a seminar in economics, an evening seminar, which was held at Evans's house. Evans was there of course. Also Larry Klein, who later won a Nobel Prize for economics. One of the students in my course was Julia Bowman [later Robinson, the first woman to be elected president of the American Mathematical Society]. I think our first official student from 'outside' was Dantzig. We really had quite a group."

With all this activity Neyman did not have much time for other things.

"However, I am attached to the idea of the book," he reiterated to Egon, "and, if there are realistic prospects for it being eventually published, I shall certainly find time to work on it."

In December he heard from the State Department that Karol Neyman and his family had remained in Gdynia until the beginning of November. Then, abandoning all possessions which they could not carry, they had set out by train ("third class carriage") for an unspecified destination in the interior of Poland. Karol's only son, Jerzy, a military surgeon, had been imprisoned.

"It appears that it is much easier to talk about [my countrymen] than to give them any help," Neyman lamented to Egon. "…In Warsaw the Germans are carrying out some thing like a systematic program of extermination of Polish intellectuals. An old man…, whom you met in Mądralin but probably do not remember, was shot. Several others died in prisons and concentration camps. The whole equipment of physical & chemical institutes were shipped to Germany. Well, you probably have heard all about it from the papers.

"Now what shall we do with the book? Shall we try to finish it up…."

The mimeographed copies of *Lectures and Conferences* had been exhausted in a year and a half. It was exciting to think that hundreds of people had been interested in his ideas. Even a paragraph in *Science Progress*, over the initials R.A.F., summarily dismissing the contents—"not enough original material to justify publication as a book, and too much that is trivial"—did not faze him.

"It is a very nasty feeling when one sees that somebody powerful tries to keep him down, but what a feeling when this keeping down does not succeed! The gain is worth the price.—Life is beautiful and all—or at least some—ladies desirable!"

He thought of putting Fisher's review beside the *Zentralblatt* review of William Feller when a second edition was issued.

"The point of departure for the author is always actual practical problems," Feller had written, "and he never loses sight of the applications. At the same time his goal is always a truly rigorous mathematical theory. He appears to insist on absolute conceptual clarity and rigor, not only as a sound foundation, but also because it is really useful and necessary, particularly where the practical problem goes beyond the mathematical aspect. This attitude and the lively combination of theory and practice are especially apparent in a collection of conferences, where the point of view and problems of methodology play a more prominent role. Particularly pleasing is the clarification of the boundaries between purely mathematical schemata and everyday conventions. The lectures (all very lucidly presented) deal with subjects only loosely connected….Some are not easily accessible in contemporary literature, but they are nevertheless of the utmost significance."

In Neyman's letters of this period, the modest problems of the Statistical Laboratory are a continuing counterpoint to the unknown but suspected tragedies of his colleagues in Poland. The second semester of 1939–40 he increased his voluntary teaching program to twenty-five hours a week. Several students were helping on simpler questions which came from other departments. One of these was Mark Eudey, who had graduated from Berkeley in 1934 with a degree in mathematics. In 1939 Eudey had decided to return for his doctor's degree. When he needed an extra course to fill out his program, his friend Ronald Shephard suggested that he take statistics. Eudey

169

objected that he had already taken a semester course in statistics and another in probability; but Shephard assured him that Neyman's course would be different.

After a while Neyman put Eudey to work on a project concerning voting patterns in Hawaii. The question was whether people were voting by race, and the solution involved fifteen or twenty different equations which Neyman himself had set up. No matter what Eudey did, the answer never seemed right. Finally he examined the equations one by one and found that not all of them were independent. Neyman was delighted.

"I think he was just tremendously pleased that a student had taken the trouble to really look at the problem and figure out what was wrong. It didn't seem to bother him at all that I'd found him in a mistake."

With more people working in the lab, there was not enough room on the first floor of the old divinity school.

"Having in view an arrangement which would be satisfactory in the long run," Neyman informed Evans, "we must think of the possibility of nearly doubling the space of the Laboratory."

According to Dresch, Neyman seemed on occasion to feel that Evans had "gone back" on some of his initial promises.

"One of the things he seemed to expect was to be an independent department right off, and Evans seemed to think it would be a kind of subdivision of the mathematics department. That was a source of conflict. I would occasionally hear complaints from Neyman. I don't think I ever talked to Evans about them."

Documentary evidence seems to indicate that Evans, as chairman of the department, consistently and strongly supported Neyman's requests to the administration in regard to statistics. In the spring of 1940, for example, Neyman was asking for an additional $300 "to prepare in advance a collection of suitable applied problems together with their solutions..." A letter from Evans to Sproul, dated less than a week later, recommended this expenditure "if it is at all possible to be granted."

It began to appear that there might be some ways in which Neyman could give help to Polish colleagues. He had received proofs of the *Biometrika* paper of Przyborowski and Wilenski—the Jewish cobbler's son who had become Przyborowski's assistant and co-worker. "Poor fellows, nobody knows whether they are dead, in a concentration camp or in exile, but it is certain that they do not feel comfortable." Would Egon be inclined to share the cost of reprints until such time as the authors themselves might be in a position to pay for them?

A countryman whom Neyman had been successful in helping was Antoni Zygmund, the young friend of his days at the University of Warsaw, a professor at Wilno before the war.

"I owe him very much my life," Zygmund explains shyly to me. "Retreating from the Russians, our regiment had to cross Lithuania so I was able to return to Wilno, which was essentially Lithuanian. It was Neyman who organized things—there was a Kosciuszko Foundation, and they gave me a stipend in March of 1940. Just before the Russian offensive, I was able to

170

escape with my family. Had I stayed in Poland, probably I would have died....I don't advertise it, but my escape to America I owe primarily to Neyman. So I have personal reasons to feel very deep friendship."

At about the same time that Zygmund arrived in the United States, Neyman heard that Karol Neyman was in Warsaw with his family "in distressing circumstances." A Polish businessman was not so welcome in the United States as a distinguished mathematician, and there was nothing he could do to help Karol and his family except to arrange food packages to be sent to them through the Danish Red Cross. In time he also heard from Wilenski, who was in Biolystok. Wilenski described an enormous number of schools being built and organized—"well, this is not so bad after all," Neyman wrote Pearson—but Przyborowski, threatened with capture by the Germans as he and Wilenski had fled Krakow, had committed suicide.

Throughout the spring of 1940 disaster followed disaster in the headlines and the broadcasts. In April the Germans invaded Norway and in May, the Low Countries. In June—the day after Paris fell—Neyman cabled Pearson and invited him and his little family—there was a baby, Judith, by then—to come to America and stay with him and Lola until the end of the war. Pearson gently refused.

"For me there could be no question of moving as I have a definite job of work to do—unless at some future date the centre of the British Empire were to migrate to Canada! And though we might both like Judith to be in a safe place, as Eileen doesn't want to leave me, the wish is hardly realisable. Of course the future is entirely unpredictable but I certainly don't see why the Germans should win for they have a pretty tough nut to crack in this country." The English reaction to the French collapse had contained an element of relief—"We are now alone by ourselves...there is no longer that uncertain feeling of reliance on foreigners...."

In the United States, "War Preparedness" replaced "Defense." In response to a notice circulated by the American Mathematical Society, Neyman volunteered his services: "I have had various contacts with mass production, control of quality, etc., and it is possible that I may be useful in cooperating with the present efforts to arm."

"My own desire for action," he wrote to Egon, "took the form of a course in statistical methods for engineers in industry. It was advertised a few days ago, and the response was immediate and inspiring."

Under the threat of invasion, Pearson fretted whether Neyman, planning to work all summer in his lab, was going to get a much needed rest. He himself reported devoting a Sunday (with three other statisticians) to Local Defense Volunteer Duty: "I was assistant cook this time & turned over frying sausages & tomatoes & washed potatoes. At night we sleep on mattresses on the floor in our clothes & do two hours patrolling." He was still hoping that he could issue *Biometrika*. "At present nothing is coming along....If you have anything to spare I should be glad of it, but probably you will prefer to publish in America."

He had earlier asked Neyman to review Jeffreys's new book (in which Bayes's theorem was described as being to the theory of probability what the

Pythagorean theorem was to geometry). Neyman had written some thirty pages, trying—unsuccessfully—to write "nicely." He now offered to return the book to Egon, since he "[did] not like to engage in controversies," and send him instead a paper on a subject they had been discussing for some time: the connection between the fiducial theory of Fisher and his own theory of confidence intervals.

"My efficiency now is very low and the paper is not ready. But your letter is so nice and cheerful and I do want you to have plenty of M.S.'s and have the Biometrika out at the usual time. So I will set out to work and, maybe, within a week or so will send you something."

Another American publisher had indicated that he might be interested in *Whither Mathematical Statistics?*

"You might ask why I didn't collect everything, put it in order, and have the book out. Well—this is what being a spectator of European happenings means. I just let the time go and did not manage to do anything."

In mitigation, he pointed out, there was still one aspect of Egon's lectures which he felt would have to be remedied before publication.

"As I have already written to you, I do not like your history of Fisher's work.... If it [were] a question of a review in a journal or the like, I would not mind letting it go. But our book is intended—at least in my mind—to suggest to people the necessity of more accuracy in statistical work. 'La critique est la vie de la science' and that kind of thing and it seems to me that it would not be proper for us to praise in it the works which are so outstanding for their vagueness and inaccuracy combined with a pontifical highbrowness...."

He suggested that they drop the sections of the book which would require much discussion between them. He would give up the chapter on "bad habits," which Egon had always objected to, if Egon would agree to the omission of his lecture on Fisher's work.

"They say in the paper this morning that the Blitzkrieg on England is approaching. Good luck, and remember that if circumstances change and there is the necessity of sending Judith over alone or with Eileen, we shall be glad to have them. What I am afraid of is not bombs and parachutists but poison, bacteria, and hunger."

All that August of 1940, as students began to return to the Berkeley campus, the threat of a channel invasion, as well as a blitzkrieg from the air, hung over Britain; but by August 24, when Pearson responded to Neyman's letter, neither had materialized.

"I see that I have ascribed a little bit too much to R.A.F.," he conceded, "but that is a matter of detail. My writing was to some extent autobiographical, and therefore you may say not suitable for an accurate book; nevertheless from my point of view the fact remains that I personally (and therefore possibly you?) would never have got on the idea of testing hypotheses if Fisher had not written his paper of 1922 and Rhodes and K.P. theirs of 1924.

"I remember somewhere some remarks of K.P. about Galton or Darwin, of how they probably made all sorts of faults, but nevertheless the fact remained that because of their work other things happened and things developed. This is somewhat so with R.A.F.—he was the cause of much progress, and however objectionable he may be, one gains nothing by denying it."

As to publishing, if Neyman had the time and energy for it, he was perfectly willing to let him handle the matter as he saw fit. Maybe after the war, when they could talk again, there could be a second, revised edition.

"The only point I want to urge is that you remember that, however inaccurate and pontifical R.A.F. may have been, he did—no doubt unwillingly—play a part in our education. He stirred up our minds, by making it clear that chi-square with appropriate degrees of freedom was necessary, i.e., that we must control the first kind of error. Further, whether we should have hit on the likelihood ratio without his likelihood development is not clear; at any rate I certainly reached the likelihood ratio from the idea of his L. Again it was enormously useful having available his analysis of variance tests; would the linear hypothesis have been formulated without? Perhaps, but one just doesn't know."

He concluded in another vein:

"All goes on here very steadily. From extracts we see of speeches and articles in your papers, it is clear that you are more anxious about our fate...than we are. Of course this is natural; you can't feel the determined confidence that is deeply rooted in most of us over here...."

1940
—
1942

In September 1940 the quiet little village of Hanover in the foothills of the White Mountains—the seat of Dartmouth College—seemed as far away as one could get from the past year's bloody subjugation of Poland and, now, the nightly bombings of London.

What was to be for Neyman the most memorable paper of the IMS meeting being held there was one presented by Harold Hotelling. In it Hotelling had created a character he called "Young Jones" to illustrate the low level of the teaching of statistics in America. Jones—his personality "all that might be desired...white, Protestant, native-born American"—is drafted to teach statistics by the "Department of X," which has just become aware that some very important work in the field of X involves the use of statistics. Eventually he learns that Karl Pearson is the great man of statistics, and that *Biometrika* is the central source of information. Unfortunately most of the papers in *Biometrika* and Pearson's writings, while not lacking in vigor, trail off into mathematical discourse of a kind with which young Jones feels ill at ease. After Jones has taught the statistics course for several years, "[it] expands, takes on a settled form, and after a while crystallizes into a textbook....He becomes a Professor of Statistics and perhaps an officer in a national association. His textbook...has a large sale and is used as a source by other young men writing textbooks...."

To Neyman, returning to Berkeley after Hotelling's talk, it seemed that a simple, sound textbook on probability and statistics was as much needed in America as his and Egon's proposed examination of the logical foundations of the subject. He would write a *First Course in Probability and Statistics* for the young Joneses. He began to work on such a book almost immediately, but it was not to be published until 1950.

In London, Egon, bombarded nightly by the Luftwaffe, no longer thought of books, joint or otherwise. He still managed to carry on with *Biometrika*, editing and rewriting at night, sometimes in a bomb shelter.

"...rather thin I am afraid," he commented apologetically as he sent the most recent volume to Berkeley, "but no one writes over here now. What about that contribution of yours...."

The contribution to which he was referring was the paper which Neyman had presented at the Hanover meeting: "Fiducial argument and the theory of confidence intervals." In it he had explained that although he had initially found Fisher's conception of fiducial argument difficult to reconcile with what he considered essential in his own theory of confidence intervals, he had been inclined to think that the difficulties were "more or less, lapsus linguae"—hard to avoid in the early stages of any theory. It had come as a "great surprise" to him that, "far from recognizing them as misunderstandings, [Fisher] considered fiducial probability and fiducial distributions as absolutely essential parts of his theory." As a result he, Neyman, had begun to doubt whether the two theories—his and Fisher's—were really equivalent. Now, seven years later, he had concluded that they were not.

The theory of fiducial argument had developed from ideas similar to those underlying the theory of confidence intervals.

"These ideas, however, seem to have been too vague to crystallize into a mathematical theory. Instead they resulted in misconceptions of 'fiducial probability' and 'fiducial distribution of a parameter' which seem to involve intrinsic inconsistencies as described....In this light, the theory of fiducial inference is simply non-existent in the same sense as, for example, a theory of numbers defined by mutually contradictory definitions...."

Neyman sent this paper off to Pearson on February 11, 1941, "and had a good sleep—nearly missed my 9 a.m. lecture." Rested and refreshed, he decided that perhaps he had been too pessimistic in thinking that he could not help with other contributions to *Biometrika*. He mentioned to Egon a number of subjects on which people at Berkeley were working.

Preparations for eventual entry into the war were intensifying in the United States. The preceding September the Selective Service and Training Bill had been signed by the president. While its effects were not yet felt on the campus, they would be soon. The young men whom Neyman had been training were very much in demand by government and military. Dresch and Dantzig were shortly to be on their way to Washington, one to the Navy and the other to the Army Air Forces. There was even a possibility of a job for a woman with only the most elementary upper division statistical course behind her!

Hours for "good work on theory" had been few since Neyman had arrived in Berkeley. They now became a thing of the past. His bibliography lists only one paper published in 1942 (and that was the Geneva talk of 1939), none in 1943 and 1944, and one in 1945. He worried constantly that, with his classes and without competent help, he was "lagging behind" in his answers to questions put to him by other departments. He listed some of these for the Committee on Statistics: relationship between the density of soil and such factors as stone content (Department of Soils), mutations produced by irradiation (Department of Genetics), rate of egg laying and egg eating in connection with population density of *Tribolium confusum* (Department of Entomology), estimation of timber stands (Forestry Experiment Station), estimate of contamination of paper milk bottles (Department of Hygiene), analysis of experiments with new varieties of wheat (Department of Agriculture).

The departments which he was assisting were appreciative of his efforts. Five members of the Committee on Statistics—including Evans—called upon Provost Deutsch "especially to urge that everything possible be done for the work of the statistical laboratory which has proved to be of enormous benefit to their departments and many other departments." Later Professor E.B. Babcock, who had not been a member of the group, told President Sproul that he felt "impelled" to write to express his personal appreciation of the "fine work" being done by Professor Neyman in helping to make statistics "more serviceable" to biology.

By the end of March 1941, Neyman had still not received any response from Pearson regarding the paper on fiducial argument and confidence intervals or the other papers he had proposed to send to *Biometrika*.

"I wanted you to tell me whether you like the subjects so that, if I do get a manuscript, there would not be any special danger of its being refused."

Egon replied that he had read Neyman's paper "with great pleasure" and expressed interest in all the suggestions for contributions to *Biometrika*, especially those having to do with the Berkeley cyclotron. "I feel a little hesitant...but of course [the authors] are free men & if they like to do this it would be very welcome. If you are satisfied with the statistical treatment, then I think it will be quite safe to send them & there will be no risk of refusal...." There might of course be other risks. The Cambridge University Press (which published *Biometrika*) might be bombed, or all paper for scientific publication might be withheld by the government. "Still probably if the authors feel well disposed to us, they will be prepared to accept a little risk."

For some reason Neyman did not immediately send Egon any of the papers he had suggested. He was occupied with the problem of finding an appropriate successor to Dantzig, whose new job with the Air Forces would pay him twice the salary he had been receiving as a research assistant in statistics. The ideal replacement would be someone like J.L. Doob, an American probabilist for whom Neyman had already developed great admiration; but obviously the position was not a suitable one to offer to a man of Doob's stature. He wrote for other suggestions to a number of people, including Stanisław Ulam, whom he had known in Poland, and his former London student, Churchill Eisenhart. Both of these recommended Dorothy Bernstein, a young mathematician at the University of Wisconsin; and, in the early summer of 1941, Neyman pushed through an appointment as a research assistant for her. She was concerned that, trained as a mathematician, she was not really suited for work in statistics; and he went to pains to reassure her in several long letters. But, although it was not easy for a woman to get a job in mathematics at that time, she decided not to accept the Berkeley offer.

Neyman, desperate, wired an offer of a lower grade position to another woman who had attended two of his courses during the summer of 1940. Evelyn Fix, thirty-seven years old, had studied mathematics at the University of Minnesota, obtaining a bachelor's and a master's degree in 1924 and 1925. Since then, she had taught mathematics in junior and senior high schools and had obtained a degree in library science. She had come to Berkeley for the summer session because she knew Evans and his family and, as Neyman liked to say, "had caught the bug." She came to the Lab as a "technical assistant" in August 1941.

At the beginning of the 1941–42 academic year, Neyman was at last able to send Egon a manuscript for *Biometrika*, this one connected with forestry experimentation, a subject which he had not previously mentioned as a possibility. The author was A.A. Hasel of the California Forest and Range Experiment Station.

Several months passed without a response. Neyman was busy. Although he and his laboratory had not yet been "cleared" for war work, he had been retained in an advisory capacity by Warren Weaver, who now headed the Fire Control Section of the National Defense Research Committee. When, however, still another American publisher expressed an interest in the joint book, he wrote again to Egon to say that if he could firm up the offer, he was prepared to use the time between semesters to get the manuscript in shape for publication. He recalled that they had not agreed on certain points. Would Egon leave alterations and omissions to his discretion "on the understanding that I shall do my best to follow your intentions....

"A co-operation is difficult these days and it may be reasonable to drop it altogether if a satisfactory form of it is too difficult. However, the years 1927–1934 meant a great deal to the development of both of us and it would be nice to mark the closing of that period with a book summarizing the results of our effort."

This letter from Neyman crossed a letter from Pearson in which the latter, explaining that he had been ill during the summer, enclosed a negative report by his referee on the manuscript by Hasel.

"On the decision there is little more to be said than that I cannot feel that the results so far attained justify so much space. It seems to me that from the Forestry Service point of view more evidence is wanted before the various possible methods are set out in this rather elementary way for the ordinary worker. However, you will feel sore about it, I fear...."

He then went on to say that, in regard to the book, if Neyman was not going to make any distinction between their separate contributions, it was only right that he should combine them with whatever modifications he thought necessary.

While this letter of Pearson's was on its way to Neyman, the Japanese attacked Pearl Harbor. Neyman was in the mountains at the university's experimental forestry station when he heard the news.

The United States had been at war for ten days—on the Atlantic as well as the Pacific front—when he received Egon's letter refusing the Hasel paper, but in his ten line reply he made no mention of Pearl Harbor or of the war.

"I have just seen your letter refusing Mr. Hasel's paper and the report of your referee.... This again indicates that now you and I attach different importance to things. I did not think this to be the case and therefore recently offered to formulate a joint book of ours. I think now that it will be safer for me to withdraw the offer."

"Of course I knew that the refusal of Hasel's paper would be painful," Egon replied at the beginning of 1942. "...I may be clumsy in expression, but I cannot see unjust. Our difference of opinion as to what is the importance of things no doubt exists. As to the book...I can only point out that it is really here a question of my taking the risk of accepting your judgement, which I have done. If I was wrong, I shall be the loser, but I did not contemplate any serious risk—unless of course your outlook has completely changed since you wrote the Washington lectures.

"As to refusals, I have had to make 4 in the past 6 weeks: (1) led to a charming letter, (2) to some distress, followed by a Christmas card, (3) a quite unjustified attack on your confidence theory based on erroneous assumptions—no answer—probably a new enemy & the revival of an old one, (4) (Hasel's paper)?"

He signed himself, "Yours ever."

1978   In September 1978 I am in England with a letter from Neyman introducing me to Egon Pearson as someone who is interested in the history of statistics—he finds it embarrassing to say that I am writing a book about him.

The Pendean Home, where Pearson lives, is a large old house set in the lovely Sussex country around Midhurst. People speak pleasantly as I follow a middle-aged woman down the hall, but I am aware of old people, established in rooms which they will probably not leave again.

Although the door is open, my guide knocks: "Professor Pearson—"

I am startled because the old man in the chair has a face so much like that of Karl Pearson, which I have recently seen in a publication of the Royal Society. He looks now more like his father than he did in his younger years. Although he has just returned from a stay in the hospital ("an imbalance of the body's fluids," he explains, "which made my brain woolly") and is wearing a surgical collar to treat some other ailment, he is a far from decrepit figure. His voice is strong, of course very British, and there is an eager, almost boyish excitement about him as he talks. He is a busy man. Papers and books are everywhere. I take a chair and glance at the shelves behind him. The first title which catches my eye is *The Grammar of Science.*

When I comment that I feel as if I am in a time warp—he says, "Oh, yes." He knows the feeling well. He tells me he has just subsidized the publication of his father's lectures on the history of statistics in the seventeenth and eighteenth centuries. (These were given at University College between 1921 and 1933.) He produces a couple of flyers advertising them. Currently he is absorbed in editing his correspondence with Student. He is fascinated by the project, for it is also his own history that he is writing.

"I have no idea how long it's going to be, but I'm going on writing. I'm putting all the correspondence, about a hundred letters, into an appendix. So that is free from what you call 'opinion.' That is fact. Then I have a long introduction about how I came into statistics, and my father, and how I began to doubt my father. He was a very powerful man. It was terribly difficult to shake off. And he didn't—it upset him, you see—but I finally did. Well, that's all put out. My article at present covers only letters of Gosset and myself from 1926 to 1929, but of course they did go to Gosset's death in 1937. But the important part to Jerzy and myself is this."

I ask what started him on the project.

"What Jerzy had written and published I was fairly clear was wrong," he says, referring to an article by Neyman in *Synthese* in 1977, which describes the development of the theory of confidence intervals. "And it was partly this that made me think, 'Oh, I must put down my memories, not critical of Jerzy in the least; but just the facts as I remember them, because otherwise both Jerzy and I will be written up by historians of science who don't know the facts.'"

Later he says, "They have already begun on that. There was an article on statistical inference in which the writer said that at a certain point—he asked me to look at his paper, you see—that at a certain point, he said, 'I think yours and Neyman's views switched across.' This annoyed me very much. I said, 'How can you know what we thought?' And this was a warning—that this is the thing that historians of science will do and, therefore, let me put facts on paper and, if they don't like them, nevertheless they are facts."

One section of the book is titled "Historical Note on the early development of Interval Estimation." He says that he has sent a copy to Neyman, and I say that Neyman has shown it to me.

"...I am giving particular attention—for example—to Jerzy's, Fisher's, and my simultaneous, independent thinking about interval estimation. You see, Jerzy doesn't know. It didn't arise. We were so absorbed in testing hypotheses that we didn't discuss interval estimation, and I have the feeling that he has perhaps quite forgotten that I was thinking about it, too; and, unconsciously, he was perhaps doing me an injustice. I tried to make this out to him in a letter recently.

"I wasn't sure whether he liked it. I just have this feeling that he doesn't like being contradicted. But I think it is quite possible that I am wrong. I mean, this has happened over a good many years that I've known him, but I haven't—I think—taken offense. I would tend to say to myself, 'Oh, that's dear old Jerzy.' I may remember it, but I'll let it slip, and I suppose he lets it slip."

I have seen much of the recent correspondence of Neyman and Pearson, and I recognize the reasons for the feelings which Pearson expresses. He has been trying for the second time to piece together and fill in the history of the collaboration, especially his own share in it, since his letters to Neyman have not survived. The first time was in "The Neyman-Pearson Story," which he *ch* wrote in 1964 for a festschrift honoring Neyman on his seventieth birthday.

"Clearly I could not push my priority in certain ideas in an article written in Jerzy's honor!"

It is a Saturday afternoon when I see Pearson for the first time, and I arrange to talk to him again on Monday. Early that morning, even before breakfast, the phone rings in my room at the Spread Eagle in Midhurst.

"Constance, this is Egon!"

He is exhilarated. He has spent almost all of Sunday writing his recollections of his early contacts with Neyman. Seventeen long handwritten pages! He has got as far as March 1934, when he offered Neyman the temporary appointment in his department at University College. Later that morning, when I come again to the Pendean Home, the director, who is working in the front garden, says to me, "It has done Professor Pearson a world of good to have you here!"

Pearson and I are more relaxed at our second meeting. He confides to me that John Hammersley has sent him a copy of the written "portrait" of Neyman which he composed at the time of Neyman's eightieth birthday on the basis of a week of taped interviews, "lubricated," according to Hammersley, "with warm whiskey specially flown in from the duty-free shop at London airport."

Pearson is deeply hurt.

"Neyman in this thing says some very harsh things about my father. That's one of the things—I wasn't going to comment on it—but he did."

He is also hurt by Neyman's apparent deprecation of their joint work and the implication that the only things of real importance to come out of it were the joint 1933 paper and his own 1937 paper on the theory of estimation. It does not seem to be the old "Jurek" who is talking into Hammersley's tape recorder—he seems to have forgotten completely the exciting period of comradeship and discovery which Pearson described in "The Neyman-Pearson Story."

"It *is* quite possible that at some later time when I was editor of *Biometrika* I felt I must refuse one of Jerzy's papers because it was too long—and perhaps too mathematical—and this *may be* subconsciously rankling."

I am aware from my interviews with Neyman that there *is* always an underlying bitterness when he speaks of either of the Pearsons. I can only say to Egon Pearson that Neyman himself has consistently refused to read what Hammersley has written about him.

As I prepare to leave, Pearson asks me if Lola is still alive. I tell him that she is and describe my meeting with her.

"I always liked Lola," he says.

He gives me another copy of the flyer for K.P.'s lectures on the history of statistics; "to take to Jerzy," and then a photograph of himself which he has located. It shows him in formal morning dress with one of his daughters, about to be married, on his arm.

Although Pearson is now "on the shelf," as he puts it, he feels that he has had a varied and eventful life: three memorable summers sailing in the northwest waters of Scotland—"affairs of the heart"—interesting trips to Italy and to the United States—"fears of K.P. and R.A.F."—and for nineteen years "a fascinating sheltered garden" in Hampstead.

"Mathematical statistics was not my life as I guess it has been for Jurek."

Back in Berkeley I report to Neyman on my visit to the Pendean Home and give him the items that Pearson has sent with me.

He looks a long time at the picture.

"Handsome, isn't he?" he says admiringly.

Then he carefully examines the flyer advertising K.P.'s lectures on the history of statistics.

"I think I would like to order a copy of this book," he says.

I say, "That will please Egon."

"I am not doing it to please Egon," he objects, "and that will please Egon more."

1942
—
1944

By the beginning of 1942 the war had spread to all the countries in which Neyman had lived for any period in his life. There were rumors (false as it turned out) that Sierpiński had fled Warsaw and that such well known Polish mathematicians as Hugo Steinhaus and Stefan Banach had been executed. Leningrad, where Bernstein had gone from Kharkov, was under siege, and Kharkov itself had fallen. Paris was occupied. Lebesgue was dead, and Borel had been imprisoned by the Germans on suspicion of subversive activity. London, bombed night after night, continued under threat of invasion.

In February, security clearance having at last come through, Neyman received a sub-contract from Princeton University's NDRC contract for bombing research. The problem on which he was to work, according to Weaver, was considered by "the highest Army officials" to be "of very great importance."

This new assignment gave a degree of autonomy, but in most situations Neyman still had to go through Evans. Everyone who was in the Berkeley mathematics department at that time remembers well the constant pressure which he exerted on its chairman. There were, however, to be only two real disagreements between the two men. One was scientific, the other more personal. Evans firmly believed that statistics was a part of mathematics and should remain in the mathematics department. Neyman believed with equal firmness that, although statistics should always remain "close to mathematics," it should become a separate department. The other "big disagreement" (Neyman's words) occurred in 1942 when an old acquaintance of Neyman's from Warsaw days was trying to place himself in the United States.

"It is one of the major ironies of the time that [Alfred Tarski] should be out of a job," the mathematician W.V. Quine wrote from Harvard to the philosopher William Dennes at Berkeley. "In my estimate he ranks with Gödel as one of the two leading logicians. Also his hundred odd publications include enough in other branches of mathematics to make him a distinguished mathematician even in abstraction from his logic. And there are his few but crucial contributions to philosophy...."

When Neyman heard that Evans was planning to offer Tarski a year's appointment at Berkeley, he pressed for the appointment to go to Zygmund instead. Evans, quietly smoking his pipe, remained unmoved by argument. Thus it happened that from 1942 on, Berkeley had in its mathematics department two outstanding Poles so different in their personalities, interests, and views that they came to be known among some of their mathematical colleagues as "Poles apart."

Neyman is still "regretful" that Evans took Tarski instead of Zygmund, who was at the time "inappropriately" established in the mathematics department of a small women's college.

"At some moments I think he thought that I am kind of too much trying to tell him what to do," he says—a remark that brings a wry smile to the faces of the older members of the mathematics department. "I think that motivated him in not bringing Zygmund."

When it is pointed out to him that as a result of Tarski's presence, the University of California has become the strongest, most active center of research in logic and foundations in the world today, he is silent. After a moment he says, "Tarski—that was modern mathematics. Zygmund—classical. Different but just as good."

The first people whom Neyman himself employed were Evelyn Fix and young Elizabeth Scott. He thought it necessary that they be informed about the whole setup of the NDRC problem, he told Weaver, "since this is likely to make them to work more intelligently and also add to their enthusiasm."

Some mathematically well trained person was still desperately needed to help with graduate teaching and theoretical aspects of the consulting work which the Lab continued to do for other departments—someone to fill the post which Dorothy Bernstein had turned down the previous summer. As before, Neyman would have liked to get Joe Doob. Perhaps, he thought, a suitable substitute within the range of the Lab's budget could be found among Doob's students.

"I had only three students at Illinois who would be suitable," Doob wrote from the Institute for Advanced Study, "and two of them are already placed. The third is a Negro.... If not for his color, this student would be just the man you want, but I don't suppose you could use him. The democracy we are to fight for abroad could stand some extension at home in some directions."

The idea of a black mathematician immediately appealed to Neyman. Since his youth he had been fascinated by the phenomenon of the Negro in America, his interest fed by reading *The Adventures of Tom Sawyer* and *Huckleberry Finn* as well as *Uncle Tom's Cabin.* He tried to ascertain the reaction of his colleagues to the hiring of a Negro. After several weeks he reported to Doob that it seemed to him that the people in Berkeley "would like to do something, but expecting some opposition, are inclined to be very careful in selecting a rather strong case."

Doob wrote that David Blackwell was an excellent student—"our best student last year...a quiet likeable chap who has always been popular with his fellow students."

Blackwell, who was currently at the Institute with Doob, offered to come to the west coast at his own expense; but Neyman, dismayed at the idea of a young person's spending so much money, suggested an interview in New York instead. After talking with Blackwell, he returned to Berkeley, much impressed.

"I told Evans that in my mind there is no doubt that the best candidate available would be David Blackwell. 'Blackwell? Isn't he black?' 'Yes.' 'Well, before we go into this, we will have to consult with the president.' So, yes, he went next door—I think I told you that Evans had an office but no telephone—and he had a conversation with Sproul. Then to me. 'And you really think that he is the best?' 'Yes. I made an effort to find out. No doubt.' 'Well. So let's have him.'"

Shortly after this conversation—before a formal offer could be made to Blackwell—Neyman learned that the wife of a mathematics professor, born and bred in the south, had said that she could not invite a Negro to her house or attend a departmental function at which one was present.

"Then I don't remember who started it, but someone, one of the old fellows, said, 'Well, now, Professor Neyman, don't you think it would be a bad idea to have our department split?' And to my great regret, I said—but I was new, you see—'Well, I certainly don't want to split the department.'"

He wrote regretfully to Blackwell, "I feel that I may have been too optimistic in describing the situation to you, and that my presentation of it coupled with the present reverse may have caused you a considerable disappointment. Will you kindly excuse me and believe that whatever I told you reflected exactly the picture as I saw it at the time."

He thought again of Bernstein, who had later regretted turning down his previous offer. There would be a particular advantage to employing a woman during wartime. Already the military had taken three of the young men he had trained during his first years at Berkeley. He offered Bernstein the position, and she accepted.

He did not forget David Blackwell but added his name to a mental list of people he would like to bring to Berkeley after the war. The list already included, in addition to such emigré luminaries as Feller and Wald, a native-born white American named Henry Scheffé who had passed through Berkeley the previous summer on his way to Princeton. Trained as a mathematician, Scheffé had had only one formal course in statistics when, at thirty-four, he had decided that that subject would offer him more opportunity for interesting research than would differential equations.

That spring of 1942, shortly after receiving the assignment from Weaver, Neyman visited the Air Forces' Proving Grounds at Eglin Field, Florida. "...for a satisfactory performance of a statistician's duty," he explained, "it is...necessary that he or she fully understands the circumstances of experiments, whatever their nature, to which statistical methods are applied."

He returned to the Lab with copies of the Air Forces' Technical Manual, which seemed to him "to suggest the importance of properly trained statisticians and possible efforts to obtain them...[but] the way in which certain data that we had occasion to consult are handled suggests strongly that the number of properly trained statisticians serving the Air Corps could be usefully increased."

By March 1942, computing for train bombing probability studies had started in the Laboratory "but on an admittedly unsatisfactory basis," as Weaver conceded. There were six people and only five machines. The lights in the Lab were on far into the night, and usually it was Neyman himself who turned them off. By the end of April what had started out as a slight cold had developed into pneumonia. Confined to bed, Neyman suggested that D.H. Lehmer of the mathematics department could replace him temporarily. Although Lehmer, a number theorist, did not know any statistics to speak of, that would be no problem. He needed only to ask "Why?"

"If the others are asked 'Why?' sufficiently frequently, I am sure they will get the right answers."

Everyone in the Laboratory struggled loyally so that no time was lost because of Neyman's illness. In June, with nine reports having been sent to Weaver and Dorothy Bernstein on her way, Neyman thought things were again in pretty good shape. He had received an invitation to spend the summer at Ann Arbor, and he was eager to accept. Weaver wired an unenthusiastic approval: NEARNESS TO NEW YORK PERHAPS OFFSETS DISADVANTAGES OF DISTANCE FROM CALIFORNIA.

In Ann Arbor, Neyman began to prepare his Geneva talk for the *Journal of the Royal Statistical Society*. In a long appendix he took up, among other things, some remarks by Fisher, who the year before had virtually accused him and Kołodziejczyk of plagiarism.

"The solution given in the British Association Tables (Fisher, 1931)...was adopted without acknowledgment by a pupil of Dr Neyman, a certain S. Kołodziejczyk, who published a note (Kołodziejczyk, 1933) in the *Comptes Rendus*," Fisher had written. "As I had previously rather pressed this solution on Neyman's notice, owing to its important industrial applications,

I was led to inquire why no acknowledgment was given of the origin of the solution, but acknowledgments only to Neyman's writings. Dr Neyman assured me, however, that in the original form of the note, reference to my introduction had been inserted, but had been cut out by the editor...in *shortening* Kołodziejczyk's note."

Kołodziejczyk was by that time dead, killed during the first days of the war in an attack by Polish cavalry on German tanks; and Neyman was by no means inured to Fisher's criticisms. (A student of the period remembers that whenever he referred to Fisher's work in his lectures, he always, with deliberation, ground out his cigarette.) Fisher's parting remark especially rankled: "So far as I know, neither Neyman nor his follower has done anything to rectify the invidious position in which they have been placed."

To begin with—Neyman pointed out in reply—before 1934 Fisher had had scarcely any opportunity to press a problem on his attention. In the second place, although he no longer had any memory of the references in the original manuscript, authors who were not members of the Paris Academy were strictly limited in the number of pages permitted them, and a cut had indeed been made in Kołodziejczyk's paper, which contained only a part of his tables. In the third place, the tables appeared in full in the paper on agricultural experimentation which had been presented in 1935 to the Royal Statistical Society with a complete acknowledgment of Fisher's contribution in a footnote. Professor Fisher certainly had had an opportunity to see this footnote since he had "honored" the paper by a "lengthy and interesting" discussion and "in so doing [had] registered a number of protests, duly recorded at the end of the paper, but the adequacy of the footnote was never questioned."

In Ann Arbor, Neyman also wrote a friendly note to Egon. Without referring to the half year of silence between them, he suggested that Pearson should publish a lecture he had written up earlier for the now abandoned *Whither Mathematical Statistics?* There was quite a bit currently being published on a somewhat similar subject and what the other people wrote— "with the possible exception of R.A.F."—seemed to Neyman "sheer rubbish." Pearson replied promptly, also in a friendly fashion. The correspondence, taken up again, continued, although with less regularity than in the past. When Pearson, who was working on fragmentation, complained about the lack of communication between British and American scientists working on similar problems, Neyman wrote to Weaver to see if something couldn't be done to improve the situation. He also offered to arrange a visit to the United States for Pearson, but the latter declined on the grounds that a necessarily hurried trip, under war conditions, would be too fatiguing.

During all the time Neyman was in Ann Arbor, he and Weaver debated by mail various ways of speeding up the computations being done at Berkeley. The most efficient method of processing such extensive calculations would be by punch card. Columbia had such equipment, and Neyman energetically explored the possibility of acquiring the equipment at Berkeley as well; but, as Weaver told him, it was "obvious nonsense" to set up two laboratories to do a job that one could already do. Neyman, "beaten," wrote sadly to Fix and

Scott—"My dear friends"—all his efforts had proved useless, and it was his conviction that the whole Berkeley project would soon be liquidated.

Weaver was reassuring. He was "exceedingly anxious that you keep available at least a few well trained people to assist you with the many exploratory and new phases of the problem which I am confident will continue to develop." But he informed Neyman that for the rest of the war there were to be no more sojourns at other universities. He was most valuable at Berkeley.

The tide of war was still running against the United States and its Allies. Faculty wives were being called in, mathematics majors employed to work in the Lab on the new assignments. Evans suggested to a German teaching assistant named Erich Lehmann that perhaps he could be more useful in physics or statistics than in pure mathematics. Lehmann, not enthusiastic about physics, decided to give statistics a try. He found Neyman in shirt-sleeves and suspenders in front of a calculating machine in the basement of the Life Sciences Building, the new home of the Lab. ("A handsome man, looking the way one imagines a Polish nobleman, romantic yet strong, with wavy brown hair, and a substantial but well trimmed moustache.") It was arranged between them that Lehmann would take the basic upper division course in statistics that year and the two graduate courses the following year—that was all there was at the time. Thus, casually, he began a career in statistics.

Neyman was convinced that people worked best when they enjoyed what they were doing. He worried whether faculty and students were interested in their assignments. He fussed about the length of time they had to spend over calculating machines and applied urgently, through Mr. Evans, for more lights and also a couch so that people working at night could lie down from time to time. He provided coffee and cake in the late afternoons and arranged weekend picnics. On occasion he brought everybody up to his house for supper. Young Micky had by then developed a great passion for all sorts of animal life; and, while guests sat on the floor with their plates in their laps, "a loving snake" was quite likely to crawl over their legs.

The workers in the Lab remember Olga Neyman as a casual hostess who considered her duty accomplished when she had placed a ham or a turkey on the table. She had found a congenial friend in Dresch's wife, Judith, who was also Russian; and the two couples were to see quite a bit of each other in the coming years. Today, in Palo Alto, Mrs. Dresch, who heads an active theater group, speaks enthusiastically of Olga Neyman's talent and displays a series of masks which she made for one of the theater's productions. "And Jerzy— you should have seen him dance," she says to me. "Such *joie de vivre!*"

By the end of 1942, work at Berkeley had shifted from probabilities of one hit on a rectangular or "long" target to those of multiple hits. Weaver was very pleased with the work and wired, CONSIDER MULTIPLE HIT PAPER AND TABLES MOST IMPORTANT....

Neyman responded, also by wire, DELIGHTED. He was, he explained, preparing material for a demonstration experiment with the trainer at Eglin Field to test, among other things, the new hypothesis that the most efficient method of utilizing a given number of bombers against the same number of ships would be for several of the bombers to take on one ship instead of each bomber's attacking a separate ship.

In December 1942 Neyman made another trip to Eglin Field to establish closer contact with the men who actually flew the planes and dropped the bombs.

"Many years of experience with applied problems have taught me to be extremely careful with hypotheses concerning observational phenomena," he once wrote Weaver. "In older days I shared the opinion of that German fellow who said if life disagrees with a good theory then 'desto schlimmer fuer das Leben' [so much the worse for life]. Since then, having received a few nasty shocks, I am of the opinion that, when trying to predict frequencies to be observed, it is imperative to start with observations."

During the war he was always eager to find out what really goes on when a bomb is dropped: the aiming error, the distortion caused by the bomb's being in a train and being released from a moving plane, the effect of winds. At Eglin Field he learned that "formation flying" in battle was nothing like what he and the workers in the Lab had imagined. He also learned that experienced pilots were woefully ignorant of anything not in their training manuals. A major, a wing commander, about to be sent overseas, "learned from me about the decrease in the trail as a means of placing the center of train on the center of target."

The commanding general at the field was very skeptical about applying probability theory to warfare. Over drinks in the officers' mess, in a discussion with "Doc"—as the men on the field called Neyman—he maintained that with an equal number of bombers and ships it would be most effective to assign each bomber to a different ship. There was a bet on the outcome of the trainer experiment—always Neyman's favorite bet—the loser to buy the winner a martini.

"In other circumstances I might have regretted [the general's] attitude," Neyman wrote in the trip diary he kept for Weaver. "As things are, however, it seems rather satisfactory. The trainer experiment is proceeding and there are reasons to believe that it will be done well. If so, we may expect it to give tangible results clearly indicating the advantage of basing bombing policy on probability tables. If the report on the results of this kind is signed by a person who is known to be skeptical, the effect may be much more satisfactory than otherwise."

The final report on the tests, which were concluded in April 1943, stated:

"Within the limits of experimental error, trials on the bomb sight trainer produced frequencies of hitting multiple targets which were in excellent agreement with the expected frequencies as tabulated in the NDRC tables for dispersionless trains.... Similar tables, computed to include the effects of dispersions of the bombs of a train, should therefore be of value to the mission planner in determining the best grouping of airplanes in attacks requiring multiple hits on multiple targets."

So he won the bet?

186

"I won, but the bum—he never paid up!"

Another result of the trip was a suggestion by Neyman for the improvement of the bombsight. There had already been some studies at Berkeley of the machinery of bombing errors, particularly those caused by the intervalometer—a mechanism that determined the spacing between bombs. After the trip to Eglin Field, it occurred to Neyman that the bombsight could be very simply modified to make the intervalometer unnecessary. He passed his idea on to Weaver who, impressed, sent it on to the military. Nine months later all new "radar bombsights" either incorporated or could be readily made to incorporate the suggested improvement.

On his way back to Berkeley from Eglin Field, Neyman talked in Washington with Colonel Charles Williamson, Chief of Status Operations Development for the Air Forces. Williamson, who kept in touch with combat through periodic trips to the front, was impressed by the Lab's work on multiple hits and suggested that in order to speed up research, the Berkeley people should make their computations in Washington, using the punch card equipment of the Air Forces, and "some of us, preferably all the crowd, [should] be resident [so as] to be in constant contact with the H.Q.'s."

Neyman immediately drew up for Williamson an eight-page plan for a statistical laboratory for the Army Air Forces—an American equivalent of the Air Warfare Analysis Group of the British—based on the premise "that statistical research on a very high scientific level is of extreme importance to the Air Force." He recommended that the proposed laboratory should perhaps be in the quiet atmosphere of Berkeley rather than in Washington.

Weaver, when he saw a copy of Neyman's proposal, was considerably annoyed. In the first place he feared that the Air Forces would never create a laboratory which would fully utilize Neyman's talents and yet his work for the NDRC would be bound to suffer. It was also "somewhat unfortunate" that Neyman's proposal had "pretty definitely" implied "that I knew just what it was before you sent it...." He was also very disturbed by the fact that, along with his proposal, Neyman had forwarded directly to Williamson a copy of an NDRC report.

"I am afraid it sounds stuffy and formal [but] the decision as to where, when, and how [NDRC reports] go to the Services rests not with you but with the NDRC—in this instance with me."

Neyman was immediately conciliatory:

"It is a human thing to disagree on various questions, but so long as some individuals decide to work together I think it is essential that the rules set for the team be observed even if someone does not think them useful. I owe you the very precious opportunity of working for defense. Apart from that I admire your organizational talent. Believe me, so long as I am on your team, I will scrupulously observe its rules.

"In the case in point I did not transgress these rules...."

Three single-spaced pages of rather complicated explanation followed.

Weaver replied that he was convinced that Neyman had acted in good faith.

"But having said this, as cordially and friendly as I know how, I have to go on and say that I pretty thoroughly disagree with most of the points in your letter."

In spite of this incident, Weaver not only extended the Lab's contract at the beginning of 1943 but also upgraded it so that Neyman was no longer "an employee of Princeton."

HAVE EVERY REASONABLE BASIS FOR EXPECTING THAT YOUR WORK WILL CONTINUE FOR DURATION ON AT LEAST PRESENT LEVEL AND VERY POSSIBLY EXPANDED, he wired. THIS OF COURSE ASSUMES ANY POSSIBLE DIRECT RELATION BETWEEN YOU AND AIR CORPS WILL NOT UPSET PRESENT ARRANGEMENTS.

But continued annoyance is apparent in Weaver's response to another suggestion from Neyman—this time for an instrument which would "instantly and simultaneously yield all the probabilities of multiple hits that we compute for a single plane."

"I trust your supply of Vitamin B_1 is sufficient to brace you for the nervous shock of reading the enclosed document," he wrote to S.H. Caldwell of the Office of Scientific Research and Development upon receipt of Neyman's suggestion. "You will note that he emphasizes that he is totally ignorant of all practical matters and assumes the engineers will make the necessary simplifications in his idea. But the difficulty is, of course, that these necessary simplifications must be absolutely heroic...or his idea...is completely worthless."

After studying the suggestion, Caldwell reported to Weaver that in his opinion Neyman had made "a rather courageous attack" on the problem which had the advantage "of presenting the real problem to an engineer in a manner which shows its magnitude. Unfortunately, he falls flat on his face on physical principles."

Throughout 1943, as a consultant for the Air Forces, Neyman was away from Berkeley with increasing frequency. During his absences Dorothy Bernstein taught the advanced statistics course with the assistance of Eudey and Lehmann.

"It was on the books as three hours of lectures and three hours of lab," Eudey recalls. "Dorothy took the lectures—they were really mathematical. Then Erich and I divided the lab hours. He lectured on mathematics as it applied to statistics, and I lectured on statistics directed toward the applications."

When I express surprise to Lehmann that he was lecturing on statistics when he had only just taken up the subject, he laughs and says, "I don't remember now, but Neyman is perfectly capable of handing you a book and saying, 'Teach!'"

"Students were getting six solid hours of lectures," Eudey goes on. "It was murderous. They had enormous notebooks." He indicates by spreading thumb and forefinger. "One of them complained bitterly to me, brought his notebook to let me see how thick it was. That was Joe Hodges [now a senior professor in the Berkeley statistics department]."

Although the war was far from over in 1943, things were looking up in Africa, Europe, and Asia. Lehmann decided that he really did not like statistics.

"Both elements that had attracted me so much to mathematics were missing. On the one hand, statistics lacked the beauty and harmony that I had found in the integers and later in certain parts of mathematics. Instead, ad hoc methods were used to solve problems which were messy and to me not very appealing. And second, although this may not have been so clear to me at the time, statistics (or any other branch of applied mathematics) does not share the reliability which I had so valued in pure mathematics. A result in pure mathematics is a fact which can never lose its validity. However, applied mathematics must justify itself through its usefulness in explaining or predicting some aspect of the real world of which at best it can be only a rather rough model."

Lehmann went to discuss with Tarski the possibility of working in algebra. Before he could inform Neyman of his new plans, Dorothy Bernstein decided she did not like statistics either and did not approve of Neyman's mathematical justification for some statistical methods. In addition, he seemed to her to be unable to accept any criticism of his work. When she abruptly resigned after only seven months, he was very angry. As in the case of Pearson, he cut off all contact with her; however, in letters which he wrote to Evans as well as in letters of recommendation for her which he later wrote, he never went beyond the statement in her own letter to him: that she was leaving the Laboratory because she did not find the work congenial.

Miss Bernstein's resignation was "regrettable," he told Evans; but if the Lab should lose Eudey or Fix, "this would amount to a catastrophe." He proposed to use the money already budgeted for Bernstein to provide raises for them. He also suggested an appointment for Lehmann. In spite of Lehmann's recently recognized distaste for statistics, he felt that as a graduate student who had scarcely begun his studies he could not afford to turn down Neyman's offer.

Today he feels that statistics has had its compensations. "It was a very new field just at the start of a tremendous development, and I was getting in on the ground floor. Perhaps more important—although I never worked on specific applications—I gradually began to like the applied aspect of the subject. It brought me in contact with the concerns of others, with ideas and questions from genetics, medicine, psychology, education, and many other fields. Since statistical thinking permeates modern life, familiarity with it has been useful and interesting. No. I really have no regrets."

He did not tell Neyman that he had been planning to leave.

During the last half of 1943 the Lab took up again the high level bombing of maneuvering ships. Neyman's file drawers began to sprout such labels as "Dodging I," "Dodging II," "Dodging III." In November he verbally reported on a new theory to the Applied Mathematics Panel, which Weaver now headed. He was never able to submit a written report on the theory, however. That same month Roosevelt, Churchill, and Stalin met in Tehran and agreed upon plans for launching the second front in Europe which the beleaguered Stalin had long urged. On November 23 Weaver assigned Neyman to a special project for the DOLO Committee. The letters of the acronym stood for *Destruction of Obstacles to Landing Operations.* According to Weaver, the DOLO Committee had not asked the Applied Mathematics Panel to undertake this project but had specifically requested Neyman's services.

Neyman dropped everything and concentrated all his efforts on the assignment. The problem was to compute—given slight enemy opposition and medium low-level bombing—the optimum spacing of bombs in a train for clearing a path across a beach of land mines. During the next four months relationships were set up among five variables: (1) the length of the mine-free path desired, (2) the number of bombs per plane, (3) the radius of effectiveness of the bombs, (4) the number of formations attacking, and (5) the probability that a mine in the corner of the proposed path (the most difficult to explode) would be left unexploded. The results were compiled in two nomograms which would give the value for any variable when the other four were known.

At one point, Eudey recalls, Neyman decided that he also needed certain information about the beach in question—the quality of the sand, the tides, the shoreline, the immediate topography. There was a meeting with someone, "presumably from the DOLO Committee," whom Neyman badgered with these and similar questions.

Finally the man expostulated.

"Have you ever taken a vacation in France?"

"Well, yes."

"And have you gone to the seashore?"

"Well, yes."

"For instance?"

"Well, Normandy."

"Then use that beach!"

1944
—
1945

In the spring of 1944, although the landing on the beaches of Normandy was still in the future, it was clear that the war had passed into a new and final phase. Weaver instructed Neyman by wire to lay out his views on the future activities of his group.

The operation at Berkeley had always been modest. Nearly all the workers in the Laboratory had been trained by Neyman. They were young, they were paid very little, and—almost as soon as they had been trained—the men among them were snatched up by the military. In his reply to Weaver, sent off the same day he received the latter's wire, Neyman proposed a number of reforms which would result in more stability, double the budget of the Lab, and considerably expand its work. He added in a handwritten note that he would also like to employ "Prof. A. Zygmund, now at Mt. Holyoke College… an outstanding mathematician and, in my opinion, in a 100 years' time his name will remain on the books with [those of] only a few of his contemporaries."

With the approval of the university administration as well as the approval that seemed implicit in Weaver's request, he had high hopes for the expansion of the center at Berkeley. The response which came a few weeks later found him quite unprepared.

During the past weeks, Weaver wrote, the Executive Committee of the Applied Mathematics Panel had been giving careful and extended attention to its future program.

"It has become quite clear that the emphasis… should, from now on, shift rather rapidly from long range fundamental studies (of admitted importance but dubious application to this war) and even from analytical studies of specific mechanisms and devices… to activities of the 'operational research' type, related as closely as possible with the practical day-by-day conduct of the war." This activity would of necessity have to be carried out in the closest possible contact with the military. "Thus the stage has been reached in our work at which it no longer seems to us efficient or justified for us to extend further support to any activity at a great distance."

The blow could not be softened by complimentary phrases in the last paragraph of Weaver's letter: "sincere appreciation"—"the completely unselfish and devoted way in which you and all of your group have served"—"a large amount of work, often under great pressure, and frequently on problems of great importance and urgency"—"the very unusual vigor and energy which you… have sacrificially applied."

In short, there would be no further support from the Applied Mathematics Panel for the Berkeley Lab.

Weaver also wrote to Evans, who was then in Washington with the Office of the Chief of Ordnance. He would like to talk to Evans about the discontinuance of Neyman's contract, "explaining somewhat further than I can to him why it seemed necessary. We do have the highest regard for his knowledge and ability, as well as his energy and devotion. He has, nevertheless, serious limitations on the work we are doing—limitations which should in no possible way be held against him in his normal activities."

At the last moment Weaver was not able to keep the appointment, and Evans could only conjecture what "the serious limitations" were. He was already familiar with the correspondence between the two men. Although he recognized occasional flashes of annoyance on Weaver's part, followed usually by apologies, he could find nothing which justified termination of the Berkeley contract on personal grounds.

He wrote immediately to Weaver that the decision of the Applied Mathematics Panel did not in any way affect his own confidence in Neyman. Since 1938 he had made a point of learning to know and understand Neyman's character and personality "as well as one man can know another's." In some ways he was naive, in other ways sophisticated. "But for sincerity, generosity and patriotism, it would be hard to find his equal." He stated firmly, "...it is not my judgment alone that you have separated from your counsels the man most qualified in the United States, theoretically and practically, in the field of statistics."

Commenting later to Sproul on the termination of Neyman's contract, Evans pointed out that it had extended over a range of several problems and as far as he could see, "from more than a casual knowledge," Neyman and his laboratory had contributed most of the new ideas.

"I should like to add that Professor Neyman gave one of the first, if not the very first, of the courses to outside engineers on 'quality control.' Of course this subject has been known to statisticians since the days of 'Student,' but it is interesting that Professor Neyman stressed its importance during the early days of Lend Lease and expanding industry. Credit for it in this country seems to have gone elsewhere."

The minutes of the Executive Committee of the Applied Mathematics Panel for April 3, 1944, report the following discussion about the termination of the Berkeley contract:

"SSW reported that he had considered the proposals for next year's program outlined by JN in his letter to WW; that these proposals amounted substantially to the establishment of a unit of Air Warfare Analysis in Berkeley....SSW pointed out that WAW's group and EJM's now take care of studies in this field, and raised the question whether effective work of this type can be done in Berkeley where frequent consultation with Army and Navy personnel is difficult. He reminded ExCo that JN's group was originally set up to do computing for train bombing probability studies; that the computing, which had been progressing very slowly in California, was transferred to JS's group using IBM equipment at the beginning of last year; that several other requests have been sent to JN, particularly the moving target problem, which has been shelved since November because of JN's absorption in Burchard's request [the DOLO project]. There was discussion of the work done at California and of the report recently written by JN for Dolo, and of JN's interest in the dispersions of a bomb computed as a function of altitude. This problem involves the solution of a probability differential equation. After rather full discussion of the character of the problem which had been given to JN by the Dolo Committee (the assumption of no opposition in the air, essentially no flak from the ground, etc.) and of WW's comments on JN's report (failure to make unambiguously clear the extreme character of the assumptions, unrealistic factor between practice and combat bombing) and after a rather full consideration of the difficulties of remote control, it was decided that the contract would be formally terminated at the end of August."*

Mina Rees, who was Technical Aide to Weaver and from whose diary the above minutes have been extracted, writes to me, "My impression, for what it

---

*SSW = S.S. Wilks, JN = Neyman, WW = Weaver, WAW = W.A. Wallis, EJM = E.J. Moulton, JS = Jan Schlicht, Burchard = J.E. Burchard, Chairman of the DOLO Committee.

is worth, is that [Neyman] always wanted to carry out a full theoretical study that might lead to a new theorem, while the others were thoroughly indoctrinated, as I was, with the 'quick and dirty' philosophy; we had to give the military forces an answer that would help them to determine a course of action that would be better than what pure chance could be expected to provide; and that would be based on the experience of the people making the study as well as the evidence. But it would not rely on proving a new theorem."

With his contract soon to terminate, Neyman spent the spring and summer of 1944 bringing to a conclusion work already started. One by one the military took most of his remaining young men. As civilian scientists at the various fronts, they would be able to cut through the chain of command, talk to generals as well as mechanics, and bring their scientific knowledge directly to bear upon day to day problems. Eudey took a position with the Ninth Air Force after his brother, a B-17 pilot, had been shot down in Europe. Lehmann and Joe Hodges, a giant of a youth who had come into the Lab even before his graduation in 1942, shared a tent on Guam. All three wore military uniforms but without insignia. They were—as Evelyn Fix put it—the "war heroes" of the Lab.

Weaver had held out the hope that, possibly, "specific studies will arise in the future, of such a nature that they can be isolated, so to speak, from the major concentrations in and around New York" and that these might be referred to the Berkeley group. Just a few weeks before the Lab's contract was to terminate, he extended it for another six months "because of the strong possibility," he told Neyman, "that we may want to call upon you in a consultative or other capacity...." As the Americans and the British fought their way through the Low Countries and the Russian army reached the Vistula, he contacted Sproul.

"A high-powered and thoroughly experienced group of individuals is being assembled with the intention of sending this group to England for four to six weeks of specialized training, following which this group will play a very important role in connection with the planning of certain bombing operations against Japan....After the most careful consideration, the [Applied Mathematics] Panel has decided that Dr. Jerzy Neyman would be an ideal person to bring to this group the extensive knowledge of bombing statistics which has been developed over the past two or three years."

Although Weaver mentioned "four to six weeks," he cautioned Sproul that later he might have to send Neyman elsewhere and the university should consider the possibility that he might be away from the campus for a year or more.

It was not the best time for Neyman, personally, to leave Berkeley. Mrs. Neyman had just returned from the hospital after a serious operation. Teaching in statistics was "at a low ebb," no one having been obtained to replace Dorothy Bernstein. (At times Neyman posed the question to Evans whether it was even "legal" to have advanced courses taught by people who did not yet have their Ph.D.'s.) England was being bombed again in a particularly frightening way. Hitler had unveiled his long threatened "secret weapon"—the unmanned rockets which, at the time, were expected to be much more effective than in fact they turned out to be. But Neyman felt as if he had been called up for action. It was his duty to go. Evans backed him and urged Sproul to give permission.

There was again the problem of finding an appropriate replacement. Neyman insisted that it was very important for the quality of instruction to be maintained throughout the war. There was obviously no statistician of any stature free in the United States, but in China—

The year before, he had received an offprint of a paper by P.L. Hsu. After leaving University College with a D.Sc. in 1939, Hsu had spent a year in Paris with Hadamard and had then returned to his native land as a professor of statistics at the University of Peking. Hsu was, in Neyman's opinion, absolutely on a level with Wald—they were the two outstanding mathematical statisticians in the generation coming up!

During the summer of 1944 he grappled with the problem of obtaining travel money for Hsu. Conditions in China were chaotic, the country divided into three parts—one under the Japanese, a second under the Communists, and a third under the Nationalists. At one point Hsu was doing his scientific research in a cave. How his work had got printed at all was a marvel. He was eager to come to the United States.

As Neyman left on the first leg of his journey to England, Deutsch sent off a cable to Hsu (who had finally offered to pay his own fare), inviting him to lecture at Berkeley for six months with an appointment for the following year assured. Until Hsu arrived, Neyman's old friend Pólya, who held an inappropriate post as an associate professor at Stanford, would come up from Palo Alto a couple of days each week and give Neyman's lectures.

By the beginning of October 1944 Neyman and the other members of the American group of mathematicians, headed by V.B. Rojanski, were established in England at Princes Risborough—the so-called "brain" of the Royal Air Force.

In preparation for the coming attack on the Japanese homeland, the American Air Forces needed to know the relative effectiveness of certain bombs. Data relating to these and their use against German industry were available in England but not sufficiently organized for transmission to the United States. It was the assignment of Neyman and the other members of the American team to study the problems of target-weapons analysis and the methods of treating such problems as they had been developed at Princes Risborough.

Neyman was delighted to become better acquainted with Jacob Bronowski, a lively, articulate Pole who was ultimately to become most widely known as the author and narrator of the popular "Ascent of Man" series on television. Although Bronowski's prewar interest had been algebraic geometry, he had quickly mastered the methods of treating weapons problems and had invented a number of new ways of dealing with them. He participated in a series of indoctrination lectures for the Americans, and Neyman found his delivery and insight "distinctly above all the others."

There was a sharp difference of opinion between the group at Princes Risborough and a group in London on the relative importance of primary fires (those started by bombs) and the fires spread by primary fires. Neyman's first assignment was to prepare a critical evaluation of this controversy. In his view the weak point of the theory of damage by incendiary raids, as developed at Princes Risborough, was that it did not take into account methods of delivering the bombs. Discussion with Bronowski and others followed, and it became apparent that a satisfactory treatment would require a combination of the Risborough theory and certain ideas which had been developed earlier at Berkeley. The result was a joint paper by Bronowski and Neyman outlining a new theory of what was referred to as "vulnerability"— as Bronowski said, "big bombs versus little bombs."

Also at Princes Risborough was another old acquaintance of Neyman's. This was Florence David, K.P.'s last assistant. All of Neyman's admiration for the English character revived when he visited Miss David and the two other "ladies" from the old laboratory with whom she lived and observed the difficult conditions under which they were living and working without complaint.

By the middle of November there was a long cable from Rojanski urging the still further extension of the contract of the Berkeley group.

"Dr. Neyman is apparently the key man of the project," H.C. Hottel told Weaver. "Princes Risborough has provided...a Dr. David and a computer to assist [him], and CWS in England has lent...a captain and two sergeants."

Just before Christmas, after a visit to The Cell and a first meeting with Egon and Eileen's two little daughters, Neyman returned to Berkeley. The Lab contract was extended again. The work on the new data was even more frustrating than usual. Since many different kinds of bombs were used by the Royal Air Force in the same raid, it was impossible to tell which incidents of damage were caused by which type of bomb. The only solution was to ascribe the greatest damage to the heaviest bomb—a procedure which involved the obvious danger of bias in favor of heavy bombs. Neyman apologized to prospective employees, "The problems we work on are sometimes rather messy."

The liberation of Poland was by then at hand; but already a struggle which would divide Poles in and out of Poland was looming. Neyman's friend and former student, the economist Oskar Lange (who was a professor of economics at the University of Chicago) had early criticized the Polish government-in-exile as reactionary, fascist, and anti-Russian. Lange maintained that Poland's security in the post-war world would depend upon Soviet military might and that, therefore, in disputes with Russia over her eastern boundary she should yield to her larger neighbor. Neyman agreed with Lange. The year before, when seventeen Polish professors in the United States had issued a statement condemning Lange's ideas as provocative, partial and unscholarly, he had written to a local paper and, using data from Poland's last prewar census, had pointed out that, contrary to the statement of the seventeen professors, in all the provinces under dispute with Russia the number of inhabitants who considered Polish their mother tongue was in the minority—the greatest density of Poles within the boundaries of prewar Poland was in fact in the provinces adjoining Germany.

As the Western Allies raced toward Berlin in an effort to reach that city before the Russians, Neyman received orders to enlarge his laboratory so that two members—Betty Scott and Frank Massey—could be stationed full-time at the Pentagon. Lt. Commander W.W. Timmis, the Chief of the Physical Vulnerability Section of the Joint Target Group, announcing the new arrangement, noted that it was "most desirable that analysis be made of aerial warfare against Japan, since the Joint Target Group based its recommendations on such analysis of past attacks." He added that the Berkeley Statistical Laboratory would make "significant contributions" to this work. Neyman would divide his time more or less equally between Washington and Berkeley. Again it was arranged that Pólya would come up from Stanford and fill in.

At the same time D.H. Lehmer, who with his wife had worked throughout the war in the Lab, announced to Weaver and the Applied Mathematics Panel that he had constructed a version of the optical bombing device that Neyman had suggested two years earlier.

"[It] gives in 30 seconds a measure of the *average* amount of coverage of the target by 500 random sample bomb patterns under any given conditions of bombing.... Following [another] suggestion by Neyman a counting mechanism has been incorporated in the apparatus in order to obtain a measure of the probability of at least one hit or the probability of starting a fire in a given fire division, etc. This has been done with a view to immediate use in connection with the AN-23 incendiary raids [over Japan]."

The device, however, was completed too late to be put into operation.

It was obvious that the war would be over soon, and Neyman's thoughts turned toward the future. In spite of his best efforts, teaching in statistics had reached a still lower ebb at Berkeley. However, Hotelling had learned of the efforts to bring P.L. Hsu to the United States, and he now suggested that Columbia join California in making an offer which would bring the Chinese scholar to each institution for a semester. Hsu finally responded—explained what had happened in regard to the previous offer from Berkeley (a telegram had been sent but never received)—and accepted the California-Columbia offer. He indicated that he would prefer to come first to the west coast.

In June 1945 the war in Europe ended. There would still be a bloody battle in the Pacific as the Americans and their Allies fought their way toward Japan. But victory would come.

Like most scientists who had been involved in war work, Neyman had authored and coauthored many reports during the preceding few years, but he had done no research on his own and had made little progress on his *First Course in Probability and Statistics*, begun in 1940. As he flew back and forth across the country, eating dinner in Berkeley and breakfast in Washington or vice versa, he returned to an idea which had come to him on his trip home from Princes Risborough. Inspired by the memory of international intellectual cooperation which he had experienced at Geneva in 1937, he envisaged a symposium on mathematical statistics and probability at Berkeley in the summer of 1945 to mark the expected cessation of hostilities. He had already broached the subject to Provost Deutsch when victory in Europe—V-E Day—was celebrated on June 8.

"During the last several years the efforts of many mathematicians in general and of mathematical statisticians in particular have been directed towards helping to win the war. Although a number of interesting results must have been obtained in this way, it is quite certain that research activity in this country and all over the world has been deflected from its normal course, and restrained. The purpose of the proposed symposium is, then, to contribute to the revival of scientific work in mathematical statistics and allied fields."

"How much will it cost?"

"Five thousand dollars," Neyman said off the top of his head.

"Too much."

"Four thousand."

"O.K."

Although Neyman had already organized one successful large-scale affair in Berkeley—a meeting in December 1941 of the Biometrics Section of the American Statistical Association which had "interpreted Biometry in a very wide sense"—Evans was doubtful about his ability to set up an event of the magnitude of the proposed symposium on such short notice. He nevertheless gave his approval. Invitations went out immediately to statisticians all over the country, the people at Berkeley so excited that they failed to notice that no date was included. There was another round of invitations. The Berkeley Symposium would be held from August 13 to August 18.

It was Neyman's idea that each paper, in addition to presenting some unpublished results, would cover "a recently worked out chapter of mathematical statistics and probability." The preparation would stimulate their authors to bring together "into an architectural whole" a number of results which were currently scattered in a variety of journals. These would then be published in a single volume—the *Proceedings of the Berkeley Symposium*.

While detailed preparations for the symposium went ahead under the supervision of Evelyn Fix, Neyman returned with enthusiasm to teaching. In June 1945 he had a graduate class of five students—all young women. One of these was Marilynn Spanglet, now Dueker, a member of the faculty at the University of Connecticut. She had come with some friends from Hunter College (where there was even then an undergraduate major in statistics) and had obtained a job in the Lab by writing ahead. The first course she took from Neyman dealt with "stochastic processes." It was a new concept to her and her classmates—it was quite a new concept in mathematical statistics—and there was no textbook. For most of the course she did not understand what "stochastic processes" were or what Neyman was saying about them. She was, however, tremendously impressed by him. He was "a great man"—

"a genius." But especially important was the fact that "he cared about us." The day before she was to take her orals for the master's degree, she was surprised to receive a long letter from him, telling her not to worry but to relax and go to a movie with some friends—everything would be all right.

For his own contribution to the symposium, Neyman prepared a paper entitled "Contribution to the theory of the chi-square test." As a result of his consulting work for other departments, he had become increasingly involved in studies of a practical nature for which new statistical procedures were frequently needed. One large class of such procedures was introduced in this paper—what he called for short BAN estimates: Best Asymptotically Normal estimates. The theory, which provides computable answers in many complex situations, has since been described as "an asymptotic equivalent of the theory of least squares."

Seven years after the productive decade of 1928–38, he often expressed the feeling to his students that he could no longer contribute creatively to statistics: that all he could do was to teach and turn out good sound statisticians. In this connection there was one particular subject which he was eager to have treated during the symposium. He had never forgotten the character of "Young Jones" sketched by Hotelling in 1940. When he sent a symposium invitation to Hotelling, he suggested that in addition to presenting a theoretical paper, Hotelling should also deliver a paper on "The Place of Statistics in the University."

On August 13, 1945, the Berkeley Symposium on Mathematical Statistics and Probability (the word *first* was not yet in the title) opened with Hotelling's talk. It was exactly one week since the United States had dropped the first atom bomb in history upon the Japanese city of Hiroshima. There were daily rumors that the Japanese had agreed to the surrender terms of the Allies. The program for the symposium, handsomely printed by the University of California Press, had had to be discarded and another hastily mimeographed to accommodate late arrivals, unexpected absences, and volunteers who had heard about the meeting and wished to contribute papers.

The statisticians present in Berkeley listened attentively as Hotelling described how the teaching of statistics in American universities and colleges had developed rapidly after the First World War and how, as the end of the Second World War approached, had reached a position of such importance in the country that various unsatisfactory features would have to be eliminated. There was a plethora of elementary courses, a dearth of advanced courses. The subject was treated in many different departments, and even the relevant books were shelved in libraries under many different headings. But the "major evil" was that, too often still, statistics was being taught by people who were not specialists—in short, although Hotelling did not mention him again by name, "Young Jones." Hotelling, who himself was still teaching statistics in the Department of Political Science at Columbia, put forward a utopian solution. In colleges and universities of reasonable size there should be a *separate* department of statistics and in larger universities *separate departments* of mathematical statistics and of applied statistics.

198

Neyman liked to remind his students that Rome "couldn't be built in a day"; and in his discussion of Hotelling's talk, he limited himself to suggestions for eliminating what he described as "maladjustments" in the university curriculum. Statistical courses for students in the experimental sciences should always be taught by people well qualified in mathematics, but sampling experiments should be substituted for mathematical proofs. Mathematics students specializing in statistics should become familiar much earlier with "conceptual mathematics." They should also have contact with as many experimental sciences as possible "because, at this stage of the development of statistics, the experimental sciences are sources of theoretical problems"—also because in almost any job they would get after graduation they would be called upon to apply theory to experimental or observational problems. He proposed that each of the sciences set up a course for statisticians (to be offered every couple of years) to give them, not details, but a general knowledge of that science.

From the "lively discussion" which followed Hotelling's talk, it was immediately clear to everyone present that the symposium was going to be a great success.

"I didn't think Mr. Neyman could pull it off!" Evans conceded delightedly.

"...the symposium was an outstanding success," he told Provost Deutsch, "and...the success was due to Mr. Neyman's foresight in seeing its possibility at this time, and to his initiative and resourcefulness in planning it. It constitutes a significant page in the history of the University of California. But what is of even more importance, it was convincing evidence of the growing importance of the relation of statistics to experimental work in many branches of science and of the service which the Statistical Laboratory is rendering to the University and the public."

The symposium, "intended to mark the transition from war to peace," had been timed more exactly than could have been achieved by any statistical method. On the second day, August 14, 1945, the Japanese agreed to the terms laid down by the United States and its Allies. The war was over!

1945 — 1946

In the fall of 1945, Neyman was looking forward to a return to teaching and to research on problems other than those concerned with questions of whether a given projectile would hit a given target. He eagerly took up again the task of building a great American statistical center west of the Mississippi. Long missives went out to Evans about the statistical program within the mathematics department and to heads of other departments about ways of facilitating cooperation between the Statistical Laboratory and their departments.

The ten statistics courses which he proposed for the fall semester 1945 would be taught by Edward Barankin (a newcomer to statistics from

Princeton), Mark Eudey, Evelyn Fix, John Gurland (another newcomer), Erich Lehmann, and Elizabeth Scott in addition to Neyman and Hsu. None of these, with the exception of the latter two, yet had a Ph.D. degree.

Neyman was delighted to have a mathematical statistician of Hsu's calibre in Berkeley. When he saw that the latter had been listed in the catalogue merely as a lecturer in statistics, he objected to Sproul about the "unpleasant mistake."

"One could, perhaps, argue that this or that listing in a catalogue does not matter very much; [however] there is a purpose of listing titles, otherwise they would not be listed. In the case of Hsu, the mistake is likely to hurt more than if it is related to someone established in this country. For a Chinese scholar an invitation from two important American universities to act as a visiting professor means considerable honor. In fact, if his services were properly described in the Catalogue, it is not unlikely that a copy... would have been kept in his family and shown in pride to his grandchildren...."

Shortly afterwards he himself circulated a "Statement Concerning Dr. Hsu." It described Hsu's distinguished background, the "very tough" conditions under which he had been conducting his scientific work in China, and the extended efforts that the University of California, and later Columbia University, had made to bring him to the United States.

Letter after letter now came from Poland. *Szanowny i Drogi Panie Kolego! Szanowny Panie Profesorze. Drogi Jerzy. Drogi Jurku.* Poland was worse off than any of the victor nations, worse off even than defeated Germany. The *New York Times* reported: "800,000 Poles Living in Dugouts. Food, Clothing and Fuel Insufficient for Present Population. Repatriations Threaten More Suffering."

Indirectly, Neyman heard from a Swedish friend of his brother that Karol had had "a very hard and cruel time" but that he still appeared to be in good spirits. All Karol's family, with the exception of his old mother-in-law, had survived.

Neyman's friend Oskar Lange relinquished his American citizenship and accepted the post of ambassador to the United States from the new Polish government. When Neyman, who became a citizen at the beginning of 1944, is asked if he ever considered an action such as Lange's, he says no. He felt that he was committed to the United States—"There was a moment. When was it?"—but he goes on to explain that since he had not grown up in "Poland Proper"—as had Lange—he probably had less feeling for the geographical land itself than did Lange.

In addition to helping relatives and friends in Poland, Neyman was concerned about what could be done to speed the recovery of the intellectual community all over the world. A little money had been left after the symposium in August, and a modest sequel was planned for January so that statisticians and probabilists who had been unable to attend in August could present their papers. The banquet concluding this "post-symposium" should have as its theme "International Intellectual Cooperation."

200

"It is possible that the opinions expressed will be entirely divergent, but I do not expect it," he wrote to a potential participant. "On the contrary, it seems probable that on several important points there will be an essential unanimity. In that case somebody might move to adopt a resolution which might be published. Small as it will be, it will be another drop intended to carve a rock."

This symposium session was just a fortnight away when, in the middle of a Saturday afternoon in January 1946, a garbled telegram for Professor Neyman arrived in Berkeley.

PRESIDENT TRUMANS SPECIAL MISSION TO OBSERVE ELEC-
TIONS IN GREECE URGENTLY NEED SAMPLING STATISTICAL
WITH YOUR EXPERIENCE TO ASSIST IN THE DESIGN AND
CONDUCT OF OPERATIONS...

—DESSEN

Neyman knew nothing more about the political situation in Greece than any American who read the newspapers with attention. During the past decade there had been four and a half years of dictatorship, four years of occupation by Axis powers, more than a year of civil war between the Right—represented by the Royalists—and the Left—represented by the EAM, or National Liberation Front, a coalition of which the dominant element was the Communist Party. The day after the publication of the Yalta Declaration, an agreement ending the civil war had been signed. The final article had provided for a plebiscite to determine whether the king would return. Parliamentary elections were to follow. For various reasons, it had been decided to hold the parliamentary elections before instead of after the plebiscite on the monarchy. Representatives of both sides agreed "that for the verification of the genuineness of the popular will, the great Allied Powers shall be requested to send observers."

When Neyman received the telegraphed request to join the Allied Mission to Observe the Greek Elections, he was torn. If he accepted, he would have to leave Berkeley immediately and miss the symposium sequel. Would the guests he had invited be offended? What about his efforts on behalf of international intellectual cooperation? Who would present the speech on that subject on which he had spent so much time? But—PRESIDENT TRUMANS MISSION URGENTLY NEEDS....It seemed both an honor and a duty. There was also the possibility of a very interesting statistical problem. Even though the telegram had arrived on a Saturday afternoon, he managed to contact the university authorities. Saturday evening he wired his acceptance.

Initially he was under the impression that he was to be the only statistician, but at the beginning of the next week he learned that the "Dessen" who had signed the telegram was the American statistician Raymond J. Jessen and that there would be something like a half a dozen other statisticians from Great Britain, France, and the United States. (The Soviet Union had declined to participate on the grounds that it was opposed in principle to the supervision of national elections by foreign states.)

The elections would not take place until March 31, but Neyman was to leave immediately for a period of preliminary preparation in Washington. A "hasty and haphazard" letter to Evans laid out arrangements for the teaching of statistics and the running of the Lab in his absence. He was especially concerned about obtaining the services of Hsu after his semester with Hotelling at Columbia, and an enclosure of four single-spaced typewritten pages repeated the points in favor of Hsu "on which we have already agreed." He had had a communication from Hotelling, who was by then planning to leave Columbia the following fall for the University of North Carolina (where he would at last head the separate department of mathematical statistics which he had so long advocated). Hotelling had said that Chicago, Yale and Columbia were all after Hsu.

"I am afraid that if we delay our offer," Neyman warned Evans, "we shall lose an excellent addition to our staff."

By the time his plane reached Omaha, he had worked himself into a fret.

"The weather is nice, I have my favorite seat, etc.," he wrote to Mrs. Neyman on the "In Flight" stationary, "but I am not very happy."

"Loose ends" ranged from the beetles involved in some experiment, about which he had had "an inadequate 10-minute talk with Erich," to the conviction that there would be some disturbance in connection with a talk by Oskar Lange for which he had been acting as a "go-between"—he envisioned tomatoes and eggs being thrown at the new Polish ambassador to the United States. Down around "6" on his list was the question of whether arrangements had been made to serve coffee and tea to symposium participants during the morning and afternoon breaks.

From Washington he wrote unhappily to Lehmann:

"The symposium does not appear to have a good prospect. Lange just phoned that he is ill. Somehow or other I am scared that Dr. Hsu may fail to appear at the last moment. My own contribution is haphazardly prepared. Who knows what else may happen.

"However, one must carry on as well as possible."

A few days after he arrived in the capital, he received a soothing letter from Evelyn Fix, to whom Mrs. Neyman had turned over his letter. Everything was under control except that Hsu had saddened everyone by announcing that after his semester at Columbia he was following Hotelling to the University of North Carolina.

Across the continent Neyman was plunged into gloom: "Again it is a case of too little too late."

He was spending his days in the capital attending lectures and studying documents relating to Greece and the coming elections. This activity did nothing to alleviate his depression. It was all politics and "dirty politics."

"Dear Lab," he wrote, "if you only knew how much I regret that I agreed to go to Greece.... I am convinced that nothing good will come out of it."

He thought unhappily of the symposium he was missing and demanded details from Miss Fix:

"Who discussed what? What did you discuss personally? How were the people seated at dinner? I suppose you talked with your neighbors—who were they and what did they say?"

But he had to leave Washington without receiving her report.

Arriving in Athens in February 1946, he and the other members of the Technical Section set up a meeting with the three Greek members of the International Statistical Institute. One of these was a government statistician; the other two, professors. Although in their political points of view, the three represented Right, Center and Left, the meeting went satisfactorily enough with the Greek statisticians coming up with almost the same estimate of the size of the electorate as the Allied Mission. All agreed it was substantially smaller than that indicated by the election rolls. Afterwards there was a casual gathering for cocktails in the hotel suite of one of the Americans. It was then that Neyman and his colleagues discovered that their Greek guests did not speak to one another.

The objectives of the Technical Section were to determine the number of eligible voters in Greece; to ascertain whether any were being denied an opportunity to register; to discover whether there were names on the electoral lists which did not belong to valid voters; and to evaluate the general state of readiness for the elections. The first weeks in Athens were devoted to developing sample surveys to obtain this information. It was a difficult assignment but, to Neyman's disappointment, not one of any great theoretical interest.

To get a sense of the general situation and the problems which would be faced by the personnel who would do the actual surveying, he and the other members of the Technical Section did a considerable amount of traveling during February in the particular areas to which they had been assigned. Neyman's area was that around Salonika, the seaport town which is the capital of Greek Macedonia. It occurred to him that there was a professor of international law at the University of Salonika who had been a Rockefeller Fellow in Paris in 1926; and, although they had not met at that time, he decided to look up Georges Tenekides.

"My contacts with Professor Tenekides were purely social," he later explained to Henry F. Grady, the chief of the American Section. "We dined together twice and discussed a variety of questions. These questions naturally included the elections and also our efforts to check their fairness. All the time Professor Tenekides was speaking very cautiously and I could not gather what was his political orientation.

"On my own initiative I asked him whether he thought that the Electoral Lists are worth investigating. He thought that they are and that, in some places, they are likely to be in disorder. When I asked about particulars, Professor Tenekides said that, if I think that this might be helpful, he would make inquiries and, perhaps, give me some information."

The next day, when Neyman dined again with Tenekides, the latter suggested that he should investigate the electoral lists of the village of Khortiatis.

With the help of an interpreter, a university student employed by the Mission, Neyman conducted his survey from February 20 to 24 and reported the results to the head of the Technical Section on February 26. While there were virtually no irregularities in Salonika itself and only a few in Gefiras, a village which had been suggested to him by the official to whom he had had to apply in order to gain access to the electoral lists, there were in the village of Khortiatis "a substantial number."

The day after Neyman made his report, an angry editorial entitled "The Enemies of Democracy" and referring by name to Tenekides and Neyman appeared in *Nea Alithia*, a Salonika newspaper which favored the return of the monarchy. A long paragraph at the beginning dealt with "various intellectual personalities—professors, scientists, doctors, lawyers, etc., who have enrolled in the ranks of the E.A.M. and who feel the necessity to dissociate their responsibilities from those of the Communist Party."

The writer then presented his own version of Neyman's visit with Tenekides.

"An American Observer comes to our town with the task of examining the charges of falsification of the electoral lists, and of errorism [sic]; and he comes with the recommendation to meet a certain person—

"It is the E.A.M.-ite propaganda in America that had recommended him to meet the Greek Professor. An intellectual to an intellectual. It is understood that there shall be good faith on either side. The one asks and the other answers. But how does, indeed, the other answer?... Does he feel the moral obligation to get away for once from the fetters for the good of his country? Unfortunately, the answer is NO...the *Greek* professor falls from his intellectual pedestal. He thinks and acts as a party-man."

A "conversation" between Neyman and Tenekides was then described:

"'Elections,' says the Greek professor, 'cannot take place under conditions of falsification and terrorism.'

"'Do you have concrete facts...'

"'Endless. Look, here are two electoral lists. These fifty names in this one, in the district of St. Sophia, all belong to dead persons, who are here given as voters. And in this other one, all these names have the same address...'

"The Foreign Observer takes the lists and busies himself with them for two whole days.... And he ascertains that from the first list of the fifty dead men the 49 are still very much alive, and only one did in fact die, but even then only last week.... And as regards the second list, he finds out that the 38 voters with the same residence were registered through their Party and that it is the address of the Party Officers that was put down on the lists but nevertheless, from personal visits to the home of each and every one of them, he is satisfied that none of them has a second [voting] booklet."

The American Observer was then described as going back to the Greek Professor.

"'They must have given you,' he says politely, 'false information.'

"'Yes,' answers shamelessly the Greek Professor, 'they must have given me false information.'"

Neyman did not immediately see the article in *Nea Alithia*. His travel in Salonika had been by jeep and small open cockpit airplane, and he had caught a cold which had developed into a severe case of laryngitis. Upon his return to Athens he was hospitalized for a week and then sent to the Red Cross Convalescent Home, where he was finally able to respond to the various reports he had received from Berkeley about the success of the January symposium. Still not having seen the article in *Nea Alithia,* he wrote an informal account of his trip to Salonika, apparently intended eventually for the members of the Lab.

The canvassing of the selected district in Salonika was uneventful, he wrote in this account—he and Dmitri, his interpreter, were able to locate nineteen of the twenty registrants they sought. But the canvassing of the villages, especially Khortiatis, was "quite thrilling." Khortiatis was magnificently situated on a steep slope—in utter ruins as a result of retaliatory bombing by the Germans. After instructing the driver to open the hood of the jeep so as to distract any curious bystanders, he interviewed—through Dmitri—one of the inhabitants and ascertained that almost all the thirty-eight people on the sample list were either dead or unknown in the village.

"However, I wanted the information confirmed. So we went outside and wandered to the local tavern. By this time, a good proportion of the village's population had gathered so that we had a regular town meeting in the tavern. . . . I wish I could reproduce this scene adequately. . . ."

Dmitri read the first name on the list and also the age of the man, 25. After a moment of complete silence, someone spoke "with inflections of doubt." Others spoke with more conviction. Dmitri explained to Neyman that the man, a pilot in the Greek Air Force, was killed in 1941 during the war with Italy. Since registration for the election took place in 1945–46, this name was obviously fraudulent. The second name was genuine, that of the local shepherd.

"The real fun began with the third name. It started a long series of dead or entirely unknown persons. . . . The meeting was in an uproar, partly angry and partly hilarious."

It seemed that thirty "irregular" registrations out of thirty-eight checked could not have been the result of unavoidable random error.

". . . [yet] with everyone in the village knowing everyone else, the appearance of strangers on election day or multiple voting by some citizens under several alternative names, etc., does not seem to be feasible, unless the village is thoroughly terrorized. . . . They speak of terrorism around here, but up to now I have not witnessed any."

He concluded, "Even if all [the 500 listed voters in Khortiatis] were fraudulent, one such village in the whole of Greece would not matter. The question is, how many villages are like Khortiatis. I hope that there are but few. However, the existence of even one such village seems to justify our presence here."

It was three days later that he first saw a copy of the editorial from *Nea Alithia,* which had been sent to him by Tenekides. He immediately wrote to Grady as the head of the American Section of AMFOGE:

"Professor Tenekides finds that [this] article is both unfair and detrimental to him (with which I perfectly agree) and requests that I publish a formal denial."

He wanted to circulate to the Greek press, through AMFOGE, a personal statement that he had contributed nothing to the article—that it was at variance with the facts and that the passages quoting his conversations with Tenekides were "pure inventions." The statement which he enclosed with his letter to Grady explained that he had contacted Tenekides as an outstanding scholar and had found his personality "in perfect harmony" with his high reputation. He had not asked Tenekides for his party affiliation, nor had Tenekides volunteered it. He had not discovered that anything that Tenekides had said to him was untrue.

Grady (at one time the dean of the School of Commerce at the University of California) did not respond personally. Instead, James M. Keeley, the secretary general of the American Section, informed Neyman that it was contrary to policy to reply to press attacks, no matter how unjust. Neyman objected that the policy could hardly apply to a "non-political" employee like himself who wanted only to deny "charges of non-gentlemanly and non-scholarly behavior."

Keeley insisted that the policy did indeed apply.

Neyman responded that he would send a statement to the press on his own.

"However, following somewhat heated arguments by Mr. Keeley, I finally gave up [that idea]," he reported later to President Sproul. "On the other hand, since I felt it my duty to do something in defense of my host, I insisted on writing a letter to his University authorities...."

He explained to Keeley that since it appeared that the party supporting the return of the monarchy would triumph in the elections, he wanted to protect Tenekides's future position at the university; but Keeley informed him that a letter to the administration of the university would also be "inappropriate." A day or two later Keeley called him to his office and inquired whether he had sent the letter. He replied that he had.

"Thereafter I was summoned to Dr. Grady's office and Mr. Keeley explained that I had disobeyed his orders. This, of course, was exact. However, I felt and still feel, that what I did was correct and tried to explain it to Dr. Grady."

Keeley urged Neyman's immediate departure so that, if there was another newspaper attack in connection with his letter to the rector of the University of Salonika, AMFOGE could release the information that he was no longer in its employ.

Neyman objected again.

"It seems to me that this statement of Mr. Keeley is inconsistent with the principle of not replying to press attacks...."

As soon as transportation could be arranged—it took several days—Neyman was hustled off to Rome. On the plane he found himself in the company of Helen Crosby, a political analyst who has since become the wife of his long-time colleague, Hans Lewy. She explained to him that she had resigned from the Mission because she felt that the minimum conditions for public security and free elections simply did not exist at that time in Greece. He had earlier nicknamed her "Mata Hari" upon hearing that she had worked for the OSS for four years, and now he became convinced that she was a spy sent along by the Mission to keep an eye on him. Later, however, he heard her talking to a newspaper reporter. "So—ah!—I was unjust!"

On March 28, three days before the elections—Neyman was by that time in Paris—a communist paper in Athens alleged that he had resigned because, traveling through Greece, he had become convinced that a genuine election could not be held and he did not wish "to be party to an immoral coup."

Grady had a wire sent to Washington which he later paraphrased as follows to his old friend, Provost Deutsch:

"Since Neyman was ill in the hospital about three out of six weeks' stay in Greece, he visited only one District (Salonika). Therefore, he was in no position to evaluate situation as a whole. Also, his views on the elections wholly unconnected with Mission's termination of his services. He did not resign."

In a personal note to Deutsch, Grady added: "If you know Neyman, and I assume you do, you will realize that he is somewhat difficult. He would not take instructions in a situation which is extremely delicate...."

In Paris, Neyman had to wait again for transportation. It was one of the few times he had been in the French capital since 1936, when he had delivered a series of lectures on his then new ideas on confidence intervals at the Institut Henri Poincaré. Now, as soon as he could, he went to the institute. To his delight, it happened that at the same time both Fisher and Cramér were giving a series of invited lectures and, without being informed of each other's subject matter, they had decided to talk on fiducial argument and confidence intervals, respectively. When Neyman entered the hall, Cramér, who had been lecturing, had just left. Some members of the audience were still sitting around and discussing Neyman and his theory when Georges Darmois, glancing up, exclaimed delightedly, "Here he is!"

Neyman still enjoys telling the story.

"And so, 'How long are you here, this and that, and will you give a lecture?' 'Well, all right! It will be a pleasure.' 'When can it be?' 'Well, tomorrow!'"

The invitation, the coincidence, and—most importantly—the description of Cramér's "vigorous defense" of the theory of confidence intervals did a great deal to raise Neyman's spirits. In his own lecture he presented a recent result—the last paper in point of time which appears in his *Early Statistical Papers*.

When the Greek elections took place on March 31, Neyman was back in

Washington. As had been predicted, a royalist majority had been returned to the parliament, the communists abstaining from voting as they had threatened.

Neyman's considered stand on the Greek elections, which he expressed in talks and letters to colleagues during the spring of 1946, was that "it was a miscarriage of justice and that the American Mission played a role in it which I regret." Three days after his return to Berkeley he spoke on a local radio station. He was asked, "What do you think of the Greek elections?" He answered by citing "a few items of common knowledge," such as the fact that prior to the elections the Greek prime minister had repeatedly complained of royalist terrorism and further the fact that the royalists had indeed won the elections by a comfortable majority.

"The key to the question whether the elections were fair or not and whether their result represents the popular will of the country, seems to be provided by the fifth news item, published only two days ago. This item reports that the victorious Royalists favor the continued presence of the British army in Greece.

"To my mind, if the elections were really fair and if the Royalists did have behind them a large majority of the population, there would be no need for them to seek the continued support of a foreign army."

On his return he had immediately compiled a detailed report on the circumstances surrounding his dismissal for President Sproul; but he went to the Faculty Club for the first time with a certain amount of trepidation. "Dismissed for insubordination, you know." He was always to remember with gratitude how across the dining room came the powerful voice of Robert Gordon Sproul:

"WELCOME HOME, PROFESSOR NEYMAN!"

During the coming year Neyman followed events in Greece with interest and made efforts to obtain positions in the United States for Tenekides and another professor, also not a statistician, whom he had met during his stay there. In the first few months after his return, he gave several talks on the Greek situation; but when, in May, he was asked by the Greek American Council to be the principal American speaker at a mass rally in Hollywood honoring two visiting representatives of the EAM, he refused.

"The reason for this decision is that my appearance at the meeting might be attributed to a desire to support the E.A.M. against their political opponents," he explained to Provost Deutsch. "This would be contrary to the actual situation." Although he thought that most Greek intellectuals supported the EAM, he was not interested in *supporting* either party.

An odd postscript to the adventure in Greece came when the plebiscite on the return of the monarchy was being set up for the following summer. Neyman was asked again to serve as a member of the American team of observers.

"...in spite of (or because of) the unsatisfactory conclusion of the first one," Evans mused.

Neyman declined.

1978   Neyman has a wedding invitation in his hand when I arrive for our weekly interview. Two students are getting married.

"I must send them something."

Jumping to the conclusion that he does not know what to get them as a gift, I make a few suggestions; but he shakes his head, he knows exactly what he wants to get.

"I always give towels. They have gorgeous towels in this country." He likes the word and rolls it in his mouth. *Gorgeous.* "There is a store in Oakland—do you know it?—where they sell gorgeous towels. I can't remember the name."

I suggest Capwell's. No. But Betty will know. He goes to the phone and calls.

"Betty—what is that store with the gorgeous towels?"

He comes back from the phone, pleased.

It is I. Magnin's.

I offer to stop by Magnin's after our interview and pick out some towels for him, send them to the young couple with his card.

But no—

"I would like to *participate,*" he says. "If you would take me in your car—?"

He goes to the closet and gets the new beret Betty has bought for him—his old one, he explains, has "left"—and we drive to downtown Oakland. It is some little distance from Neyman's house in the Berkeley hills.

At I. Magnin's he makes his way slowly through the mannequins and the glass and gold of the main floor.

"Where do you have your gorgeous towels?" he asks the young woman at the elevator.

"Fourth floor. Gifts, sir."

A pleasant, middle-aged clerk brings out an assortment of pastel colored towels, pale green, delicate mauve, soft yellow, all appliquéd with satin flowers and butterflies.

I observe that they are very pretty but may not launder well.

Neyman rubs the material between his thumb and forefinger.

"I like *thick,* you know."

The clerk leads him to shelves where rows of plain, deep colored towels are stacked. Mustard. Turquoise. Emerald. Wine red. Mulberry. Salmon.

He feels the material.

He takes down the salmon, then the emerald, puts the latter back. Examines and rejects mustard and turquoise. Debates between wine red and mulberry, laying them beside salmon, which he likes best.

After consultation with me and with the clerk, he finally decides on the mulberry with the salmon.

Further consultation. A bath towel, a face towel, and two washcloths in each color. He carefully writes a check, then writes a card, arranges that they be sent to the home of the bride's mother.

"I would like to take them myself," he explains to the clerk, "but it is difficult."

"I am sure she will be very pleased with your gift," the clerk assures him. Later, again in the car, he settles back with a cigarette.

"I *like* to give gorgeous towels," he says.

1946
—
1949

The Second World War had established the usefulness of statistics. The years that immediately followed saw a scramble for qualified statisticians in government, industry, and academia. Competition was "fierce." So Neyman described it to Evans.

Hotelling's decision to leave Columbia had struck a blow for the cause that he had so long advocated—separate departments of statistics in colleges and universities. Columbia, through its committee on education, recommended to its budget committee "to take steps to make it plain that the departure of Professor Hotelling does not mean an eclipse of the subject at Columbia." It also recommended that mathematical statistics be set up as an independent department. Abraham Wald should head the department as a full professor, and provision should be made for a distinguished visiting professor at the same salary as that of Wald.

Thus it happened that in the spring of 1946 Neyman received an invitation to spend the fall semester of that year at Columbia. Scarcely back from Greece, he was nevertheless eager to accept. As a consequence of the presence of Wald, Columbia had become—in his opinion—the outstanding statistical center in the country. He saw Wald as "the ideological successor" of himself and Egon and spoke and wrote of him in glowing terms. Working conditions at Columbia would also be incomparably better than those at Berkeley—four hours of lectures as compared to nine! There would be time for research and time also to finish the elementary book which he had been trying to write for the last several years. When, however, he applied for a semester's leave to accept the Columbia offer, he found Sproul and Deutsch reluctant—Berkeley, they felt, had seen remarkably little of Mr. Neyman in the past few years.

True, Neyman conceded—he had had several leaves, but these had been, not holidays, but chores "in the service of the country and therefore of the University." Since 1941, he had not had even so much as a single week's vacation! As to the more general interests of statistics at Berkeley, he agreed that his absence would create something of a problem. With the current demand for statisticians in the United States, it would probably be necessary to go abroad for a semester's replacement. But finding a semester's replacement was not the real problem.

He recapitulated the activity of the past years with a shade of bitterness.

"...after I arrived in Berkeley, eight years ago, the Statistical Committee decided that the teaching program in statistics should aim at training future research workers.... On my remarking that in that case more than one man was needed, it was suggested that I might train additional men.... Since an extended program seemed attractive to me, I offered to teach it all by myself.... Dresch and Dantzig were assigned to attend these courses so that, later on, they could teach some of them in my stead.... Unfortunately the war came and, prior to its actual outbreak, both Dresch and Dantzig and, later on, Shephard were claimed by the armed forces and, already engaged in war research, I had to begin the training of junior personnel anew."

This practice was simply not fair, and he was no longer willing to continue it.

The alternatives, as he saw them, were clear: either the university should give up the idea of developing a great statistical center, or it should embark "on broad and bold reforms, giving us, the statisticians, both the freedom to revise the programs of teaching and [the opportunity to obtain] more men of high standing...."

Under the force of these arguments, Sproul granted the request for a leave. He also encouraged Neyman and Evans to discuss widening research and instruction in statistics and making a major appointment.

The first choice of both men for such an appointment was Pólya. Time and again, during Neyman's many absences, Pólya had come up from Palo Alto to give his lectures and conduct his seminar. Half a dozen years older than Neyman, he was an exceptionally broad mathematician with a worldwide reputation in analysis, also a most effective teacher—his little book *How to Solve It*, which was to become a popular classic, had just been issued by the Princeton University Press. In addition, while filling in for Neyman, Pólya had himself become interested in mathematical statistics and had successfully tackled an important problem arising out of Wald's new theory of sequential analysis. His position in the mathematics department at Stanford was, in the opinion of both Neyman and Evans, wholly out of keeping with his stature as a mathematician. Although Berkeley and Stanford had a gentlemen's agreement not to "raid" each other's faculties, Neyman made discreet inquiries as to whether Stanford would consider releasing Pólya to accept a higher appointment at Berkeley. Thus approached, Stanford belatedly promoted Pólya to the rank of full professor.

Neyman and Evans were also disappointed in their efforts to obtain the services of a young French probabilist named Michel Loève, the "intellectual son" of Paul Lévy, who had greatly impressed Neyman during his brief stop in Paris on his way home from Greece. Loève was a Palestinian-born Jew who had received his education in a French lycée in Alexandria. He had migrated to Paris and had been incarcerated by the Germans during the war. Although he had indicated an interest in coming to the United States, by the time he received the Berkeley invitation he had already accepted an invitation from the University of London.

Late in the summer of 1946, when Neyman went east for his semester at Columbia, he left Lehmann, who had only just received his Ph.D., nominally in charge of the work in the Lab. In spite of many disappointments, he had not abandoned the Berkeley development. During his semester

211

at Columbia he outlined an extended program for basic research in hypothesis testing and estimation for which he hoped to obtain funding from the newly established Office of Naval Research. He envisioned young men and women carrying on the work which he himself had been so instrumental in developing twenty years earlier. There would be a systematic study of the performance characteristics of several current tests the properties of which were still obscure; the development of tests and methods of estimation for several problems recently encountered in applications; and the working out of the concepts of optimum tests for cases where known concepts were not applicable. In addition, efforts would be made to extend the recently developed theory of stochastic processes in which the French and the Russians had done such spectacular pioneering work. For this last part of the program he intended to make another try for Michel Loève.

The semester which Neyman spent away from Berkeley was a pleasure, although he was again without his family. He was stimulated by the group around Wald and by the presence in the east of so many important academic centers. He found the preparation of the undergraduate students at the great private university superior to that of comparable students at the University of California, and he was gratified to find on the list of those in his advanced course a name which he had already noted in the *Annals of Mathematical Statistics*.

"So when I am at my first lecture, 'Is Charles M. Stein here?' And he speaks with kind of hesitant voice. 'Yes?' 'Are you the author of this paper?' 'Yes.' Very softly. Gentleman and *gentle man*. 'Then what are you doing here?' 'Well, I am studying. I want to get a Ph.D. degree.' 'So how did you write this thing? This paper deserves a Ph.D. without any change!'"

Soon Wald informed Neyman that "wheels were turning" to offer him a permanent appointment at Columbia. Neyman wrote immediately to Birge, who, he felt, knew better than anyone else at Berkeley how much time and energy he had devoted to trying to build up statistics there.

"So, when I think of leaving Berkeley, my heart is bleeding. On the other hand, the conditions of work in Berkeley are so unconducive to research in statistics that, if I stay there without any reforms, I shall have virtually to accept that my research work is concluded."

A separate department would be a solution to many of the problems. He knew that Birge and several other prominent professors, including Charles Morrey of mathematics, had previously gone to Provost Deutsch with the suggestion that a separate statistics department be created at Berkeley. He understood that Deutsch had appeared favorable to the idea, but nothing seemed to have come of the suggestion.

"I would hate to leave the Lab. Equally I would hate to repeat the story of Hotelling who, after years of struggle, prepared the ground in Columbia for the creation of a Department of Statistics, but just failed to complete the thing, and the Department was created after, or perhaps because of, his resignation. On the other hand, I feel that I still can produce and, that being my main ambition, I would not like to have it buried in minor bickerings."

The formal offer from Columbia, when it came in January, was in every respect—salary, teaching load, retirement benefits—an improvement over the position in Berkeley. Neyman informed Sproul and Evans, laying out again for them the development and achievement of the statistical group since he had come to Berkeley.

"It may be presumed that the graduate students coming to us from abroad, and also their respective teachers, think that actually we are now one of the most effective centers of statistical instruction. Unfortunately they are mistaken."

The recent advances in mathematical statistics had been so great that no single man was any longer capable of giving adequate instruction to students. It was his opinion—he was very explicit—that the only way in which Berkeley could be brought up to the level of such places as Columbia would be through the organization of a Department of Statistics entirely independent of the Department of Mathematics.

Columbia wanted his answer by February 21. Since he would be returning to the west coast by automobile, there would not be much time for discussion after he got back to Berkeley.

"Luckily the problem is not new...."

Evans continued to stand firm against the creation of a separate department of statistics. Systematic instruction in statistics belonged in mathematics—not in economics or political science, as it had been at Columbia, and not in any other subject. This view had been the motivation for his original proposal to add Neyman to the mathematics department, and it was one which he still considered valid. In Evans's opinion the pressure from Neyman for a department of statistics came from "an overestimation of the administrative position of a chairman." He hoped that some way could be found "which will suitably gratify personal ambition without destroying the essential character of the organization."

Since the end of the war, Evans himself had been actively working toward the development of applications in the mathematics department, "both for the sake of the applications and for the sake of mathematics itself." This was a pioneering effort in the face of the American penchant for pure mathematics. He conceived of a "Mathematics Center" which would ultimately include an Institute of Statistics, a Bureau of Computation, an Institute of Actuarial Science, and an Institute of Applied Mathematics and Mechanics. Eventually, he expected, there would be at Berkeley a postgraduate School of Statistics and Insurance. In the immediate future it might be possible to add dignity to the Statistical Laboratory, and to its director, by having its advanced functions taken over by an Institute of Statistics of postgraduate status.

The Budget Committee endorsed this last suggestion and expressed its belief that being director of such an institute would satisfy Professor Neyman's desires for development and autonomy appropriate to his distinction. Other points raised by Neyman in his letters to Sproul and Evans, including the matter of salary, could be arranged—in the opinion of the committee—to the satisfaction of both Neyman and the university.

While these decisions were being reached, Neyman was driving across the country. He dropped south to visit Hotelling at Chapel Hill and to see Hsu again. He still had hopes of enticing the latter to Berkeley—or to Columbia—but he found the Chinese scholar miserably unhappy, disappointed in love, and desiring only to return to his native land.

Scarcely a week before Neyman was to give his decision to Wald, he arrived back in Berkeley. He wired for an extension of time, and a new deadline was set for March 6. In the next few weeks he had several talks with Sproul and Deutsch, first alone and then with Evans joining in. After the last of these meetings—it was March by then—he and Evans sat down together in the latter's office and hammered out a letter to Sproul.

It was agreed between them that in the future Neyman, as director of the Statistical Laboratory, would submit his budget directly to the administration without the previous approval of the chairman of the Department of Mathematics. Appointments of research personnel would be completely in his hands, but appointments of teaching personnel would continue to be made in consultation with the mathematics department chairman. Revision of courses in statistics would be worked out by a small committee composed of representatives of the Statistical Laboratory and the Department of Mathematics—"with the understanding that in case of difficulty of agreement, recourse may be had to the judgement of other representatives of the University as a whole."

There would be some new appointments.

"We understand," the two men wrote to Sproul, "that you are favorably inclined to the addition at this time of two persons to contribute to the work in statistics, one a position of tenure grade...and another of the rank of instructor."

In the first draft of the letter the proposed candidates for these appointments are mentioned by name: Michel Loève of the University of Paris and Charles M. Stein.

Neyman disdained the offered sop of the directorship of a postgraduate Institute of Statistics. The Lab would remain the Lab.

Asked to give the reason for his decision, he replies, "Past motivation is difficult to reconstruct. I think I felt that in some ways it might separate us too much from mathematics."

On March 6, 1947, Neyman and Evans put their names to the letter to Sproul. The same day Neyman wired Wald: SORRY MUST DECLINE YOUR KIND OFFER....

With the signing of what Neyman was always to refer to as the "Magna Carta Libertatum," the Statistical Laboratory entered upon one of those happy periods which occur close to the beginning of any great group. Youth, mutual interest, and commitment to the growth of the institution seemed to make for perfect congeniality.

The Lab was by then installed on the upper slope of the campus in what is now Minor Hall, the third of its many homes. Most of the active members had been trained by Neyman: Erich Lehmann (Ph.D. 1946), Evelyn Fix (Ph.D. 1948), Joseph Hodges (also 1948), Elizabeth Scott (1949). Edward

Barankin, who had joined the group toward the end of the war, remained and got his Ph.D. at Berkeley. In the fall of 1947 Charles Stein came, and the following year Michel Loève. Of these, only Stein was to leave.

The survivors remember themselves as a happy few, sharing cramped offices, playing tennis on the nearby faculty court, working nights and Saturdays, lunching together on Saturdays at George's, a little restaurant in Oakland no longer in existence, often collaborating on work. Hodges and Lehmann, Fix and Hodges, Fix and Neyman, Neyman and Scott. ("It is a firm rule in this laboratory," Neyman would explain to a job applicant, "that there is no discrimination on the basis of age, sex, or race...authors of joint papers are always listed alphabetically.") Visitors invariably commented on the spirit of the group. ("As always you and your staff were kindness itself," Bronowski wrote after a trip to Berkeley with a contingent from the RAF, "and my thanks are mixed with envy at the delightful relations in the laboratory.") The group was compared to a family. Neyman liked the idea. He assumed the role of father with assurance and delight. He loved to do things for the young people—plan parties and picnics and theater-goings, provide cake and tea in the afternoons, buy gifts, promote careers, participate with zest in their lives and loves.

For those who had their own families or those who had recently won their freedom from their own fathers, he was a difficult person to work under. One young man welcomed the news that he had been awarded a scholarship for which he had not applied—"I applied for you," Neyman told him—but another had a different reaction when he learned that, without ever having been consulted, he had been "volunteered" to deliver a paper at an important meeting.

Neyman wanted things, even sometimes very minor things, done his way. Once, when a student at the blackboard stubbornly persisted with a demonstration of which he did not approve, he took the student by the hand and forcibly wrote what he thought should be on the board.

A conflict on a more serious scale had its beginnings in 1948–49 when for the first time he entrusted Lehmann with the basic graduate course. Notes on Lehmann's lectures were taken by Colin Blyth, one of the students in the class; and since no other systematic treatment of hypothesis testing was available at that level, they had a fairly wide circulation in mimeographed form. A summer class that followed resulted in a comparable set of notes on estimation, also taken by Blyth. Lehmann conceived the idea of converting the notes into a book; and a great deal of his research effort during the next few years was devoted to filling in gaps in his knowledge for the project.

"It would have been very helpful, and also quite natural to try the work out in class. However, Neyman never let me teach the course again. It appeared that I had not followed his script. I had introduced some new topics and approaches. They were absolutely along his lines, but just extended a little bit and carrried a little bit further here and there.

"After several years, during which courses based on my notes had been developed at a number of universities, I asked Neyman directly why he wouldn't let me teach the course again. He answered that he would consider it if I would let him see the manuscript of my book so that he could determine how suitable it was. At this demand I bristled. He already had a copy of the notes, and there really did not exist a reasonable manuscript of the book at that time. It was all bits and pieces. So I never taught the course again until many years later when Neyman was no longer chairman of the department."

Such incidents, Lehmann points out, are balanced by others of great kindness to young colleagues.

"Neyman has always been generous in acknowledging priority, putting a student's name on a paper as a co-author, or omitting his own name."

Behind his back he was "Our Glorious Leader," a benevolent but ironhanded father-figure who was able to arrange that things always turned out right—and his way. Shortly after the war, for example, there was a wide range in ability among the statistics students, a number of whom were paying expensive "out-of-state" tuition fees which would be forgiven if they made the honors list. Aware of their difficulty he proceeded to "hand-tailor" a final. Each student was given a problem geared to his individual abilities; if he solved it, he got an "A." Other people on the campus, looking out their office windows and seeing Neyman with his young colleagues, would on occasion turn to each other and murmur, "Jesus and his disciples."

(Neyman, told of this old campus nickname, shakes his head. "Jesus?" he murmurs, bemused. "No.")

For the statisticians at Berkeley and for statisticians in general, the period following the war was a gratifying one. Their value seemed at last recognized, the relationship between mathematical theory and applications appreciated, even by the government. Money became available. At first Neyman had been doubtful about government-sponsored research in peace-time; but Mina Rees, traveling across the country, had managed to convince him (and other scientists as well, for he was not alone in his reluctance) of the pure intentions of the newly established Office of Naval Research. Grants were made without strings. "We do what we want to do," Neyman explained to Deutsch when the ONR funded his proposal for a long-term project concerned with hypothesis testing and estimation—a proposal which had been energetically backed by Evans as of importance for the development of statistics in general. ("This is the way we are able to compete with Columbia," he had reminded Sproul.)

There were a large number of visiting lecturers and speakers during the academic year and, beginning in 1947, high-powered summer sessions as well "for graduate and post-graduate students, perhaps for those young men (read "Young Joneses") all over the country who [have] recently obtained high University positions in statistics and feel that their educations could be usefully improved by being introduced to new fields of study."

Colleges and universities jockeyed for the leading statisticians. Columbia had tried to lure Neyman from Berkeley, other universities tried to lure Wald from Columbia. So it was, all down the line. Everyone was on the lookout for promising young people.

"Both Lehmann and Hodges are with us," Neyman wrote in response to an inquiry, "and it is my intention to do everything I can to keep them."

The "Magna Carta Libertatum" had contained the promise of a series of statistical publications at Berkeley or, alternatively, the joint sponsorship of a journal in cooperation with some other institutions. For a long time Neyman had pushed for the Lab to have its own journal. He envied the "unlimited access to the printing press" which people like Fisher seemed to have and smarted under the seemingly arbitrary decisions of various editors. (The year before he had withdrawn from the editorial board of the *Biometrics Bulletin*—"[leaving] a trail of smoke behind," according to Evans—because references to the work of two specific persons had been omitted from a short article of his: "[they] work hard and well and, owing to just such tricks..., cannot get recognition.")

Indeed, he has never lost his strong feelings about what he describes as the "bossiness" of journal editors.

"And so this is disgusting," he tells me. "It was recommended that I should be editor of the *Annals*. But I declined. I declined to participate in this bossiness. To be on the board—to review papers. That's a different story. But I didn't want to participate in the category of individuals who for some reasons not relating to scientific things reject this, accept that."

In August 1948 he brought up again in a letter to Egon the idea of reviving the *Statistical Research Memoirs* in the United States. Pearson was agreeable, but nothing came of the plan. The following year the first number in the series, *University of California Publications in Statistics*, appeared (the doctoral thesis of Joe Hodges). The series was a "journal" after Neyman's own heart. There were three conjoint editors: himself, Loève, and Jacob Yerushalmy of the School of Public Health—and no referees.

"...it is my experience," Neyman once remarked, "that whenever a generally decent fellow is asked to act as an anonymous referee, he is apt to acquire hateful qualities: presumptuousness, quarrelsomeness, and bossiness."

Yerushalmy's presence on the editorial board was a consequence of a developing interest on Neyman's part in problems of medicine and health (stimulated by his friendship with Joe Berkson). Although their initial contact had been marred by "unpleasantness," Neyman and Yerushalmy cooperated extensively in the ensuing years and in 1954 joined to propose a degree in biostatistics to be administered by the School of Public Health and the Statistical Laboratory.

Neyman directed several of his students into public health. One of these was Chin Long Chiang, who with Elizabeth Scott now heads the Biostatistics Group in the Berkeley School of Public Health.

Chiang tells me that he considers the paper by Neyman and Evelyn Fix, "A simple stochastic model of recovery, relapse, death and loss of patients," Neyman's most important single contribution to biostatistics.

"Up to that time only length of survival was considered. One state. Neyman and Fix introduced a second state. Health. But their model is actually far more general. For instance, in engineering—not just how long does a light bulb burn but whether it burns brightly."

The paper, according to Chiang, is also another example of how Neyman has never received credit for many of his contributions. His ideas and methods have passed into the common body of statistical knowledge and methodology—"the way it is done"—without his name being attached to them.

"I tried to correct this in my book on stochastic processes in biostatistics by referring to 'Neyman-Fix method,' but most authors still call it 'zero-crossing process.'"

Another student whom Neyman directed into public health was William Taylor, who joined Neyman's friend Berkson at the Mayo Clinic and later succeeded him as head of the Section on Biometry and Medical Statistics.

Neyman's own research continued to suffer from the distractions of his many administrative duties and the extensive amount of time which he gave to students of all levels of ability. In 1947 he was elected vice president of the American Statistical Association with the duty of arranging an International Statistical Conference the following year in Washington, D.C. The next year he was elected president of the Institute of Mathematical Statistics.

He was increasingly aware of the "isolation" of west coast statisticians and the consequent dangers of "inbreeding." It occurred to him that another symposium might be at least a partial solution of the problem. At the beginning of 1948 he submitted a request to the Budget Committee for $4000 to be allocated for compensation of participants and publication of the proceedings of a Second Berkeley Symposium on Mathematical Statistics and Probability to be held in 1950. The committee recommended the expenditure as justified in view of the fact that the Statistical Laboratory was in the process of development.

The capital of statistics, which had once been London, had moved to the New World. The mathematical approach now dominated the subject. The membership of the American Statistical Association, a basically applied group, had doubled between 1935 and 1945; but the membership of the Institute of Mathematical Statistics, a society which had been organized by a couple of dozen men in 1935 to promote the theoretical side of statistics, had multiplied by almost one hundred.

The impetus for this great growth in mathematical statistics in the United States stemmed rather directly from Neyman and his work: the Neyman-Pearson paper of 1933 on the testing of statistical hypotheses and the 1937 paper on the theory of estimation; the popularity of Neyman's 1937 lectures in Washington, D.C., and their subsequent wide circulation; the work of Neyman and Pearson and their students published in the *Statistical Research Memoirs;* and, finally, the presence of Neyman himself on the west coast of the country and on the east coast Abraham Wald, in many ways—as Neyman said—the "ideological successor" of himself and Egon. Nearly all the papers published in the *Annals of Mathematical Statistics* concerned themselves with issues raised by Neyman's early work.

It must be said that after the war the Americans were inclined to feel superior to many of their foreign colleagues. This was something that it was easy to do when the adjective "American" could be applied to such men as

218

the Russian-born Neyman, the Romanian-born Wald, the German-born von Mises, the Yugoslavian-born Feller, and the Chinese-born Hsu (who was still considered, hopefully, "on leave" from North Carolina). The committee in charge of nominating foreign speakers for the first International Mathematical Congress to be held on American soil (in Cambridge in 1950) hoped that the Russians, whom they did admire very much, would be permitted to attend and said privately among themselves: "After all R.A. Fisher is no longer the man to show off as the best that mathematical statistics has to offer...."

Fisher, making his first postwar visit to the United States, found a disconcerting change in the general attitude toward himself and his work, according to his daughter-biographer.

"...new influences made themselves effective. J. Neyman emigrated to America...and A. Wald...; in teaching, theoretical and mathematical aspects of statistics received increasing emphasis. Changes took place in the mood of American statisticians which extinguished the enthusiasm that had charmed Fisher in 1936. When he returned to the United States in 1946 he was welcomed by the younger statisticians as a great originator and authority certainly, but also as a foreigner whose ways were not always their ways, nor his thoughts their thoughts."

Perhaps—she speculated—much might have been different if Fisher had accepted the position offered him in 1936 in America.

In America, however, there were apparently no regrets—certainly not at Berkeley—as to the choice made in 1938.

"...the University of California is now one of the two real Statistical Centers in this country, the other being at Columbia," proclaimed Charles Morrey, who had succeeded Evans as chairman of the mathematics department. "...That our university is so fortunate is due in the first place to the foresight of our Professor G.C. Evans in inviting Professor Neyman to come here at a time when Mathematical Statistics was only just beginning to be recognized in this country as a 'respectable' branch of Mathematics. However, since then, Professor Neyman has become more and more responsible for the development of the Statistical group and should therefore be given a large share of the credit for the high regard in which his group is now held all over the country....In spite of the difficulties caused by [the demand for statisticians], Professor Neyman has been able to keep his group together and even add [very able people]. This is, of course, due in part to the reputation of the University and the advantages of being here....But also Professor Neyman has fostered very strong ties of loyalty by means of almost continuous informal personal contact....I therefore believe that it is only fitting and proper that the statistical group form a separate department and that Professor Neyman receive recognition for his stature and services by being appointed its chairman."

In 1949 the future looked very bright for statistics at Berkeley, but—as Neyman says—"Life is complicated...."

1949
—
1951

The academic year 1949–50 came to be known at the University of California as "the year of the oath" after the title of a small book by George R. Stewart, professor of English and a well-known novelist. The "oath" in question was a disclaimer that the signer did not belong to and did not believe in any organization which advocated the overthrow of the government of the United States. The initial objection by the faculty was to the regents' imposition of a special oath upon the members of the university community. Later, questions of tenure and academic freedom became paramount; finally the issue became one of power—who was going to run the university?

Attitudes toward the oath varied from campus to campus and department to department. Berkeley had more non-signers than Los Angeles, and the Berkeley mathematics department had a particularly large number. These, however, did not include Neyman. Many of the people who were closest to him would have expected him to take a stand against the oath with all the intransigence of which he is capable. Instead he described himself as being "color blind" on the subject. To him, "not believing in the oath and not signing it" were two different things. His attitude was that everyone who objected should sign under protest and should keep on protesting. Nevertheless, William R. Gaffey, who was for a time chairman of the organization of teaching assistants who opposed the oath (and another of the students whom Neyman directed into public health), recalls that Neyman never tried to influence him. "What he did do was to offer to arrange with Harold Hotelling to take me on as a graduate student and teaching assistant, if it became necessary. In effect, he removed the threat of unemployment so that I could decide without economic pressure whether or not to sign."

Neyman himself felt considerable economic pressure to sign. As in England, he still lived always beyond his means, generously buying and spending for visitors and students out of his own pocket. Even during the war he had insisted—to Warren Weaver's annoyance—that he had to teach an evening course at fifty dollars a month "to get by." His own decision in regard to the oath was to sign but, as he said, to protest. All through the long months of the struggle he consistently supported those who felt that they could not sign.

The debate over the regents' oath simmered during the summer of 1949, but by September the situation on the campus was tense. In October, Neyman flew to Paris to take part in the International Congress of Philosophy.

One of the topics proposed for discussion was the philosophy of the theory of estimation—a 150-year-old problem for which he had been "lucky enough"—he explained to Sproul—to produce what seemed to be a solution.

"As is usually the case with novel ideas, my theory met with almost unanimous scepticism. Later on, however, it took root and at present seems to be generally accepted in this country. However, on the continent of Europe it is still hotly debated...."

It was no surprise, therefore, when after his talk—"Foundation of the theory of estimation"—a letter attacking it was circulated by C. Gini and G.

220

Pompilj. The former, Neyman pointed out in his response, "pendant la decade 1938–1948,...à lui seul, a publié dix sept mémoires, tous consacrés à la 'démolition' de la théorie anglo-américaine de statistique mathématique." How he envied such "unlimited access to the printing press"!

He returned from Paris in November to find that the oath had not "gone away." The Berkeley non-signers had organized themselves into an informal group. They were convinced that the imposition of a special oath was unconstitutional; and they intended, if necessary, to take their case to court. The regents had divided as a result of the unexpectedly violent reaction to the oath on the part of the faculty. The influence and authority of Sproul as president—previously unassailable—was seriously threatened. The alumni struggled to put together a compromise which would restore peace.

In February, Neyman left Berkeley again for a semester's sabbatical abroad, this time accompanied by his family. He had high hopes that by the time he returned he would be chairman of his own department.

The semester abroad—his first sabbatical—was in a way a reverse reprise of his prewar career. His first stop was University College, and he was surprised that he still felt at home in its halls. "I wonder how many years must pass for me to lose this feeling." The atmosphere, however, was much different from that in the old days. Fisher was no longer upstairs, having been appointed during the war to the Arthur Balfour Chair of Genetics at Cambridge. He had been succeeded by the geneticist L.S. Penrose.

Pearson had asked Neyman "to tell my staff and post-grads and E.S.P. just where, after war-time experiences and years at Berkeley, you [have] got from the joint foundation of 1936." But Neyman chose instead to lecture mainly on theoretical aspects of the medical problems on which (under a grant from the Air Force's School of Aviation Medicine) his laboratory had been working. And, as the disappointed Pearson later reminded him, all the while he was very absorbed in problems of the oath "—just, old boy, as some 20 odd years before, you had for short periods been 100% absorbed in problems re Pilsudski!"

After the first cycle of lectures in London—there were to be two—Neyman went to Paris, where Mrs. Neyman and Michael were living while the family was abroad. He found a great and satisfying change in the French capital. There was a lively group of young people who seemed eager to hear what he had to say. One of these was a tall, very thin young man of twenty-five. The son of a farmer, Lucien Le Cam had had a haphazard scientific education. Currently he worked at Electricité de France but, having flexible hours, "hung around" the Institut Henri Poincaré, where it was his duty to see that there was a speaker each week for Darmois's seminar. Although he had no degree, he had already published several papers in the *Comptes Rendus* on theoretical ideas that had developed out of his job. Neyman, "impressed," offered him a Junior Lectureship at Berkeley the following semester. Le Cam thought, as he says, "Well, why not?"

"It seems that people in Paris begin to appreciate our work," Neyman wrote happily to "Dear Joe, dear Erich." Borel had invited him to contribute a volume to the famous "Borel Series" on the theory of probability. This was a notable honor, since all authors in the series to date had been French. "I think of this thing in terms of our group in Berkeley," he told Hodges and Lehmann. "As you know, the main text of Borel books is frequently followed by 'Notes' written by other persons. In this spirit I invite you two...to write a 'Note' summarizing your results either obtained separately or jointly. I already invited Charles [Stein]...and am writing to Betty [Scott]....It will be nice to have a nice volume published representing the 'Berkeley ideas.'"

From the French capital, Neyman went to Poland, arriving in Warsaw—for the first time since the war—on the eve of his fifty-sixth birthday. His friend Lange, by then Poland's ambassador to the United Nations, arranged a room with a private bath at the Society of Polish Economists, and he was much more comfortable than he had been in France and England.

"There appears to be a hell of a lot of people who like to see me, and I am taken from dawn to late at night," he wrote to the Lab. Warsaw had been virtually leveled. "...rubble—rubble—rubble and plenty of people swarming over the rubble, practically day and night, clearing space and building—building—building."

Under the rubble, he knew, there was a graveyard. His own personal losses as a result of the war are listed "reverently and affectionately" in the dedication of the second edition of *Lectures and Conferences,* on which he was working while he was abroad. He searches for the book in his study and reads the names aloud to me. They come in a dreadful litany. Lost. Murdered. Missing. Killed. Lost. Murdered. Starved. (This was his old mentor from Kharkov days, Antoni Przeborski.) Committed suicide. Murdered. Missing. He closes the book without comment.

And what of Sierpiński? The founder of the great Polish school of mathematics had survived the war but felt neglected because he was no longer consulted by the Rockefeller Foundation. Neyman wrote to Warren Weaver to see if some little attention could not be paid to the old man.

While he was in Poland, he also saw his brother again—for the first time in a dozen years. Before the war it had been Karol—an impressively successful businessman, carrying on a lively trade with Sweden in iron ore—who had been the "father" of the family. Now the roles of the brothers were reversed.

In addition to the cycles of lectures which he delivered in London, Paris and Warsaw, Neyman also gave talks in a number of other European cities, traveling particularly extensively in Poland. When he returned to England, he went for a week to Cambridge. Coincidentally, Fisher was not there but in America; and Neyman instructed Loève, whom he had left in charge during his absence, that if R.A.F. should come to Berkeley the Lab should by all means join in inviting him to lecture. "Also, if it is convenient to formulate the invitation so as to mention that I, Neyman, am one of the persons inviting him...I should be grateful."

While Neyman was abroad, the subject of a department of statistics had been thoroughly debated at Berkeley. Finally, in May, Dean Alva R. Davis, a colorful botanist familiarly known as "Sailor," made his recommendation:

"Because of the [many] conflicts and because of a definite lack of any enthusiasm for the proposal except that presented by the Statistics group itself, and furthermore, because I believe that the functioning of the Statistical Laboratory as a quasi independent unit is not being hampered by its present status, I recommend against the establishment of an independent department at this time."

He was under no illusion, he added ruefully, that this delay would "diminish in any degree" the efforts of Professor Neyman to gain the independence of his group.

In the early summer of 1950, Neyman and his family returned to the United States. Despite the stimulation which he had received from his travels, there were regrets. He and Egon had hoped to produce another joint paper but, as he wrote, the "old resonance" between them had seemed to be lacking.

"From my side, I think the explanation...is not hard to give," Pearson responded. "...I don't really enjoy being head of a Department nor managing editor of a journal; having a slow working mind, I can't deal rapidly with all the administrative business nor quickly get to the bottom of people's papers...the result is that my power of original work gets pretty well stifled and whatever I may have hoped before you came, it at once became clear that with the burden of things like exam papers & proofs and daily correspondence, I hadn't the energy left to contribute to any joint work....one lives in hopes of finding a solution to an impasse, which probably appears much easier to solve to others than to myself!"

In addition, although Pearson did not refer to the fact, Eileen Pearson had died in 1949 of a viral infection; and, along with his other duties, he was struggling to be both father and mother to two small daughters.

Back in Berkeley, with a Second Berkeley Symposium on Mathematical Statistics and Probability only weeks away, Neyman found that the problems of "the oath" remained. The non-signers had, however, been granted an extension of time and the opportunity to petition the president of the university for a review of their individual cases by the Committee on Privilege and Tenure.

Some members of the faculty had already left. Among these was Charles Stein, who was looked upon by many as the most original young statistician in the country. The extent of Neyman's "high admiration" for Stein even at this early date is indicated by the fact that for his inaugural lecture at University College he had chosen to talk on Stein's work and, in revising his *Lectures and Conferences* while he was abroad, he had devoted a whole section to "the brilliant work of Charles M. Stein."

A few days after Neyman returned to Berkeley, the regents voted, 10 to 9, to accept Sproul's recommendation to retain the majority of the non-signers who had applied for hearings before the faculty committee. Immediately after this result, Regent John Francis Neylan—who had voted against retention—announced that he was changing his vote from "nay" to "aye"— a parliamentary maneuver which would permit him to move for a reconsideration of the vote at the next meeting of the regents.

Such was the situation at Berkeley on the eve of the Second Symposium. In spite of Neyman's absence, arrangements for the affair were well in hand. Then, unexpectedly and at the last moment, came a number of regrets "on account of illness." Neyman raised an eyebrow. An epidemic among statisticians and probabilists? It appeared that there was a move to boycott the symposium in order to show support for the faculty against the regents. A letter arrived. How would Neyman and his people feel about such a boycott?

The regular members of the Laboratory agreed that they themselves were not clear about the best answer, and so they limited their reply to describing their own unanimous attitude:

"(i) We deeply regret the recent happenings and, especially, the threatening dismissals of our colleagues. [Six of the forty-five non-signers had not been cleared by the faculty committee.] We find that the atmosphere created by the action of the Regents is not conducive to research and to normal academic activities. We believe that, if this atmosphere, and what goes with it, spreads to other Universities, this would be to the great disadvantage of the country as a whole.

"(ii) However, while we are at this University, we continue our usual functions, including the Symposium."

There was no boycott. The Second Symposium was satisfactorily successful, a much larger affair than the first, with two foreign visitors—Paul Lévy and Bruno DeFinetti—and thirty-five from outside California. Early during the meeting, however, it was proposed that the participants join in a resolution expressing support for the faculty against the regents. Neyman decided that he should consult Sproul.

"I think I have a tendency to be loyal to people with whom I work. Sproul—president—and Evans brought me here. And who knows what Sproul's reaction would be? *Welcome home, Professor Neyman!* I remember." The president was not on the campus at the time but at his summer home on a lake in the mountains. "And I took my dog, so-called 'Socks,' and drove to this lake. The lake was composed of two lakes essentially, with a little bit of a passage between them, a narrow thing. I had to take a boat to get to the other side where Sproul's house was. Currents from one lake to another were strong, and I remember—unexpected—my dog sitting next to me and pressing against me. Scared. So then I told Sproul what we were thinking of doing and asked him whether he would welcome a resolution on the part of the symposium. And he told me, 'I would welcome a resolution anti-communists but not anti-regents.' I said, 'Sorry. That is impossible.' So I said goodbye and left. We didn't have the resolution, but it wasn't that our feelings weren't—well, I went to ask him and he implied that it was kind of appropriate for me to follow his preference."

Neyman pondered how to get the Lab—in its entirety—away from Berkeley. During the symposium he broached the subject to several friends at universities in the midwest and the east. Would their institutions be interested in acquiring, ready made, an effective statistical center? One of those to whom he spoke was an old friend, the astronomer Otto Struve, who had just left the University of Chicago for the University of California—partly, in fact, because Neyman was there. Struve wrote to Robert M. Hutchins—the president at Chicago—about the Lab's desire to leave Berkeley.

"Now, it all seems very funny that I should be writing to you about this! My own attitude toward the oath is one of quiet resignation: having fought against communism on the battlefield, I can only smile when I am asked to assert that I never have been and never shall be a communist... it would be a blow to me, personally, if Neyman and his associates should leave.... But the statistical laboratory has become an important cultural achievement, independently of any institutional connection. If these men cannot work here and be satisfied, then, I suppose, a change would be justified."

Hutchins's reply was prompt but noncommittal.

"You may hear from him directly," Struve noted as he passed the letter on to Neyman, "or this may mean he has no way of solving your problem."*

Overtures to other universities were, for various reasons, equally unsuccessful.

At the end of August 1950, on the eve of the regular monthly meeting of the Board of Regents, the non-signers, by then represented by an attorney, were prepared for the regents to reverse themselves on the question of accepting the president's recommendation to retain them. The regents met that month in Berkeley—it was their custom to move from campus to campus. After some complicated parliamentary maneuvering, they took up Regent Neylan's motion and voted, 12 to 10, to reverse their previous decision and *not* accept President Sproul's recommendation to retain the non-signers who had been approved by the Committee on Privilege and Tenure.

For Neyman, as for his colleagues, this was "the culminating point of our conflict." Although the faculty moved energetically to provide financial assistance for the eighteen non-signers (among whom were five members of the mathematics department) while their suit—immediately filed—was being argued, they were by no means unified. To further confound the confusion, Governor Earl Warren, the ex officio chairman of the Board of Regents and one of those who had consistently supported the faculty position, called upon the legislature to adopt a special loyalty oath—stronger than the regents' oath—to be signed by every public employee. It was not immediately clear what effect the Levering Oath, as the new oath came to be known after the assemblyman who had introduced it, would have upon the university.

During the preceding half dozen years, in the turmoil of the times, Neyman had let references to himself and his work in Fisher's writings pass without notice. When questioned by E.B. Wilson as to what R.A.F. meant by certain statements, he had responded merely, "I presume that explanation of

---

*Struve taught at Berkeley until 1960, when he became director of the newly established National Radio Observatory at Green Bank, West Virginia. His presence delighted Neyman. "What is the probability," he demands, "that two 'random walks' begun in Kharkov should meet in Berkeley, California?"

these and similar remarks could be expected only from Fisher himself. On my own part I only have the feeling of his disapproval." In the fall of 1950, however, a comment thrown off by Fisher in a discussion at a meeting of the Royal Statistical Society thoroughly provoked him.

"The implication of this passage is that, some time before 1935, I failed to distinguish between cases of continuous and discontinuous observable random variables, that R.A. Fisher himself saw this distruction [sic] and tried to put me right...."

It seemed to him that "...if an individual systematically indulges in misrepresenting facts and in making unjustified claims that he is the author of results or ideas actually due to other members of the Society...a brief historical note exposing such practices is in order." The society, however, refused to publish the "Historical Note" he submitted on the grounds that the objectionable remark had not been part of an accepted paper.

The rejection coincided with the arrival at Berkeley of a review copy of Fisher's *Contributions to Mathematical Statistics,* in which, in some new material, R.A.F. again referred to Karl Pearson in a particularly unpleasant manner. ("If peevish intolerance of free opinion in others is a sign of senility, it is one which he had developed at an early age. Unscrupulous manipulation of factual material is also a striking feature of the whole corpus of Pearsonian writings....") Neyman was delighted to be given an opportunity to write about Fisher's book. Although he customarily had a member of his staff read his letters and, under the guise of "correcting his English," remove objectionable remarks, there was apparently no such "censoring" of his review. In it he described the distinguished English statistician as "a very able 'manipulative' mathematician...[who] also made frequent attempts at the conceptual side of mathematical statistics [in which] he was much less successful." The prefaces with which Fisher had introduced each of his papers received special attention. Although not all were so "brutal in form" as the one pertaining to K.P., "many of them are more insidious because, in a skillfully hidden manner, they involve unjustified claims of priority...."—a statement which he illustrated with examples relating to the work of Pearson and Edgeworth. Lancelot Hogben hailed Neyman's review (in a personal letter) as "[a] fine specimen of English prose style in the manner of Gibbon and Bertrand Russell"; but it caused L.J. Savage to remark (sometime later) that Fisher did not always emerge "the undisputed champion in bad manners."

Throughout 1950–51 "the oath" continued to cast a pall over the several campuses of the University of California. The situation at Berkeley was so chaotic that Neyman never followed up on the invitation to contribute a volume on the "Berkeley ideas" to the Borel series.

At the beginning of 1951 the Committee on Academic Freedom (Northern Section) published a report on the harmful effects of the Board of Regents' action in reversing itself and dismissing the non-signers. The damage, which was laid out department by department, included loss of staff, disruption of program, negative reactions in the profession, refusals of offers of appointments, and resolutions of condemnation by learned societies. The most important consequence to the statistical group was, of course, the loss of Charles Stein and the inability to obtain an appropriate successor. The person Neyman wanted more than anyone else was Henry Scheffé. In the years since their first meeting at the beginning of the war, Scheffé had published outstanding papers on the subject to which he had come so late and had also shown exceptional abilities as an administrator. But although Scheffé had always wanted to be a part of the group at Berkeley, he had refused even an offered summer position in 1950—specifically because of the situation resulting from the oath controversy.

In the spring of 1951 the United States District Court handed down a decision favorable to the non-signers. Neyman, exuberant, proposed printing one hundred thousand copies and sending them all over the world. But it seemed certain that the regents would appeal. The future looked very bleak. The word in academic circles was that the University of California was "through."

A visitor that summer was Grace Bates, who had been teaching probability and statistics "despite woeful lack of background" at Mt. Holyoke and had expressed to Zygmund a desire to go to Berkeley and learn what the subject was all about.

"Jerzy set me to work reading some of his early papers. He also invited me to sit in on various consulting sessions. One day he brought me the manuscript of a paper on which he was working concerning accident proneness models and asked me to read it. I was able to follow details of going from step 1 to step 2, etc., because of the lucid exposition typical of Neyman's writing, until I got to a certain point and couldn't arrive at the next step however hard I tried—in fact I got a contradictory conclusion. I went to Jerzy almost in tears to explain that I had hit a snag and didn't know what was going on apparently. It turned out that I was right due to a fairly trivial slip-up, and Neyman thanked me wholeheartedly. Nothing would do but that 'we' publish the paper jointly—a source of great pride but considerable embarrassment to me in that I had at that point not even read Part I (a disjoint part) of the paper."

Bates's name appears as well on Part I of "Contributions to the theory of accident proneness"—which Neyman dedicated to the memory of George Udny Yule, whose fundamental memoir on the subject is cited in the first line of the paper.

In the fall of 1951, as classes took up again, Neyman noted that while enrollment at Stanford was essentially the same as it had been the previous year, enrollment at Berkeley had dropped significantly. He recalled that at the last meeting of the Board of Regents one of the regents—according to a newspaper report—had described the Berkeley faculty as a "disreputable" group, "100 per cent lacking in integrity." Wasn't there a relationship between the drop in enrollment and such disparagement of the faculty? He decided that when "New Business" was called for at the next meeting of the Academic Senate, he would propose a resolution authorizing a committee to investigate whether that body had grounds for a libel suit against the offending regent—and also whether the university could sue for the financial loss it had suffered as a result of the drop in enrollment.

"Then after the meeting," Neyman recalls, "President Sproul said to me, 'Mr. Neyman, the next time you have new business, I wish you would speak first to me about it.'"

He enjoys telling the story. It is a pleasant memory from a dark time. He says he didn't expect his resolution to pass. He wanted only to direct the attention of the senate to the remarks of the regent and to provoke some discussion.

"There was an element of humor there. Sproul with his hand over his mouth—like this—to keep from smiling."

There was not a great deal of humor in life that fall. Mathematical statistics had suffered a severe blow in December 1950 when Abraham Wald and his wife, Mary, had been killed in an airplane crash in India. The death of Wald, 48, had brought to an end the career of the one man who, since the great pioneering days of the nineteen twenties and thirties, could rightly be called a *giant* of mathematical statistics. Editing the *Proceedings* of the 1950 symposium, in which Wald had participated, Neyman placed Wald's paper first in the otherwise alphabetical arrangement of papers.

"Professor Wald's activity in the field of mathematical statistics proper was very brief," he wrote. "His first papers in the *Annals of Mathematical Statistics* appeared in 1939, barely a dozen years ago. Yet he has left behind him an indelible imprint on the thinking of generations that will follow. In fact, the fruits of Wald's work, especially sequential analysis and the general theory of multiple decision functions, are with us to stay and it would hardly be an exaggeration to say that they cover the theory of statistics in the modern sense of the word."

The death of Wald drastically changed the configuration of mathematical statistics in the United States. In the fall of 1951, Columbia, eager to retain the eminence which it had achieved as a result of his presence, offered Michel Loève a professorship at a salary substantially higher than that which he was receiving at Berkeley. Loève was—as Dean Davis explained to President Sproul—an outstanding man in the field of probability and his leaving, hard upon the loss of Stein, would undoubtedly affect the type of men whom Berkeley would be able to attract in the future. In addition, two of Mr. Neyman's most outstanding young people, Joe Hodges and Erich Lehmann, both currently on leaves of absence, had received extremely attractive offers from other universities.

"Mr. Neyman," Dean Davis informed the president, "is in a very pessimistic frame of mind."

228

1951
—
1954

Scientific work did not cease during the years of the oath. It was in fact during these tumultuous years that Neyman was fortuitously presented with a "big problem" of the type which Bernstein, in Kharkov, had encouraged his students to seek out—a problem which would fascinate him for more than a decade and result in a great series of Neyman-Scott papers.

The problem was brought to the Statistical Laboratory by the astronomer C.D. Shane and arose out of a project he had inherited upon becoming director of the university's Lick Observatory. A complete record of that part of the sky which was accessible to the telescopes of the observatory was being compiled. The photographic plates thus obtained were then to be stored for fifty years when another similar record would be made. The results of the two surveys would be compared, and the exact displacement of the stars in half a century revealed. Shane had decided to try to get, as well, some more immediate data. He would count the galaxies.

"I was of course interested in what the distribution was going to look like, so after my assistant, C.A. Wirtanen, and I had counted in an area where the plates were adjacent, I proceeded to plot the numbers per square degree and draw the contour lines. It had been thought before—and Hubble had rather insisted—that, although there were occasional clusters, the distribution of the galaxies was, on the whole, pretty uniform. But it certainly looked to me as though there was clustering everywhere."

Shane, who had been chairman of the committee that had recommended the hiring of a statistician at Berkeley, had felt for some time that not enough high-level statistics was being applied to astronomy; and it was he who had suggested to Betty Scott ("a very, very able astronomy student") that she take some work in that subject. Now he attempted an elementary statistical test on his data—"That was about as far as I could go!"—and then took the problem to Scott, who tactfully suggested that he present it directly to Neyman.

Neyman was not entirely unfamiliar with astronomy. Ever since Betty Scott, with her attractive enthusiasm for the subject, had come to work in the Statistical Laboratory, he had exerted himself to learn a little of the modern rudiments. He had consulted Otto Struve for recommendations of recent literature; and, after the war, when Betty Scott had begun to work on a doctoral dissertation which involved a statistical problem in astronomy, he had pressed Struve even more urgently for advice.

"The trouble is that what we [statisticians] call modern statistics was developed under strong pressure on the part of biologists. As a result, there is practically nothing done by us which is directly applicable to problems of astronomy."

What were the problems of real interest to astronomers? From his reading he had got the impression that it might be helpful to apply modern statistical methods to test the "two drift hypothesis" against an alternative. But Miss Scott had protested that the two drift hypothesis was about as out of date as the theory of phlogiston and had insisted that instead he learn "galactic dynamics and such."

In spite of his developing interest in astronomy, Neyman—emotionally absorbed in the constant debates and interminable meetings over the oath—felt "a little bit pestered" by Shane and his photographs of the sky and somewhat skeptical that he would ever complete such a tedious project. It was perhaps the word "clustering" to which he responded. Since he had first been introduced to that mechanism in connection with the egg-masses of moths and their developing larvae, he had found it also relevant in the Lab's war work on the dispersion of bombs. Here, it seemed, was another opportunity—one which would take the interesting mechanism of clustering from the petri plate to the outer reaches of the cosmos.

He and Scott began to work on Shane's data; and as they did so, his enthusiasm for the problem grew. He was later to describe the collaboration as one of the great "emotional involvements" of his life. For him—he wrote—an essential factor in determining the direction of work had always been "the presence of congenial companions, whether next door or a thousand miles away, who are actively interested in the problems concerned." A pleasure that was not shared was not a real pleasure. He and Scott began to spend many of their weekends at the observatory with Shane and Wirtanen, whose enthusiasm and interest also added to Neyman's emotional involvement.

In their first work on the clustering of galaxies, Neyman and Scott used a few basic assumptions to create a chance mechanism which ("applied repeatedly inch by inch over all space and hour by hour over all time") would recreate a universe that could be tested "in the large" against the astronomers' data. This approach of "model building" arose out of the new—largely French and Russian—work on stochastic (or chance) processes and tied in happily with the idea which had impressed Neyman in his first reading of *The Grammar of Science*—the idea that there are many possible different mathematical models of natural phenomena, their value to be judged only by how well they coincide with observations of the actual phenomena. It is this principle of specifying a clear probabilistic model and then testing it against the data which runs through all his work.

In the next dozen years Neyman was author or coauthor, almost always with Scott, of some twenty papers on the subject of galaxies. He joined societies and subscribed to journals in the new field, attended astronomical meetings, proposed collaborations with physicists who could teach him "something" about relativity theory, invited famous astronomers to Berkeley to lecture on hypotheses regarding the expansion of the universe, since (as he explained to them) he and Scott were "stopped by uncertainty on a number of particular points." Among the topics which he took up in the later papers were the separation and interlocking of clusters of galaxies, the expansion of clusters, the theory of fluctuations and its inapplicability to galaxies, the expansion of galaxies, the dispersion of the redshift of field galaxies, the instability of systems of galaxies.

"During the 1950s the most extensive statistical program was that of Neyman and Scott...," P.J.E. Peebles has written in *The Large-Scale*

Structure of the Universe. "An important motivation for their work was the Lick survey: Shane emphasized that the large amount of data coming from this survey was a considerable challenge and opportunity for the statistical approach. Neyman and Scott devised *a priori* [i.e., before seeing the data] statistical models of clustering and then adjusted the parameters to fit model statistics to estimates from the data. Their mathematical methods were remarkable and perhaps have not yet been fully appreciated and exploited."

During these same "years of the oath," as Neyman was becoming increasingly "obsessed" (his word) with galactic problems, the Division of Water Resources of the California State Department of Public Works came to him with a request that he and the Statistical Laboratory evaluate claims of commercial cloud seeding operators in the state. This request was also to develop into a satisfactorily "big problem," but the story of the work in connection with weather modification will have to be postponed until a later chapter.

Between January 1951 and June 1953, representatives of forty-some different departments of the university came over to Dwinelle Hall, the Lab's fourth home, to consult about their problems. Twenty-one institutions outside the university, ranging from the Internal Revenue Service to the Weather Bureau and from the Mayo Clinic to Moore Business Forms, also sought help.

Neyman was indefatigable in his efforts to bring in generous grants, largely from government and military agencies, for the support of students and faculty. (Salaries of the latter were still low in relation to those paid at the big private universities, and this was one way he could supplement them.) He was always aware of the financial situation of students. When Chin Long Chiang's small savings from China were nearly exhausted, he called him in, asked if he could draw, and arranged a job for him lettering charts. On one occasion, stopping at a gas station, he came upon Charles Kraft, the father of four, manning the pumps. The next day a "research assistantship" materialized for Kraft, now a professor at the University of Montreal. On another occasion, hearing that Julius Blum, struggling to support a wife and a small baby on $120 a month, was considering giving up his studies, he called Blum into his office. "I understand, Mr. Blum, that you have recently acquired a microorganism. Of course you know that graduate students are not supposed to have microorganisms....But how much do you need to get by?" Blum, who became associate dean of the Statistics Division on the Davis campus of the university, thought he could live handsomely on $300 a month. The next day he learned that he had been appointed a Research Statistician II with that salary. At times, when Neyman could not obtain the funds he needed to support students from any other source, he took the money out of his own pocket.

The genealogical tree of Neyman students had begun to spread in 1950 when Colin Blyth, Erich Lehmann's first Ph.D., received his degree. Neyman, however, continued to be thesis adviser to almost half the students working for degrees and knew all of them from his courses. They found him a demanding taskmaster. In the case of one young man, he even went so far as to ask that he postpone his wedding to put the finishing touches on his thesis. Nothing seemed too small for instruction to his students in "the right way" to do it. They learned to hold chalk so that it wouldn't squeak, to make their small x (two half circles tangent) so that it could be distinguished from their capital X (two intersecting straight lines). They were expected to be in the Lab at all times. As he was. Only "A's" were acceptable, and even a single "B" on a transcript was noted year after year with a displeased "What's this?" When he asked a student to do something, he went over what he wanted patiently; but he always ended with an admonition which they remember as *"Do it!"* They quickly learned that this meant he wanted "it" done by the next day. Gaffey, trying to describe the feeling engendered by Neyman's constant demands "to do impossible things," recalls a time in the army when "as a result of a rather improbable set of circumstances, I ended up going through a combat course (where you crawl along the ground while they shoot machine guns at you and blow up explosives all around) at night while I was drunk. For the first year or so as a graduate student with Neyman I felt exactly the same way."

Results were what counted. When there were results, Neyman would evaluate them precisely—"nice but rather limited" is a phrase which remains in the mind of one student. When there were no results, he would clearly but politely "give a guy hell."

Once he accepted a thesis with an anecdote from his youth.

"Have you ever heard gold coins being dropped on a marble topped counter? They have a peculiarly pure ringing sound. I used to go with my father to the bank in Russia, and I remember that sound. The thesis of Markov was like that. It sounded like pure gold. Now your thesis is more like modern day Russian money. It is good enough, but it doesn't have any ring to it."

He was deeply concerned about all the students in statistics and counseled them frankly as to their scientific strengths and weaknesses. It was said in the Lab that if Neyman thought you should get a degree, even if you weren't quite sure about it yourself, you would get one! If Neyman did not think that you had the ability to become a theoretical statistician ("The goal of all of us was to become a theoretical statistician!"), he would steer you to some field of applications. At first you might be bitterly disappointed. Later, successful and happy in your field, you would realize that he had been right. He was also, I am told, a great salesman; and a number of students from other departments, coming to Dwinelle Hall to take a statistics course, found themselves ultimately getting a degree in that subject.

In the course of preparing this book, I ask Neyman's Ph.D.'s and a number of other former students to send me their recollections of him. A lively description of his long-term professional impact comes from Joseph Putter of the Israeli Agricultural Research Organization:

"When I try to evaluate what Jerzy Neyman gave to us on the professional level, I find that it can be related to two main sets of ideas. The ideas seem (now) so obvious to me that it takes an effort to recall where I got them. (They are not quite so obvious, though, to some of the people to whom I try to sell them.) All the efforts to recall lead me in the same direction. I'm not sure which it was, but it was always either a paper (by JN), or a lecture (by JN), or a critical remark (by JN).

"The first set of ideas concerns 'good' statistical procedures. It isn't enough for a procedure to be 'natural,' or intuitively correct. It's got to be proved good. And to prove it's good, one has first to define precisely just what 'good' means in the particular context of the problem which the procedure is supposed to solve.

"The second set of ideas concerns the role of mathematical models in investigating natural phenomena statistically. A model should not be just a *deus ex machina* formula assumed or shown to be a (more or less) correct description of the data we get. Rather, it should be a meaningful guess about the process behind the data, assumed to be (more or less) an oversimplification of the complex truth. And our purpose should be to find out (by comparing the data we get with the predictions of the model) just where our guess is wrong, and to go on to a better guess.

"Come to think of it, maybe it wasn't only statistics he taught us. You surely have been told the story about the student at the blackboard who (after a while) stopped in his tracks. Well, says JN, do you know what to do? Y-y-yes, says the student. Well, says JN, then do it! I can't recall whether I witnessed this or just heard about it from someone who did; but at a couple of important points in my life I have found myself saying, You know what to do, don't you? Well then, do it!"

The student for whom Neyman had the greatest expectations during these years was the young Frenchman Lucien Le Cam who, although he had not planned to remain more than a semester in Berkeley, had ultimately stayed and taken his Ph.D. There were a number of things about Le Cam which recommended him to Neyman. He was highly intelligent, he was French, he was "theoretical" in his point of view, and he was one of the "underdogs" of the world—a farm boy who had become "a French intellectual." As soon as Le Cam received his degree in 1952, Neyman saw to it that he became an assistant professor and a "regular member" of the Statistical Laboratory. On a number of occasions in the future he was to express his great admiration for the abilities of Le Cam, whom he describes as the "mathematical" leader of modern day statisticians—"What one can do with a problem depends on the mathematical tools which he has, and I envy Le Cam the tools in his mathematical tool chest!"

By the time Le Cam became a faculty member, new appointments and new ex officio members of the Board of Regents had changed the composition of that body in relation to the oath controversy. The pro-faculty faction clearly outnumbered the anti-faculty group, whose leader, John Francis Neylan, had resigned. The State Supreme Court, which had decided on its own to take the case of the non-signers under submission, had not yet announced its decision; but everyone at the university, from president to policeman, signed the Levering Oath required of all state employees.

Neyman had long held the hope that Charles Stein would return to

Berkeley when the oath controversy was settled, but in the spring of 1952 he regretfully concluded that Stein was permanently lost. He again approached Henry Scheffé. This time Scheffé responded positively. He considered "the active group of scholars at the Statistical Laboratory... the strongest center in the world" and, as he told Neyman, "...nothing would please me more than joining your group at Berkeley!" He asked only that his appointment be delayed until July 1953, when he would have completed his three-year term as executive director of the statistics group at Columbia.

It was Neyman's plan that Scheffé would be, not only a professor and a research worker, but also "vice director" of the Statistical Laboratory. For the first time in years the personnel situation seemed eminently satisfactory. Sproul had matched Loève's offer from Columbia, and Hodges and Lehmann had turned down the attractive offers which they had received from other universities during the oath controversy. Writing to a colleague in Michigan, Neyman remarked happily on the great change since 1939:

"At the present the personnel of the Laboratory consists of 16.8 full time equivalents and includes two full professors [When Scheffé arrived, there would be three.], two associate professors, four assistant professors, two instructors and one lecturer plus several assistants...."

During the fall semester of 1952–53 he accepted an assignment of several months' duration in Bangkok in the service of the Food and Agriculture Organization of the United Nations. This was at the instigation of P.V. Sukhatme, his student from University College days, who was the chief of the statistics branch of the economics division of the FAO. At first Neyman had been skeptical about Sukhatme's plan to organize training centers on sampling "in all the unheard of places in the world"; but after several weeks in Bangkok, where he had students from Thailand, Cambodia, Laos, Japan, Indonesia, Pakistan, Nepal, Korea, Ceylon and the Philippines, he was enthusiastic about "Dr. Sukhatme's initiative, foresight and the execution of the project."

Because of the assignment in Bangkok, Neyman was not on campus when the California Supreme Court announced its decision in the case of the non-signers. The view of the majority of the court was that the state legislature had fully occupied the field by the enactment of the Levering Oath and the Regents' Oath was, therefore, unconstitutional. The non-signers were ordered reinstated and were later awarded the salaries they had lost and the privileges they would have had if their employment had not been interrupted. It was, however, "a hollow victory," as David P. Gardner pointed out in his book, *The California Oath Controversy:*

"Not only was their reinstatement conditional on their swearing to an oath more offensive than the one they had fought earlier... but the principles for which they had been willing to be professionally injured, financially harmed, and personally hurt had been utterly disregarded by the court....

"The controversy had been mostly a futile interlude in the life of an otherwise highly productive intellectual community."

But the long years of the oath were over. With the arrival of Scheffé in the summer of 1953, Neyman began to lay plans for a Third Berkeley Symposium on Mathematical Statistics and Probability.

In 1947, when the Budget Committee had set aside funds for a second symposium, Dean Davis had commented that if the Statistical Laboratory was to be singled out in the future for support beyond that given to other units of the university, it must be because of some unique claim which statistics could make as to its general value. He had doubted that such a claim existed.

Now, again applying for funding for a symposium, Neyman put forth an argument from a book on *Indeterminism, Probability and Statistics* which he was planning to write for the general educated reader. In recent years interest in probability and mathematical statistics had grown at an unprecedented rate. This growth had been the result, not of some academic fad, but of a revolutionary change in the point of view of a great many of the sciences. The old view had held that, knowing the state of the universe or some lesser phenomenon at a given moment and knowing the appropriate formulas, one could compute the state at any point in time.

"The search for relations..., connecting the present and future..., was frequently fruitful [and this] confirmed the faith that they always exist," he pointed out. "However, early in the twentieth century, certain physical phenomena were found that defied all efforts at the discovery of [such] relations...and the idea was conceived that, perhaps, in addition to familiar 'deterministic' phenomena, there were others, to be termed 'indeterministic,' in which the occurrence or non-occurrence of certain events does not depend on any cause."

The calculus had been the mathematics of determinism. The mathematical tools of the new indeterministic studies were probability and mathematical statistics. Who then could say that these subjects did not have "a unique claim...with respect to their general value"? A further argument was that statistics was the servant of all the sciences, and the sciences were in turn the indispensable sources of interesting problems for mathematical statistics.

Approval thus obtained for the holding of a third symposium, Neyman began to bombard President Sproul and Clark Kerr, who held the newly created post of chancellor of the Berkeley campus, with detailed letters and requests. This time there would be a winter session on applications and then a six-week summer session on theory. It was his desire to have two outstanding scholars on the campus for the whole period. These, he explained, would serve as "drawing cards" for statisticians and probabilists from all over the world. He had not forgotten the 1937 meeting he had attended at Geneva when exclamations of delight at the magnificent scenery had been "mixed with unfinished sentences concerning point sets, measure, Bayes, etc." He wanted to show his visitors some of the many natural wonders of California. But what concerned him most in all his planning was the publication of the symposium papers. Because of the extra session the amount of material would be considerably larger than it had been in the past. Although Sproul and Kerr had willingly given their approval to the holding of another symposium, which they now saw as bringing worldwide distinction to the university, they were concerned about finances. The publication of the *Proceedings* of the Second Symposium had run well in excess of its budget.

There were other financial problems in connection with Mr. Neyman and his Laboratory. Statistics had an extremely low average class size and a large proportion of tenured faculty who would soon be coming up for sabbatical leaves, often in pairs. Although it was university policy that replacements in such cases should be made at a lower grade, and hence a lower salary, Neyman pressed for replacement by highly qualified scholars. At the end of 1953 he was asking for David Blackwell to take the place of Michel Loève, who would be spending the year 1953–54 in Paris. Blackwell had fulfilled the promise Neyman and others had early seen in him. He was a professor at Howard, a leading Negro university, and currently a visiting professor at Stanford. Neyman rated him as one of the outstanding men in the country although Blackwell's view of statistics, which inclined toward the neo-Bayesian, was diametrically opposed to his own.

"I do not know whether I am eloquent enough in my present pleas," he wrote Dean Davis in regard to the proposed invitation to Blackwell. "In order to achieve a high level of scholarship at an institution, years of care, planning and hard struggle are necessary. But it is so easy to go down. Now the Laboratory receives the unmistakable marks of recognition as *the center* of statistical research and instruction. Up to the present we did not have to bother about gaps created by sabbatical leaves. Now, however,...I am most uneasy about inadequate replacements that may lower our general standing."

Davis passed this message on to Kerr.

"I must confess that despite being irritated from time to time by the nature of Professor Neyman's pressing requests..., there is no doubt but that we have here the most distinguished center for mathematical statistics in the United States and perhaps in the world...."

Although the university was "on an austerity program in a budgetary sense," Davis recommended that Blackwell be invited as a visiting professor during the coming year, and Kerr recommended the same to Sproul.

In the spring of 1954 Neyman was sixty years old. In the past years there had been a change in his family life. He and his wife had grown increasingly apart. When 17-year-old Mike had left home, they also had set up separate establishments. There was never to be a divorce, and Neyman was never publicly to recognize the permanent nature of the separation—referring always, when the occasion demanded, to "Olga"—no longer Lola—to "Mrs. Neyman" or to "my wife."

For the sixtieth birthday, a special event was naturally called for, and a student who was present sent the following description of the "great party" to his parents:

"The dean gave a speech and described how the University came to invite Neyman to come here and establish the Stat. Lab. 'To this day,' he said, 'I wonder what made Professor Neyman come....' After the dean's speech, Neyman rose...and promptly addressed himself to that question. He was visibly moved, his voice shook, and he had to pause occasionally to hold back the tears. His explanation was simple.—'In 1921,' he says, 'I finished my studies in Russia, and I was an alien. I came to Poland, and found myself an alien. I went to Britain, and found myself an alien. By 1938, I was feeling, all around, like a professional veteran alien. And then I came to California,' he says, 'and here I stopped being an alien.'"

There was no doubt about the sincerity of Neyman's emotion. The impact on the student writing to his parents was very strong, but he had to add, "...one must of course remember that Neyman is the kind of man who will feel alien wherever he is not chief."

1978   Neyman and Scott have been invited to attend a conference on mathematical statistics to be held in Poland during the second week of December, at which time Neyman will receive an honorary "diploma" from the Polish Mathematical Society. Is he looking forward to going back to Poland? He fingers the synopsis of a monograph on weather modification which he is planning to write during the coming summer. He is very busy, he says—there is much to do for the Lab. An important grant from the National Institutes of Health which ran out in September has not yet been renewed. He murmurs something about "our so far unsuccessful efforts to obtain funds for our research." Also Zygmund, who has been in Poland more recently than he, has reported many new "gaps" among colleagues and friends. But at the end of November the airline tickets appear on Neyman's desk.

"Life is complicated," he murmurs when I comment on their presence, "but not uninteresting."

He has been having a recurrence of an old trouble with his leg, and he apologizes for wearing sandals to the Lab, explaining that the only time his leg doesn't hurt is when he is working at his desk. In spite of the discomfort he is obviously suffering, he insists on getting to his feet when I enter his office and walking down to the floor below to let me into the mailroom. By the end of the week his leg is much worse, and his doctor orders him to bed.

The trip to Poland has to be canceled.

Confined with a very uncomfortable leg, he is still eager to have visitors. When I call to inquire how he is, he demands, "Are you thirsty?" and urges, "Come and have a little drink." He hastens to explain, "It will have to be after five o'clock, because from nine to five I must work for the Lab."

He is constantly on the phone to his secretary, who drives up the hill twice a day with freshly typed, revised manuscripts for him to read and letters for him to sign. An already scheduled Ph.D. examination is held in his bedroom.

The refrigerator is well stocked, thanks to Betty Scott—a roast chicken, a leg of lamb, an almost inexhaustible supply of the *bayalda* which he so loves.

A morning visitor must stay for lunch, one in the afternoon must have a drink and some crackers and cheese. "A pleasure which is not shared is not a pleasure."

By the middle of December, when he had planned to be in Poland, he has been in bed for three weeks. He is neatly shaved and barbered, lying flat under the covers, fully dressed and wearing an old sweater with holes in the sleeves. On the bed beside him is his open briefcase, on top the most recent version of a paper entitled "Public Health Policy and Basic Research" with the notation "Subject to further modification." The table beside the bed is neatly organized: a small radio and a copy of the previous day's *New York Times*, cigarettes, lighter, ashtray. A rented hospital bar holds the blanket off his leg, which is propped up on a pillow.

Christmas is approaching. I ask if he will take a vacation from the Lab. "No no. There is too much to do!"

Accused of being a "workaholic," he quietly considers.

What is work? What is play?

"Work is something you have to do because you are paid to do it," he says slowly. "But if you are interested?"

I suggest that perhaps he should consider retiring after his eighty-fifth birthday in April.

He does not respond directly to the suggestion. Instead he says, "Loève— you have heard the name Loève?"

He expounds for a few moments about the great success of Loève's book on the theory of probability. Many editions. Translated into many languages. He himself has never had such a success. He is a little envious. It develops, however, that Loève, who is some years younger than he, is now retired.

"And I see him in the hall—he still comes to the university sometimes—to the library—and then I have the feeling when he sees me—that he is a little bit ashamed in front of me. Do you understand?"

Sometime later, after talk of other things, he reopens the subject of retirement, again tangentially.

He knows a young poet who is halfheartedly studying statistics at one of the nearby state universities. "On one occasion I have expressed an interest in his poems, and he has recently sent me some." He has not read all the young man's poems. "There is so much to do at the Lab!" But he has had time to read a few. There is a line in one of them that has impressed him, and he quotes it now "because it is relevant to our earlier conversation":

Why does one write poetry?
One has to.

Photographs

The photographs are the result of the generosity of many people. Anna Neyman-Żukowska took apart her family album to provide me with early pictures. Ingram Olkin lent photographs of David, Feller, Gosset, Lévy, K. Pearson, and Scheffé from the collection he has made at Stanford; Konrad Jacobs lent those of Bernstein, Borel, and Lebesgue from his collection at Erlangen. Sarah Pearson located appropriate pictures of Neyman and her father during the period of their collaboration, and Michael Neyman made available the portraits done by his mother. I was very fortunate to have at hand the extensive group of photographs of statisticians which Shanna Swan took during the Fourth Berkeley Symposium in 1960.

Salon of Neyman's aunt and uncle, the Osipovs, in Kharkov

Karol Neyman, his wife, Marta, and his son, Jerzy

Wacław Sierpiński

S.N. Bernstein

Self-portrait of Olga Neyman

Drawing of Jerzy by Olga Neyman

Olga and Jerzy Neyman

Karl Pearson (a pencil drawing from 1924 by Miss F.A. de Biden Footner)

R.A. Fisher

W.S. Gosset ("Student")

Egon S. Pearson

Henri Lebesgue

Emile Borel

Paul Lévy

Jerzy Neyman

Neyman and Pearson at The Cell

W.E. Deming

Shanna Swan

Evelyn Fix

P. J. Bickel

Elizabeth Scott

Angela S. mola

Mark Eudey

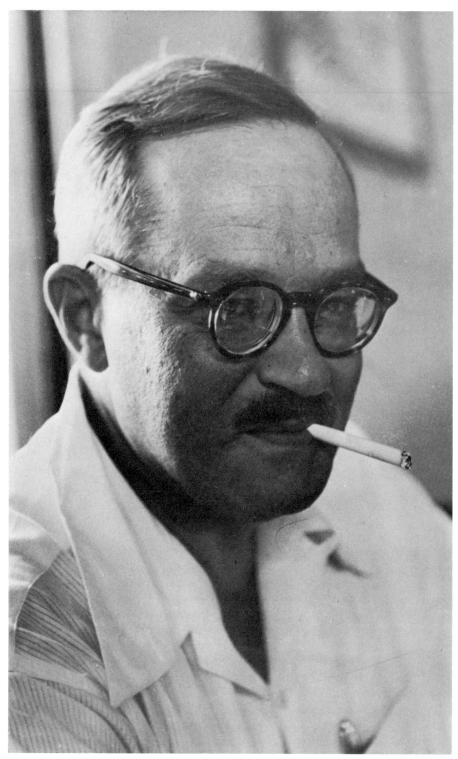

Neyman in the Berkeley Statistical Laboratory

Portrait of Griffith C. Evans by Erle Loran

Top: Joseph L. Hodges, Jr. *Bottom:* Erich Lehmann as a "war hero" of the lab, and at a later and happier time with daughter Barbara

Hsu Pao Lu (P.L. Hsu)   George Pólya

Abraham Wald at Berkeley

J.L. Doob

Joseph Berkson

Harold Hotelling

Warren Weaver

Herbert Robbins

Michael Loéve

Charles M. Stein

Henry Scheffé

Elizabeth Scott and Jerzy Neyman

Lofti Zadeh

R.A. Fisher, P.C. Mahalanobis, and Jerzy Neyman

Lucien Le Cam

David Blackwell

A.N. Kolmogorov

Antoni Zygmund

Jerzy Zygmund

William Feller

Harald Cramér

Howard D'Abrera

Top: Neyman on his 80th birthday *Bottom:* Prime Minister Indira Ghandi dusts off Neyman's blue beret in Delhi

Neyman and Scott in Prague

F.N. David

Jack Kiefer

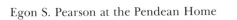

Egon S. Pearson at the Pendean Home

Peter Bickel

Neyman with Fia Lehmann at the statistics department's annual spring picnic

The dire predictions which had been made a few years earlier about the future of the University of California had not been realized—certainly not so far as statistics and mathematics were concerned. In 1954 Neyman and Tarski—Berkeley's "anti-Poles"—were two of five mathematicians from the United States who were invited to give featured addresses at the International Congress of Mathematicians being held in Amsterdam that summer. Charles Morrey, as chairman of the mathematics department, trumpeted the news to President Sproul: "Our university is the only university in the entire world from which two speakers have been invited...."

That summer, while Neyman was in Amsterdam, Chancellor Kerr formally recommended to President Sproul "that the Mathematics-Statistical Laboratory be redesignated the Department of Statistics as soon as possible in the fiscal year 1954–55." At Sproul's request an assistant summed up the long debate on the subject. The chancellor's recommendation was supported by the dean of the College of Letters and Science, by the recent chairman of the Department of Mathematics (Charles Morrey), by the director of the Statistical Laboratory himself, by the Committee on Educational Policy, and by the Special Committee on Statistics. The creation of a Department of Statistics separate from the Department of Mathematics had been steadfastly opposed by Professor Evans, now retired, and by the Budget Committee. But the fact was that in all but name the "Mathematics-Statistical Laboratory" was already a department.

"It does not seem practical to attempt to put the egg back into the shell," Sproul's assistant conceded, continuing with an impressive mixture of metaphor—"Although in this case the horse has escaped, the misgivings of the successive Committees on the Budget regarding recognition of the fact are understandable. Here, a willful, persistent and distinguished director has succeeded, step by step over a fifteen year period, against the original wish of his department chairman and dean, in converting a small 'laboratory' or institute into, in terms of numbers of students taught, an enormously expensive unit; and he then argues that the unit should be renamed a 'department' because no additional expense will be incurred....

"How, in the future, can a bureau, laboratory, or institute director be restrained from enlarging his empire against the judgment of his administrative superiors?"

Whatever the answer to this question, President Sproul approved Kerr's recommendation that the Mathematics-Statistical Laboratory be redesignated the Department of Statistics. Neyman returned from Europe to the good news. All that was still required for the achievement of the goal toward which he had pressed so long was the approval of the Board of Regents, an action that could be expected to take place before the summer session of the Third Symposium.

The probability that Neyman's seventeenth year on the Berkeley campus would be the best yet seemed very high. Soon, however, problems began to develop.

When the first session of the symposium took place at the end of December 1954, he had not yet managed to obtain sufficient money for the publication of the *Proceedings*. He could charm money off trees, as his secretaries said, and he already had in hand a generous amount of funding for the symposium. Most of it, however, had come with strings. It was not to be used for publication. Yet it was his "firm belief" that the symposia were worthwhile only if the papers presented were published—speedily and appropriately and together.

There was yet another problem.

He had begun to be conscious of "a little opposition" among his own people. He had definite ideas about replacements for Harry Hughes and Terry Jeeves, who would soon be leaving Berkeley for other positions. The men he had in mind were professionally excellent and had, for him, the added recommendation of being "underdogs." David Blackwell was a Negro. The other man had "fought his way up from a very low income group." Although Neyman often acted unilaterally, he thought of himself as democratic, and he consulted his staff about the proposed appointments.

Everyone was enthusiastic about Blackwell, who was well liked as a person and had an outstanding reputation as a scientist. The situation was not quite the same with the other man. There was opposition to him from Scheffé (who was finding his role as "vice director" of the Lab under Neyman more difficult than he had anticipated) and also from Lehmann and Hodges. Neyman, seeking consensus, suggested that Lehmann write to Loève, who was still in Paris, and explain why he and the others would be unhappy with the second appointment.

According to Lehmann's letter, he, Hodges, and Scheffé felt strongly that since the highly theoretical Blackwell was to be one of the replacements, the other should be a person capable of and interested in teaching some applied courses and in working with people in applied fields on their problems. The question was ostensibly one of the "balance" of the department, but there were also personality problems connected with the man in question. In his "struggle to succeed"—as Neyman put it—he had "developed various complexes that made him difficult to get along with."

Ultimately the opposition of his staff seemed to Neyman so strong that, "in spite of [his] best wishes," he had to yield.

He continued to have difficulties in raising the necessary money for the publication of the *Proceedings* of the symposium, the second session of which would extend over a six-week period during the summer. He again reminded the chancellor that an "excess" of manuscripts could be expected. The chancellor repeated that the university would not go beyond the $5000 already authorized and that, unless additional funds could be obtained from other agencies, publication would simply have to be restricted. Neyman examined, without enthusiasm, a variety of alternatives which could reduce costs. A cheaper method of printing. Delaying the printing of some papers. Printing some of the papers in regular journals, such as the *Annals of Mathematical Statistics*, the editor of which was Erich Lehmann of his own faculty. All these proposals were unsatisfactory to him. The whole point of the papers to be presented at the symposium was that they were *not* the type of papers which usually appeared in journals. He stubbornly argued for his point of view, but Kerr remained adamant as did Leonard Cohen of the National Science Foundation.

On April 6, in still another letter to Cohen, Neyman mentioned in passing that it might be possible for him to utilize funds already set aside by the Statistical Laboratory to furnish secretarial help to Lehmann in the editing of the *Annals*—"and thus...liberate some money out of the grants that [unlike that of the NSF] can be used for the subvention of the Press. Unfortunately," he added, "the gain so obtained amounts to only a small fraction of the funds needed." A month later, he abruptly informed Lehmann that after the end of May it would be impossible for the Statistical Laboratory to continue to provide a secretary for the editing of the *Annals*.

Lehmann found Neyman's action "incredible, absolutely incredible," as he told Scheffé, who was the president of the Institute of Mathematical Statistics, the principal sponsor of the *Annals*. Scheffé suggested that the IMS itself could probably afford to pay for a secretary; but when Lehmann went to Neyman with this suggestion, Neyman informed him that the room in which the *Annals* was being edited would also be needed for the secretary who would work on the *Proceedings* and that any secretary of the *Annals* editor would have to use a windowless storeroom in the attic of Dwinelle Hall. At this point it seemed clear to Lehmann that no matter what proposal was made, Neyman would oppose it. He felt that there was nothing for him to do but resign as editor, and he so informed Scheffé.

Even now, twenty-five years later, Lehmann is at a loss to explain Neyman's attitude.

"When I was asked to take over the editorship of the *Annals*, I had some doubts whether I wanted to spend that much time on work which was essentially administrative. Perhaps I was also afraid of the responsibility. The *Annals* was—still is—the most prestigious journal in our field. However, Neyman strongly encouraged me to accept. He gave me a rousing pep talk, as only he can. He also agreed to reduce my teaching load and to provide me with a secretary and a suitable room for her. (Most universities did this for an editor from their faculty.) When, early in 1954, I was approached about staying on for a second term, I discussed this with Neyman, who raised no objection and agreed to continue the previous arrangements.

"It may have contributed to the difficulty that as editor of the *Annals* I was rather unhappy about the symposium—symposia perhaps—since Neyman tried to persuade authors to put their best papers into the *Proceedings* rather than submitting them to the *Annals*, where one would ordinarily have expected them to appear. I personally don't like any of these things—festschrifts and things like that. Either they siphon off good work from the regular journals, which are always much more easily available to people, or else they get second-rate stuff which shouldn't be published anyway. This had been a sort of general disagreement between us, but it had never come to any big fight."

241

As soon as Scheffé received Lehmann's resignation, he fired off a copy of it and a memorandum to all the members of the Council of the IMS.

An unforgettable brouhaha resulted.

Neyman protested—at length—and with copies to everyone who had received Scheffé's mailing—about "the somewhat ungrateful tone" of Lehmann's letter of resignation. The members of the IMS Council responded "with great dismay." Joe Berkson announced to Scheffé and the other Berkeley people involved—Hodges was associate editor of the *Annals* and Blackwell was president-elect of the IMS—that he was going to cancel his plans to appear on the summer program of the symposium.

"I can anticipate that you will advise me not to do this, from considerations of mature wisdom.... But I submit to you that it might not be a bad thing all around if Mr. N., whom I love as much as anybody does and all that, learned for once that it is possible to retribute his arbitrary actions."

When Neyman received copies of Berkson's letter and those of the other members of the IMS Council, he replied with a lengthy letter of his own addressed to the Academic Staff of the Statistical Laboratory, the Council of the IMS, and the IMS Committee for the Nomination of the Editor—a total of some fifty colleagues. With great thoroughness he took up the details of employment of an extra person for editorial work, the financial situation of the symposium, and (causing some laughter among some of the recipients) the question of the storeroom in the attic. It happened, he explained, that the lady employed to do the editorial work on the *Proceedings* was "of a certain age" and had held "a substantial administrative position" on one of the university's other campuses. "Therefore I was a little embarrassed in having to put her up in the 'store room' in the attic. However, short of moving people around, this was the only possibility...." He concluded, "with regrets," that Scheffé was right: "in the present circumstances it is desirable to move the editorship of the *Annals* to another institution."

Twenty-five years later, questioned about this old quarrel, Neyman has little to add.

Why such a hullabaloo over accommodations for a middle-aged secretary? "A little more than middle-aged," he corrects mildly. And for the small amount of money that eliminating the *Annals* secretary would add to the symposium budget? He reaches across his desk for a recent copy of the *Annals* and points to the name of the Statistical Laboratory in the list of sponsors on the inside cover.

"In general I am inclined to cooperate with people," he says. "So it was not just caprice."

News of the quarrel among the members of the Berkeley group traveled with surprising speed to Europe.

From Paris, Michel Loève wrote to Lehmann, "thunderstruck":

"It is already known in more or less distorted form outside the States. It is bad for every one of us and worse for the Stat. Lab. Whatever and whose be the wrongs, pray forget them. You must take and get the editorship back. There is no other way out. I am convinced that together, you, Neyman, and Scheffé can do it.... Please do it."

Also from Paris, Betty Scott, who had just returned from ten days in England, wrote, "most distressed":

"At the very end of my stay...I began to hear all kinds of rumors....Of course, I know that some, and for all I know, all are not true, but they were recited to me as a sort of entertainment and in order to place Berkeley in a ludicrous position—absurd children who don't know how to behave like adults. But why does this get out of hand and then be advertised all over the world?....There must be some way out."

Responding to the letters from their colleagues, Lehmann and Scheffé talked with Blackwell, who agreed to reopen the subject with Neyman. It was thought that if anyone could act as mediator it would be Blackwell, whom Neyman so thoroughly admired, both professionally and personally, and with whom he had such an affectionate relationship. But in the end Lehmann had to write to Loève and Scott that Blackwell had found Neyman stubbornly and inexplicably uncooperative.

In the next few weeks Neyman continued to press for funding for the publication of the *Proceedings,* fighting off "a mistaken opinion" that the *Proceedings* were unnecessary competition to existing journals. In this connection he pointed out that the only manuscript for the *Proceedings* which he already had in hand was a joint work by Lehmann and Hodges, the chief editor and the associate editor of the *Annals.*

Thus it happened that when, on July 1, 1955, statistics became at last, officially, a department, there was a perhaps irreparable breach between the chairman of the new department and some of his closest colleagues, including Lehmann, who seemed to the other members of the group to be the one of their number who most clearly carried on Neyman's own work in theoretical statistics. Scheffé is now dead. Hodges declines to comment on the *Annals* fracas, saying only, "Neyman is my teacher. Naturally I admire him very much." Lehmann is more frank.

"The whole affair amounted to a definite break between Neyman and me, from which our relations did not recover for many years. However, there is one thing I should make clear. I have always admired very much the fact that, in spite of the hostilities being pretty severe, Neyman never took it out on my career but always tried to further it in every way. He was able to separate this personal issue completely from the administrative one, which I thought was extremely fair, and a very decent thing to do."

The members of the strife-torn department pulled themselves together to welcome their guests to the Third Berkeley Symposium on Mathematical Statistics and Probability. Neyman had ardently hoped to bring to Berkeley some of the Russian statisticians and probabilists, especially Kolmogorov—"I found him a friendly creature," he had remarked after meeting him for the first time at the congress in Amsterdam—but President Sproul had demurred at trying to get visas for a number of Russians in the current political climate.

By this time the symposia had become in fact what Neyman had always hoped they would be—a "fruitful exchange" among a large number of scholars from all over the world spending a leisurely six weeks in Berkeley with a few lectures each day and a great many hours of discussion and informal contact. A perfect example of such exchange was at hand. During the Second Symposium, a conversation between Kai Lai Chung of Stanford and Paul Lévy had resulted in Lévy's publishing a paper, "Systèmes Markoviennes et stationnaires. Cas dénombrable." In his paper for the Third Symposium, Chung now summarized the work that had been done, speaking at length of Lévy's contribution and pointing out further problems. A discussion with Chung after the presentation of this paper led Blackwell to a discovery which was in turn to be studied further by Lévy and D.G. Kendall of Cambridge University. Not only was the exchange fruitful, it was also "international"!

(The symposium was scarcely more international than the new department was to be. Today, an examination of its Ph.D.'s reveals that more than forty different nationalities are represented. Neyman always has a great deal of empathy for students who are far from home. When Israel invaded Sinai in the fall of 1956 and a major war in the area seemed a terrifying possibility, he invited Mary Hanania—now Regier, of the American University in Beirut—and her boyfriend to have dinner with him at the Faculty Club: "...and there, in a leisurely and warm atmosphere, he wanted to know all about the Middle East and its conflicts, reminisced about Poland, and somehow made me feel that we had a great deal in common. My gloom miraculously disappeared.")

Neyman's own contribution to the Third Symposium consisted of two papers, both collaborations, on widely different subjects. The first brought together his and Scott's recent work with Shane: "Statistics of images of galaxies with particular reference to clustering." (A popular treatment of the same subject appeared in *Scientific American* the following year.) The second, written with Scott and Thomas Park of the University of Chicago, treated the "struggle for existence," as exemplified by Park's extensive work with *Tribolium.*

Neyman himself had been interested in *Tribolium,* the common "flour beetle," since his arrival in Berkeley, when he had been consulted in connection with an experiment with these beetles which was being conducted there. One day, around 1950, Park learned to his surprise that following a presentation of his own research on *Tribolium* at a meeting in Ames, Jerzy Neyman of the University of California was going to discuss it from the point of view of the mathematical statistician.

"I was somewhat frightened, because I was a much younger man; but then, as it turned out, Neyman took ill and had to send one of his younger staff members. Shortly afterwards, I received a letter apologizing for his not having been able to be present at the meeting and saying that he was going to be in Chicago and would like to come and see me. I met them at O'Hare—Betty Scott was with him. I had never seen him before, but he was wearing a long heavy coat with a fur collar, very foreign looking, and I knew that must be he. Right away he wanted to see the beetles so we stopped at my laboratory on the way to their hotel. We parked and, as we crossed the street, he said, 'I would like you to come to Berkeley as a visiting professor of mathematics for four or five weeks—bring along some of your work and we can discuss it.' *I had just met him.*"

Park's "beetles" fascinated Neyman. The two species of *Tribolium, confusum* and *castaneum,* spend their entire lives in finely milled flour. If allowed to live alone, in a "laboratory-biological model of nature," each appeared capable of maintaining indefinitely reproducing populations; but when put together, one species invariably died out. Under certain environmental conditions, the identity of the surviving species varied from one replicate of the experiment to the next. The more frequent winner in the struggle for survival appeared to be the species which, when living alone, maintained a lesser population density than the losing species.

The purpose of the Neyman-Park-Scott collaboration was not so much to reach definitive conclusions as to establish a pattern for later work. Again, as in their work with Shane on the clustering of galaxies, Neyman and Scott directed their efforts toward the creation of a chance mechanism which would produce an indeterministic model providing an explanation for the observed behavior of the two species of beetles. Park feels that their application of probability theory to the outcome of inter-species competition was a real contribution to the theoretical side of biology—"and still is."

("Neyman has a rare and remarkable ability," Scott says to me. "He can see and understand complex problems in many different domains, and can make the problems precise and restate them as problems in statistics. He then develops the necessary statistical theory to solve the problems. By translating the statistical solution back into the original phenomenon, he not only solves a specific problem but often contributes, as in this case, to the basic understanding in the domain.")

Park found Neyman's personal style of working interesting.

"It was sort of a pattern. He'd stop and he would talk about something entirely different, such as a good wine, or something he'd seen, or someone he liked or didn't like—completely breaking up the work, you see. Then the obverse—we would go out to a nice restaurant for dinner and in the middle of the conversation he'd say, 'Tom, let me have a scratch pad,' and he'd start writing equations. There was something going around in his head, and he had to get to it. The other was obviously a break from the intense mental work he was doing."

He remembers that at the last moment—just before the symposium—there was a wild rush to finish the paper.

"I got Neyman a room here in Chicago, and we sat in it all day and all evening—I reading my paper to him and he reading his to me. We worked very hard. When we finally got through, he looked at me—I'll never forget this—and he said, 'Well, Tom,' he said, 'it isn't Darwin, but it's respectable!'"

The Third Symposium, in spite of the trouble over the publication of the *Proceedings,* was as successful as Neyman had hoped. After three weeks of lectures and discussions, he declared a "vacation" and loaded his many guests into buses for a week's excursion to the Sierra.

"There," he happily reported, "animated discussions of stochastic processes and of decision functions were interspersed with expressions of delight at the beauty of Yosemite Valley, Emerald Bay, and Feather River Canyon."

It appeared, however, that the total number of pages which would be submitted for the *Proceedings* of the Third Berkeley Symposium would surpass the combined total of the First and Second. The problem of appropriate publication was still unsettled; however, the American Mathematical Society had indicated some interest in taking on the project.*

On the Berkeley campus itself, during this period, relations between statistics and mathematics were strained. It was commonly felt in the latter department that statistics had developed at the expense of mathematics. There was a fear that the creation of a statistics department might have an adverse effect on its parent department—a fear that was not to be realized.

Class size continued low in statistics. Graduate students came in satisfactory numbers. In fact, statistics had more graduate students than at least half the departments on campus. It attracted, however, very few undergraduates. In an effort to correct this situation, Neyman—always a salesman for his subject—circulated among the incoming freshmen several pages of information designed to inform them of a "new kind of career as a research statistician." He emphasized that in the Department of Statistics the term "statistics" was understood to include, not only the mathematical theories of probability and of statistics proper, but also the methodology of applying those theories to the indeterministic treatment of natural phenomena:— How frequently could two galaxies subject to specified laws of motion be expected to collide? How frequently would the progeny of a given cross have a specified combination of characteristics? How frequently would a certain environment as an adolescent lead to a specified pattern of behavior as an adult? There was a "literally limitless field" of applications of probability and statistics—and a great demand for trained statisticians. The main prerequisites for the new career were an interest in research, "practically no matter what kind of research," and a good mathematical preparation.

"The above description suggests the necessity of an early, intense and persistent effort on the part of the student; however, this effort is likely to be richly repaid in future life by freedom from routine and by continuing intellectual satisfaction that invariably accompanies participation in the progress of science."

By the time statistics became a separate department, Dean "Sailor" Davis had become acting chancellor and Lincoln Constance, another professor of botany, had succeeded him as dean. From the beginning, Dean Constance was not so congenial to Neyman as Dean Davis had been, and early correspondence between the two reveals developing tension.

"I am greatly disturbed by the opinion expressed in your letter that the teaching load in Statistics [as opposed to the total work load] should be the same as in Mathematics," Neyman wrote the new dean toward the end of his

---

*The *Proceedings* were ultimately published by the University of California Press.

own first year as department chairman. "I am even more disturbed by your admission, in our conversation of yesterday, that the last paragraph of your letter implies criticism of the Department of Statistics in general and of my own activities in particular....

"It is my contention that the whole body of activities of our statistical group represents conscientious, and not unsuccessful, implementation of [the directives I received upon accepting my position on this campus]. If you do not agree with me, I would like to hear exactly what you object to."

Dean Constance apologized. Certainly no criticism had been intended. "I am well aware that we have a distinguished Department, and we want to strengthen its position." Nevertheless, as dean of the College of Letters and Science, he felt he should have more detailed information about the consulting services performed by the statistics department, since these constituted a substantial part of the work load of the faculty members and were the justification for their teaching fewer hours than their counterparts in the mathematics department.

Neyman objected strenuously. In the case of a number of consultations it would be "embarrassing" to supply the information requested. In fact, it would be "improper" for him to reveal the identity of persons requesting consultations. An exception might be made when the person concerned recognized the assistance he had received and mentioned it in a paper or, as often happened, suggested that a resulting paper be a joint one. "Even in these conditions I would feel most uncomfortable in reporting that in trying to help a given individual I spent so many hours."

Then, totally unexpectedly, he enclosed with his letter to the dean a copy of a letter sent on the same day to Kerr in which he resigned the chairmanship of the Department of Statistics as of July 1, 1956—just one year after the department had come into being.

He had come to realize—he explained in his letter to Kerr—that his continued acceptance of the chairmanship was a mistake: "The transformation of the old Statistical Laboratory into a Department of Statistics closed a period of the development of our group and opened a new phase. In these circumstances, it is only natural to have a new and younger man take over."

The members of the statistics department wrote urgently to the chancellor: "We strongly hope that you and Dean Constance will join us in making every effort to persuade Mr. Neyman to reconsider his decision."

("The ostensible reason [for the resignation] was the demand by Dean Constance for details on consultation...," Hodges wrote to Lehmann, who was in Zurich on sabbatical. "Actual reason, I think, was to elicit our support, which was of course forthcoming.")

Ultimately, at the powerful urging of Chancellor Kerr, Neyman agreed to continue as chairman of the statistics department for at least another year.

He was by then sixty-two years old. According to the rules of the Board of Regents, he would face retirement in five years.

"...from time to time I think of what will be the further development of the Department of Statistics and the Statistical Laboratory," he wrote to

President Sproul. "At present they are jointly an institution of first class importance and international recognition. The reason is that we have a supreme composition of staff, with both theory of statistics and theory of probability excellently represented. When thinking of years to come, I would like to have this composition preserved. The realization of this is not easy because of a very strong competition on the part of other universities.... In order to maintain the high level of personnel, the Department must have something desirable which other universities do not have, and something which cannot be bought with money."

What could this something be?

"To the best of my understanding of the situation, this something, attractive, durable and not purchasable by others, is the tradition of the Berkeley Symposia on Mathematical Statistics and Probability."

He was writing to Sproul to request that gifts and bequests be sought in order to insure that the symposia would continue at regular five year intervals, the next in 1960, then in 1965...

That December—six months after he had first tried to resign—Neyman wrote again to Chancellor Kerr and again submitted his resignation as chairman of the Department of Statistics. He had just received permission to accept an invitation to spend the next two months in India. He had also recently learned that he had been awarded a fellowship for the following semester at the university's newly endowed Miller Institute for Basic Research. During the next academic year he would be on sabbatical in Europe. It would not be good for a new department to have its chairman away for such a long period of time.

Kerr, still unwilling to accept the resignation as final, asked Neyman to come to his office to discuss it. Without giving any explanation, Neyman insisted that Lehmann accompany him.

"I don't know why he did that except that Joe and I were considered the leaders of the opposition to him in the department," Lehmann says. "So I was there. It was a complete surprise. I was struck dumb. Chancellor Kerr tried awfully hard to persuade him to reconsider, but Neyman was absolutely adamant."

Why did Neyman take Lehmann with him to Kerr's office?

Asked this question so many years afterwards, Neyman replies simply, "I wanted him to see that it wasn't a trick."

When Neyman's colleagues learned that this time there was no possibility of his reconsidering his decision, they had ambivalent feelings. They were well aware that their existence as a department was due entirely to his efforts. In spite of recent conflicts, they all looked up to him. "He was our leader." They were fearful that, until his retirement, there would be years of confusion and divided authority; for he had not resigned the directorship of the Statistical Laboratory, which had continued to exist, even after the creation of the department, as the university's statistical consulting service and the source of interesting problems and supplementary funds for faculty and students alike.

A quarter of a century later, it still comes as a shock that after only a year and a half Neyman would resign the chairmanship of the department which he had worked so hard and so long to establish.

He recites again the reason he gave in his letter to Chancellor Kerr. "It would not be good for a new department to have its chairman away for such a long time." Then he lifts his shoulders in a little helpless shrug.

"Mr. Turgenev," he says. *"Fathers and Sons."*

1956
—
1961
Ten days before Christmas 1956, Neyman left Berkeley for Calcutta to take part in an international meeting celebrating the twenty-fifth anniversary of the Indian Statistical Institute. The members of the department accompanied him to the airport and were shown the graceful letter from Dean Constance accepting his resignation. The next day in Dwinelle Hall they held an impromptu department meeting with Loève, the senior member after Neyman, presiding. All agreed that in the new situation it was going to be very difficult for any one of them to be chairman. The consensus was that the only person who could take over successfully was the newest member of the department: David Blackwell.

At the time he resigned as chairman, Neyman was still four and a half years away from retirement; but the question of what he would do after July 1, 1961, was much on his mind. While he was in India, he received a letter from S.C. Kleene, the chairman of the mathematics department at the University of Wisconsin. Kleene, describing that institution's plans for creating a separate department of statistics, explained that his letter was in the nature of an informal inquiry about Neyman's possible availability for such a project.

Neyman responded promptly.

"The organization of the Statistical Laboratory and then of the Department of Statistics in Berkeley has been fun, and it would be a pleasure to relive...." What, however, was Wisconsin's retirement age? "Because of the particular organizational job to be done, can one think of special arrangements insuring a continuous employment for a reasonable time, say for a decade?"

Stimulated by the possibility of ten more years of active work, he returned to Berkeley in good spirits. During the time that he was to be on leave from the department, Henry Scheffé was to take his place as acting director of the Statistical Laboratory. There was, nevertheless, a question about the Lab which he wanted to settle immediately with the administration.

The Laboratory attracted considerable financial support from outside the university; and the money thus obtained enabled the department to support its graduate students, to hire more office help, and to provide faculty members with supplements to their salaries and funds for travel to scientific meetings. All of this was good, but—not all the members of the department participated in the organized research which was the main function of the Laboratory.

"[Yet] all of us feel the present pressure on the FTE [Full Time Equivalent] ratio and all of us are aware that the current salaries in our University are, at least as far as statisticians are concerned, some $2000 lower than in other universities. In these circumstances, if the Laboratory is subordinated to the Department, it is almost inevitable that the character of its life will be changed...." It would—in the circumstances indicated—"dwindle to the status of an appendage to the Department, with its primary purpose to provide extra funds for students and personnel under the pretext of organized research."

Neyman felt so strongly on the subject that, in addition to this formal letter to Dean Constance, he also sent a handwritten note to Chancellor Kerr.

"If the independent status of the Laboratory is difficult to achieve, then I wonder whether I could be more useful to the University if I were transferred to some other campus where a vigorous statistical research unit is more desirable than it is in Berkeley."

Kerr passed the note on to Sproul—"This just arrived."—and Sproul scrawled across it, "Dean Constance: Stat. Lab.—independent until Neyman retires."

The question of his autonomy as director of the Statistical Laboratory satisfactorily settled—"I don't want to be boss, but I don't like to have boss!"—Neyman went to work as one of the first three fellows of the Miller Institute. He began to prepare for publication certain results which he had obtained earlier. These included a joint work with Scott in which the stochastic process of clustering that they had been utilizing in their cosmological studies was applied to the theory of populations. He then took off for a month "to transact certain scholarly business in Europe"—in fact to arrange a meeting, the first of its kind, on "Statistical Problems of Astronomy and the Earth Sciences" to be held in conjunction with the 1958 meeting of the International Statistical Institute in Brussels. On his return he and Scott began to modify their original theory of the clustering of galaxies to take into account the random variation of the redshift from one galaxy of a cluster to the next.

For the first time since his days in London and Paris in the nineteen twenties, he looked forward to an uninterrupted period of research. In addition, he was going to have a real vacation, his first in years, before he settled down in England for his sabbatical abroad. He described his plans with enthusiasm in a letter to a European colleague. "A nice long voyage" from New York to Naples "on a nice boat," then a little car rented in Italy and a leisurely trip up the Dalmatian coast and across the continent before taking the boat to England. "Miss Scott proposes to do the same." The two of them would work together in Cambridge, the leading center of cosmological research.

He had intended to give only a few lectures in the course of his travels, but in fact he delivered a dozen or so. He was especially active in Poland where he lectured in Warsaw, Krakow, Wroclaw, and Poznan. His brother's granddaughter—to whom he brought a complete set of the works of Jack London—recalls how in Warsaw, in front of the Hotel Bristol, such crowds gathered that he was hardly able to park.

The question of work after retirement continued to be much on his mind. Indiana now appeared interested in obtaining his services. Wisconsin was still in the picture, and Illinois shortly moved in. When he arrived in Cambridge at the beginning of November 1957, he found waiting "something in the nature of a formal proposal" from J.W.T. Youngs of the Graduate Institute for Mathematics and Mechanics at Indiana. The position which Youngs had in mind for him was that of chairman of the Committee on Statistics as well as professor of mathematics. The advantage of this arrangement was that, although chairmen of departments were required to retire at sixty-five, such was not the case with chairmen of committees, who retired at seventy. President Wells had offered to recommend to the Indiana Research Foundation that it underwrite Neyman's salary for an additional three years beyond seventy.

"...we all want you here very badly."

Neyman deliberated for an entire month before answering Youngs's letter. He had stopped at Indiana on his way across the country, given a couple of lectures there, and obviously indicated the possibility that he would be willing to leave Berkeley; however, he wrote finally, "with regret" but without giving any specific reason, that he had decided "it would be a mistake for me to accept." He sent a copy of the offer and his letter declining it to Blackwell, who in turn sent both to Dean Constance.

"We would like the University to give Neyman the strongest assurances consistent with University policy that he can continue on active status here beyond 1961," Blackwell told the dean, "and to give these assurances as soon as possible [to] stop Neyman and the rest of us from contemplating the unpleasant possibility of his going elsewhere...." But the dean could make no immediate reply, since the chancellor was away from the campus at the time and would not return for at least two months.

The sabbatical year abroad resulted in a number of small papers by Neyman alone, a joint work with Scott on "Indeterministic approach to cosmology," presented at a meeting of the Royal Statistical Society, and—for Neyman himself most significant—a contribution to the Cramér festschrift: "Optimal asymptotic tests of composite hypotheses."

Since he was contributing the paper to a volume honoring Cramér, he thought of naming his new tests after his great Swedish colleague.

"But I didn't know if Cramér would want them named after him, and so I called them *C-alpha* tests: *C* for Cramér, *alpha*—well, technical, this and that."

He was disappointed that the paper did not attract as much attention as he had hoped it would.

251

"Presumably the reason was that I made a mistake in publishing it in the Festschrift.... My motivation was that I have a great respect for Cramér and also quite warm affection [but], as is well known, not many people read the Festschrifts!"

Back in Berkeley in the summer of 1958, he learned from Dean Constance that although the chancellor did not feel that he could go as far as President Wells of Indiana had indicated that he was prepared to go, he had authorized the dean "to say that we will give sympathetic consideration to Professor Neyman's reemployment after retirement."

Neyman acceded to Blackwell's request that he take over the organization of the Fourth Berkeley Symposium and began to push the administration for "an encouraging note" and "a basic grant" to lure outside support. After a visit to Russia to attend the meeting of the International Astronomical Union, he was more eager than ever that the Russian statisticians and probabilists be invited to Berkeley for the symposium—always and especially Kolmogorov. Clark Kerr, who had succeeded Sproul as president, indicated that he would be favorable to pushing for visas for the Russians "if they were O.K. prima facie."

An aide hastened to warn the new chancellor of the Berkeley campus—the Nobel Prize winning chemist Glenn T. Seaborg—that "the Berkeley symposia have had a rather checkered financial history, characterised by extreme perseverance on the part of Neyman in gaining his ends." He pointed out that with the last symposium already underway, Neyman had still not been able to raise the $15,000 necessary to subsidize publication of the *Proceedings* and, as far as could be ascertained, even at the present date only $8,000 had ever been obtained. In the end the remaining deficit would probably have to be picked up by the university.

"Despite all this, it must be conceded (and emphasized) that Neyman has made the Symposium an event of world stature. Its continuance should clearly be ensured [and] the past success... has probably been due, in no small part, to the flexibility Neyman has enjoyed."

With retirement only a couple of years away, Neyman's career began to come full circle. In the spring of 1959 two of the professors who had examined him for his Ph.D. were in Berkeley: Kotarbiński as a guest of the philosophy department and Sierpiński as a guest of the mathematics department. Neyman remembers with particular pleasure the banquet in Sierpiński's honor. He describes how the chancellor, who was presiding, asked all the "direct" students of Sierpiński to stand; and he, Tarski, and (he thinks) a Polish visitor stood. Then the chancellor asked all the "indirect" students to stand.

"Then he asked everyone else in the room to stand. But no one got up. Because there was no one left sitting in the whole room!"

After the banquet Neyman offered to drive his old teacher to his hotel. "I thought he might be tired." But in the car Sierpiński began to hum a ballad with which Neyman was also familiar. He sings it in Russian and translates for me:

The last time I saw you, you were rather near.
At a street-crossing, you were being driven in a car.
As I dream now, in a dive in San Francisco
A blue-black waiter is helping you with your coat.

"Do you think if we went to San Francisco we could find one of those dives?" Sierpiński asked a little wistfully.

Neyman thought that they might.

"So I drove him to San Francisco, and we parked in Union Square [the fashionable shopping and hotel center of the city], and we walked around quite a bit. But I must say, regretfully, that we didn't find any dives. When I told the people in the department the next day, they laughed; and Blackwell said, 'Next time, Jerry, when you want to find a dive—ask me and I'll take you to one!'"

Cosmological questions continued to dominate Neyman's published work during these years, but he was also increasingly fascinated by the subject of weather modification. In 1950 the Laboratory had been asked by the state (as has been mentioned earlier) to evaluate the claims of success being made by commercial cloud seeders in California. He and several young colleagues had gone immediately to work, boning up on meteorology with the help of the area's chief weather forecaster. They were convinced that there could be no meaningful statistical evaluation without a properly randomized experiment, but they recognized that they could gain support for this view only by showing the extent to which the methods of evaluation currently in use were untrustworthy.

In 1957, following a report by Neyman alone and a cooperative report by Jeeves, Le Cam, Neyman and Scott, the California Department of Water Resources undertook a three-year randomized experiment to test the effectiveness of cloud seeding with silver iodide from ground based generators. A commercial cloud seeding organization was to conduct the seeding operations. The Department of Water Resources was to maintain the rain gauges and collect the data on precipitation. The Statistical Laboratory was to be responsible for the randomization of the experiment and the statistical evaluation.

Members of the Lab who participated still remember the excitement when, between eight and nine o'clock, morning and evening, they would receive word as to whether, in the opinion of the firm doing the seeding, the forthcoming 12-hour period, or "unit of observation," was a "seeding opportunity." If it was, the person entrusted with the key would then unlock the safe and take from it the slip of paper with the randomized decision: "Seed" or "Do not seed."

Unfortunately, the experiment looked better before it started than it was ever to look again. The terrain around Santa Barbara was rugged and in some of the control areas totally uninhabited. The setting up and servicing of the gauges presented difficulties. Many were damaged by hunters, who used them as targets. According to the original plan, only Santa Barbara

County was to be seeded; but in the second year, Ventura County, adjoining Santa Barbara to the east, decided to start seeding operations of its own. The combination of two counties offered "interesting possibilities" to the statisticians, who suggested that operations be subjected to factorial randomization; but, because there had been an extended period of drought in Ventura, that county—more interested in rain than in data—proceeded to conduct its seeding operations at every opportunity rather than according to the randomization procedure. The next year, operations in Ventura were combined with those in Santa Barbara, but that year—the last of the project—was exceptionally dry and there were only nine seeding opportunities.

In spite of these many difficulties, several interesting conclusions were indicated by the Lab's analysis of the data. There seemed, for instance, to be unexpectedly extreme differences in the effects of seeding from year to year. Also, when there was no seeding by the generators in Santa Barbara, seeding by generators in Ventura appeared very effective in increasing rain in Santa Barbara.

Neyman's doubts about the claims of commercial cloud seeders, expressed in print and at weather modification meetings, left a number of those individuals sputtering with indignation. One quotation can give the flavor of these reactions:

"After all this time, Dr. Neyman still fails to appreciate that the [Santa Barbara] project is not a blind statistical attack on a questionable hypothesis but an attempt to obtain a quantitative estimate of the effectiveness of a proven and accepted meteorological theory. If the statistical methods employed are unable to detect and identify the effect, this merely proves that data and techniques are inadequate, or the statisticians incompetent."

Neyman's response was most often his characteristic "Too bad!" But when, in the spring of 1959, his attention was drawn to an article by Robin R. Reynolds, the chairman of the Board of Directors of the Santa Barbara Project, stating that the data for 1957 indicated an increase in the target precipitation due to seeding of approximately 23 percent, he was indignant. He immediately took pains to make clear "that this estimate was reached and published without [the knowledge of the Statistical Laboratory] and that it bears no relation...to the preliminary evaluation [which the Laboratory] reported to the Board of Directors of the Santa Barbara Project."

The Laboratory, acting through Neyman, withdrew from the Santa Barbara Project at the end of the 1960 seeding season and returned the grant it had received for the project from the National Science Foundation. Neyman felt at the time that this abandonment of the subject of weather modification was "irrevocable."

On April 16 of that year, he was sixty-six years old. The tenured members of the Department of Statistics formally petitioned the administration that he be permitted to continue to serve as director of the Statistical Laboratory after his scheduled retirement the following year. They knew this would mean an exception to the regulation forbidding administrative titles for emeritus professors, but they offered telling reasons for their request:

"Neyman is pre-eminent among the world's mathematical statisticians, and the Laboratory and the University derive full benefit from his prestige only as long as he continues to serve as Director. Neyman continues with unreduced vigor to take the leading part in obtaining funds for, and in directing the work of, the organized research which is the Laboratory's main activity, and upon which the support of our graduate students mainly depends."

The request went from Dean Constance to Chancellor Seaborg to President Kerr. On May 26, 1960, Kerr informed Seaborg "that your recommendation for the post-retirement [employment] of Professor Jerzy Neyman in the Department of Statistics for the 1961/62 academic year, and your request that an exception to the rule be approved to allow Professor Neyman to continue as Director of the Statistical Laboratory have been approved."

Signs of recognition and appreciation began to come in. The University of Chicago awarded Neyman an honorary degree. The American Association for the Advancement of Science presented him and Scott with its Newcomb Cleveland Prize. The Royal Statistical Society elected him to honorary membership.

In June 1960 he went to Tokyo for the meeting of the International Statistical Institute. During the meeting an episode occurred which he recounts with a twinkle in his eye. A Japanese statistician, Ryoichiro Sato, who had studied with him at University College—"My 'oldest' student because he was even older than I was!"—displayed to Neyman and to Fisher a two-volume edition of *Collected Papers of Mathematical Statistics* which the Japanese, ignoring copyrights, had published during the Second World War. These contained two papers by Fisher, three papers by Neyman and Pearson, and Neyman's 1937 paper, "Outline of a theory of statistical estimation based on the classical theory of probability." It amused Neyman to see the works all in the same volumes. "But I do not think Fisher was amused."

In July and August 1960, the Fourth Berkeley Symposium on Mathematical Statistics and Probability took place. The Symposium was by that time "a notable international meeting" with "virtually all important schools of thought in statistical and probabilistic research represented." It had even been necessary to refuse some distinguished scholars because time and budget would not permit their participation. Financial support totaling well over one hundred thousand dollars had been obtained from the National Science Foundation, the Office of Naval Research, the Office of Ordnance Research, the Air Force Office of Research (which had also contributed space on military planes for overseas guests), and the National Institutes of Health. Participants came from Australia, Austria, Belgium, Canada, Czechoslovakia, Denmark, England, France, Hungary, India, Israel, Italy, Japan, Poland, Sweden, and the USSR. The guests from the Soviet Union were especially welcome to Neyman, although, to his regret, they did not include the admired Kolmogorov, who disliked travel because of an ear problem which prevented his flying.

Publication of the *Proceedings* was assured, additional financial help for the publication having come from International Business Machines, the RAND Corporation, and Space Technology Research Laboratories. This time there would be four large volumes, which would ultimately total over two thousand pages, four times the number of pages in the *Proceedings* of the first symposium. Two volumes would "provide the widest available coverage of present day research in both probability theory and mathematical statistics." A third volume would be devoted to the physical sciences and a fourth to biology and medicine.

A few years before, Neyman had spent several weeks at the National Institutes of Health in Bethesda and had come in contact with studies on carcinogenesis. He had found these closely connected with the theory of multiplicative processes and had quickly become, as he put it, "entangled." For the Fourth Symposium he had made "a substantial effort" to include, not only a comprehensive representation of the current statistical and probabilistic work on cancer, but also papers presenting various relevant empirical findings, from experiments on cellular phenomena to large-scale surveys of the possible effects of low-level radiation.

(In 1961 Neyman published "A two-step mutation theory of carcinogenesis" and in 1976 (with Prem Puri) "A structural model of radiation effects in living cells." Both papers are classics, but the approach of the latter has been particularly welcome. "It is very exactly what the people who carry out the experiments need," Le Cam tells me, "and pretty soon we will have enough data so that we will be able to check the theory. But the people who carry out the experiments are already very pleased.")

The Department of Statistics had arranged a tribute to Neyman on the opening day of the Fourth Symposium. When, however, Le Cam had compiled a few notes for the benefit of Vice Chancellor William B. Fretter, who in the chancellor's absence would deliver the tribute, he had warned that "[although] this Fourth Symposium may be considered as the crowning achievement of Neyman's regular career at the University...any implication that it is also a terminal achievement should be avoided."

Another tribute on the opening day was the presentation to Harold Hotelling, then sixty-five, of a volume of essays in his honor, published at Stanford University, where he had started his career as a young Ph.D. in 1924. At Neyman's suggestion, the editors had reprinted in the volume the 1940 talk in which Hotelling had created the character of "Young Jones."

I am interested in whether R.A. Fisher ever participated in any of the Berkeley symposia—he did not—and whether he was ever invited to do so—he was.

In the year of the Fourth Symposium, the editors of the *Journal of the Operations Research Society of Japan* called Neyman's attention to various uncomplimentary references to himself and his work in a paper by Fisher recently published in that journal. These included such comments as "some hundred years out of date," "a curious misapprehension," "this elementary error," "many young men...partly incapacitated by the crooked reasoning," "sanity and realism can be restored," and so on.

Neyman, invited to respond, was struck by the fact that it was exactly twenty-five years since R.A.F. had first attacked him.

"During the intervening quarter of the century Sir Ronald [has] honored my ideas with his incessant attention and a steady flow of printed matter published in many countries on several continents. All these writings, equally uncomplimentary to me and to those with whom I was working, refer to only five early papers…all published between 1933 and 1938."

In its early stages, the dispute had probably been useful, the issues of some interest, but "a quarter of a century is a long time for their discussion." During that time mathematical statistics had developed—there were new branches and new ideas. He titled his response "Silver Jubilee of My Dispute with Fisher."

"Because of my admiration for the early work of Fisher…, his first expressions of disapproval of my ideas were a somewhat shocking novelty and I did my best to reply and to explain. Later on,…I found it necessary to reply only when Fisher's disapprovals of me included insults to deceased individuals for whom I felt respect. My last paper…was written mainly because of Sir Ronald's particularly virulent attack…on Abraham Wald, published soon after the latter's untimely death…."

The purpose of the present paper was to observe the anniversary: "to reiterate my appreciation of the early scientific work of Fisher, to present regrets that currently it has been largely replaced by futile polemics, and to express my hope that in the years to come we shall see some more of Fisher's true research comparable to that he did in the pre-war years."

The hope was not to be realized. The year after the publication of Neyman's "Silver Jubilee" paper, Fisher died as the result of an embolism following an operation for cancer of the bowel, which he had stoically endured with only a local anesthetic.

Although Florence David and others advised against his writing anything further about Fisher, Neyman insisted on preparing "An Appreciation" for a memorial session of the AAAS in 1966. What he wrote was very personal and subjective but wholly "appreciative."

"In several earlier writings I have pointed out that certain of Fisher's conceptual developments, not mentioned here, are erroneous," he concluded. "Let there be no misunderstanding on this point. I emphasize that I continue to maintain this view. However, to err is a part of human nature and I feel that a scholar's activity should be judged by his positive achievements and, particularly, by the influence he exercised on subsequent generations. The purpose of the above outline of Fisher's work is to emphasize my personal view on his record, which is second to none."

The same month that his "Silver Jubilee" paper appeared in Japan, Neyman was in Brazil, where he was serving as the principal consultant for the development of an institute of statistics at the University of São Paulo. Elza Berquo, the young woman most ardently interested in the project, pleaded with him to come back for a year. The Food and Agriculture Organization of the United Nations was eager to have him spend a year in

India as consultant to the Institute of Agricultural Research Statistics. Other invitations—academic, governmental, and industrial—came as the rumor spread that Jerzy Neyman, retiring at Berkeley, might be open to offers from other institutions. But Neyman, concerned about leaving Berkeley even for a limited period and perhaps jeopardizing his post-retirement employment, refused them all.

On July 1, 1961—almost exactly forty years after he got his first job as a statistician in Bydgoszcz—he conscientiously began to sign his letters:

> J. Neyman, Professor Emeritus
> Recalled to Active Duty

1979 The Berkeley statistics department has been saddened by the death of Michel Loève. At our Saturday morning interview Neyman recalls again his first meeting with Loève in Paris in 1946. There will be a memorial service, and he wants to say something that will show his appreciation of the Frenchman, "the intellectual son of Paul Lévy."

He goes on to say that he does not want there to be any memorial service for him. He plans to put a specific instruction to that effect in his will, which he is going to have drawn up by one of Blackwell's daughters who is a lawyer. His executor will be Mark Eudey. He indicates the photograph of Eudey—now the president of California Municipal Statistics—which stands next to that of Evans on the mantel of his fireplace. Eudey is the one of his own "sons" to whom he remains most close.

It seems an appropriate time for me to ask him how he sees his life.

He answers my question tangentially, as he so often does, by telling me again about Thomas Park's experiments with *Tribolium*—the flour beetles.

"So the beetles, they are living in this box filled with flour, and so they walk. This way. This way. This way. And then suddenly they come to the side of the box in which they are kept. Well, and then what happens? They want to walk. But they can't. Reflecting boundary. That expression was adopted somehow. *Reflecting*, pushing away. *Absorbing*, well, is death."

So does he see his life as just such a random walk?

"Well. Yes. Many events which happened—interests that developed—might not have happened, you see. Shane, for instance, might not have come to me with an astronomical problem. But obviously there is also a Markovian element."

A Markovian element?

"So what Markov did—he considered changes from one position to another position. A simple example. You consider a particle. It's maybe human. And it can be in any number of states. And this set of states may be finite, may be infinite. Now when it's Markov—Markov is when the probability of going—let's say—between today and tomorrow, whatever, depends only on where you are today. That's Markovian. If it depends on something that happened yesterday, or before yesterday, that is a generalization of Markovian. Well, the probability of going from this to this depends on something."

One's situation at the time?

"Yes."

So when Shane came to him, not only was there the appeal of the astronomical problem, which involved the concept of clustering that had interested him for some time, but Betty was also there with her knowledge of astronomy?

"Well, yes," he agrees. "But let me tell you about my favorite poem, because it is also relevant to our conversation."

He recites from memory:

> Vers où vas-tu, toi qui t'en vas
>> Par toutes les routes de la terre,
> Homme sauvage et solitaire,
>> Vers où vas-tu, toi qui t'en vas?
> J'aime le vent, l'air, l'espace
>> Et je m'en vais sans savoir où
> Avec ma coeur fervent et fou
>> Dans l'air qui luit et dans le vent qui passe!*

He delights in the French words on his tongue; but when he translates them for me, it is clear that he loves his English version almost as well:

> Where are you aiming at, you who are wandering
>> Over all the roads of the earth,
> Man wild and solitary,
>> Where are you aiming at, you who are departing?
> I like Wind. Air. Space.
>> And I am departing without knowing quite where to
> With my heart fervent and crazy
>> Into the air that shines and into the wind that blows!

What does the poem mean to him?

"It reflects my feelings and my activities."

And his life?

"Yes. My individual life. There are some—you might call them *innate*—characteristics of my thinking. Liking this, disliking that, which were probably conditioned by genes, possibly by atmosphere at home, and so forth. So they created desires for this and not desires for something else.

---

*Neyman's first line is not quite right, and he has substituted *sauvage* for *tenace*.

Then, all right, there is wind that blows and air that sparkles and Berkeley climate which gives us fogs in the morning and fogs in the evening—for some reason I don't support heat very well—and there are also things which happen. The things that happen don't depend on me. I just make some selections. It's somewhat Markovian. You, on the other hand, will have your own impressions, conditioned by your background, you know. But, well, I am trying to go into the air that sparkles and into the wind that blows."

Later in the year he receives a letter from Chancellor Albert Bowker asking for suggestions of ways in which the different units on the campus can observe the 150th anniversary of Belgian independence. He cites to Bowker the poem he has quoted to me, the author of which was Emile Verhaeren, a Belgian who at the beginning of the First World War went to France to stir up support for his country, and expresses the hope that some way can be found to incorporate it into the celebration.

Later he arranges to have the original of the poem located and translated by a student in the French department. It is read as part of the statistics department's observation of the Belgian anniversary, following a talk by F.N. David on the Belgian statistician L.A.J. Quételet.

The student must be paid for his work. Neyman's hand goes automatically to his pocket. But he thinks he prefers his own, somewhat rougher translation:

I like Wind. Air. Space.
And I am departing without knowing quite where to
With my heart fervent and crazy
Into the air that shines and into the wind that blows!

1961
—
1964

Anyone who thought that Neyman would slow down when he became emeritus was quickly disabused of that notion.

In the summer of 1961, scarcely back from a month in Brazil, he hurried off to the Fourth All Soviet Union Mathematical Congress in Leningrad, taking a day when there were no meetings to go on to Moscow to see his old teacher, Bernstein, who had recently suffered a stroke. After two weeks in Leningrad, he flew back to California for the meeting of the International Astronomical Union, which was to be held in Berkeley; but first he must dash down to Santa Barbara for a "satellite conference"— organized by himself, Thornton Page, and Betty Scott—on the Armenian astronomer V.A. Ambartsumian's still then debatable hypothesis regarding the instability of systems of galaxies. In a few weeks he left again—for Paris and the meeting of the International Statistical Institute. (His secretaries observed that "Berkeley Travel just loves Mr. Neyman!")

On April 25, 1962, he was off to spend a month as an exchange professor at the University of Kiev. After Moscow and Leningrad, Kiev was an important center for probability and statistics in the Soviet Union, the "cell" there having been developed by B.V. Gnedenko, a student of Kolmogorov's who had since moved on to Moscow. "Regretfully," Neyman found that the professors in Kiev were careful never to be with him unless some colleague was present; but he also noted that "after a drop of alcohol" they downed their vodka with a great many toasts to exchange professorships. The rector of the university, A.Z. Zhmudski, took great care with arrangements for his comfort and entertainment. Did he have any special wishes? Neyman mentioned that he would like to see Kaniuv, where his paternal grandfather had once had an estate. "Then I had the impression—impression, you know—that this, having an estate, was not approved. But no one said anything." Later, at the end of Neyman's stay, Zhmudski arranged an excursion by fast boat to the birthplace of a national hero of the Ukraine—who happened to have been born in Kaniuv.

(Neyman's ideas for entertaining Zhmudski when he visited Berkeley the following autumn are recorded in a letter to the administration of the university: "First, I would like to take Professor Zhmudski, perhaps with a few colleagues who know a little Russian, to a restaurant in San Francisco. Second, I would like to take him to Muir Woods and show him the Sequoias. Third, I would like him to be a guest in my house. Fourth, I would like to show him something very American that is likely to impress a Russian, namely Professor Le Cam and his family building themselves their own house in a pleasant locality near the Russian River. Finally, I would like Professor Zhmudski to see the traffic boys stopping automobiles at appointed corners in order to take the little children across the street.")

While Neyman was in Kiev, Y.V. Linnik and half a dozen assistants came down from Leningrad to talk with him about recent work which had been stimulated by his talk at the Leningrad congress the previous summer ("On problems of analysis generated by mathematical statistics"). Both Leningrad and Moscow mathematicians requested that he be permitted to spend a portion of his month in the Soviet Union with them. As a result, he left Kiev after only three weeks.

On his way to Moscow he was able to stop for a day in Kharkov. There he found "the spirit...even tighter than in Kiev. In fact, of the several individuals with whom I used to be friendly since about 1912 and whom I cannot believe to be personally unfriendly to me, only two appeared courageous enough to see me during my stay...."

One of these was Leo Hirschvald, to whom he had enjoyed explaining Zermelo's axiom of choice during his student days. Hirschvald took him to see his old home on Rogatinskiy Pereulok and arranged a phone call—"I had not thought it would be possible"—to the village where Yevgeniy Butkos, the "Zhenka" with whom he had once shared his room, practiced medicine.

In Moscow and Leningrad he found that the mathematicians "treated me as if there were no cold war. I was invited to their homes and we had a number of discussions not only on scientific questions but also on other matters of more personal character."

261

An immediate outcome of the visit was "an international effort" in Berkeley the following year to translate E.B. Dynkin's *Markov Processes* into English. The four young mathematicians working on the project had come from Holland, India, and Italy as well as the United States. The publisher was Springer-Verlag, an international firm with headquarters in Germany. Neyman also played a role. "Not a single doubtful point has been settled without his active participation," Dynkin noted in his preface and added that he hoped the English translation of his book would "further the development of international scientific relations in the field of mathematical statistics."

Although the Russians led the world in the theory of probability, they were (in the words of Gnedenko) "taking only the first steps" in mathematical statistics. In Neyman's opinion, both Russians and Americans would gain much from "rubbing of shoulders." He had already begun to make plans to bring a largish contingent from the USSR to the symposium, but until 1965 seemed a long time to wait. Then one day—talking with Le Cam, who had succeeded Blackwell as chairman of the statistics department—he remarked thoughtfully, "Nineteen sixty-three. One hundred and fifty years since Laplace. We should do something about that."

Further investigation revealed that, in fact, 1963 would be a triple anniversary: 150 years since the publication of Laplace's *Théorie Analytique*, the book which had dominated probability and statistics throughout the nineteenth century; 200 years since the posthumous publication of Thomas Bayes's controversial formula—just recently returned to unexpected favor in the work of Savage and others; and 250 years since the publication of Bernoulli's *Ars Conjectandi,* the first systematic exposition of the theory of probability. Neyman immediately suggested to Le Cam that the Department of Statistics and the Statistical Laboratory join in sponsoring a series of "high class" seminars during the late summer of 1963 to commemorate these anniversaries. Invitations were sent out to a number of foreign colleagues. These included three Russians: Kolmogorov, Linnik, and A.M. Yaglom.

While Le Cam moved ahead with arrangements for the "B-B-L" seminars, Neyman took off again, at the beginning of 1963, for a six-month tour as a visiting lecturer for the Mathematical Association of America. He loved to drive, and letters began to arrive at the Lab headed cheerfully, "On the road."

Because of weather his first lectures were scheduled for the south—Texas, Louisiana, and Mississippi. The civil rights movement, under the leadership of Martin Luther King, Jr., was picking up momentum. The fall before, James Meredith had been escorted by the National Guard onto the campus of the University of Mississippi, the first Negro to enter that institution. Biracial groups of "Freedom Riders" had begun to travel from one southern state to another in an effort to desegregate buses, lunchrooms, theaters, and other public facilities. While Neyman was lecturing at Tougaloo, a private Negro college in Mississippi, President John F. Kennedy asked Congress to enact an extensive civil rights program.

A student who heard Neyman talk at Tougaloo was Shyrl Miller, now Dawkins and a senior programmer in the Berkeley Statistical Laboratory. She doesn't remember what he talked about (one lecture was titled "Indeterminism" and the other, "A Month in the Soviet Union"), but she does remember the blue beret and the long white Buick Le Sabre, also how impressed she and the other students were that anyone so famous would come and talk to them.

Neyman had been in the south before, but he had never had firsthand contact with segregation. At Tougaloo, a number of retired white professors and young white Ph.D.'s from elite eastern schools had come down to help. Mixed groups made "visits" to "Whites Only" institutions and facilities in a usually unsuccessful effort to gain admission. At the Negro state university in Jackson, where Neyman lectured next, the situation was quite different. He was shocked at the extreme secrecy necessary for a black teacher who had been a student at Berkeley to arrange for him, a white, to come to dinner in his home.

Upon leaving Mississippi, he swung north into the farm belt. From Grinnell there was a quick flight back to Berkeley to receive an honorary degree at Charter Day. Then back to Minnesota, where at tiny St. Olaf's he was delighted to find Lebesgue integration being taught and real research being carried on. He was lecturing at Kansas State when he received a telegram from Berkeley, signed "All of Us," congratulating him on his "long overdue" election to the National Academy of Sciences.

On May 15, Neyman gave his last MAA lecture at the University of Oklahoma in Norman. In the five days it took him to get back to Berkeley, he sorted out the many impressions of the last months—the disparities in educational preparation and opportunities which he had observed and the deeply affecting meetings he had had with lively, likeable young black people who saw no future for themselves in a white society. Mightn't there be another Blackwell among these? It seemed that the only hope for improvement of the situation lay in the efforts of Dr. King and his Southern Christian Leadership Conference.

"Well, I wanted to do something to help," Neyman explains to me, opening and closing the hand at his side—a characteristic gesture.

Back in Berkeley, he recounted to his colleagues—sometimes, they remember, with tears in his eyes—the pitiable conditions and the valiant struggles he had witnessed. Four days after his return he wrote a short letter soliciting funds for Dr. King's group. A copy, framed, hangs in his office in Evans Hall.

"This is in connection with the current developments in the South," it begins, "including the arrests of large numbers of youngsters, their suspension or dismissal from schools, the tricks used to prevent Negroes from voting, and the prospect of the governor of a state personally barring a few Negro students from a university entrance. These developments appear disgusting to us and we feel compelled to do something about them."

The letter concludes with a request for contributions for the Southern Christian Leadership Conference "to help the young generation of Negroes in their struggle to attain the status of citizens in the United States of America."

Neyman especially wanted colleagues from departments other than statistics to sign his letter. But time was of the essence. There are eleven signatures: Owen Chamberlain (physics), Milton Chernin (social welfare), Benson Mates (philosophy), R.M. Robinson (mathematics), Kenneth Stampp (history), and George R. Stewart (English) in addition to Blackwell, Le Cam, Neyman, Scott, and Herbert Robbins, who was in Berkeley at the time.

Four thousand copies were printed and dispatched to members of the university community and friends across the country. In the first few days more than $1300 in checks came in, ranging from $2 to $100.

"I am not sure which of the two denominations is the most encouraging!" Neyman wrote to Gail S. Young, the head of the mathematics department at Tulane, who had started a similar campaign. "As a result, I don't feel as lonely in my feelings as [I] felt before and, in fact, I feel somewhat elated."

A month after the mailing, writing to his senator to urge him to support the president's civil rights program, he reported—"on the assumption that you are wondering about the sentiments of the people at large regarding the current race crisis"—that he had collected more than three thousand dollars as a result of the letter and that checks were continuing to come in.

He also wrote to his friend Harald Cramér that in his opinion Martin Luther King, Jr., deserved a Nobel Peace Prize for his nonviolent approach to the problem of civil rights. When, in October of the following year, King did receive the prize, Cramér commented: "It is by no means impossible that your letter helped to influence the result...I forwarded it to my old friend, Mr. Jahn, who is Chairman of the Norwegian Nobel Committee and a person of great authority. In any case I should hardly believe that the matter was already settled when he got your letter, and it is certainly not unlikely that it contributed to form his decision." But Neyman took no credit for King's receiving the prize.

During the academic year 1963–64 he and a few colleagues (Blackwell, Le Cam, and Scott as well as a young newcomer to the statistics department, David Freedman) continued the fund-raising campaign with an advertisement running one week each month in the student newspaper, the *Daily Californian*, urging all members of the university to contribute to King's organization or to one of the other groups active in the civil rights movement, such as SNCC and CORE.

Such efforts as these were "all right," but still more was needed. Closest to Neyman's heart was an ambitious plan to raise the educational level of blacks and other disadvantaged groups and to create within each of these an "intellectual elite" of its own. The plan had antecedents in his experiences in Kharkov during the first years of the Soviet educational reform when N.N. Khrushchev, since 1958 the Soviet premier, might have applied for admission to the University of Kharkov. ("I say 'might,'" Neyman reminds me.) Under the plan Neyman had in mind, below-standard high schools across the

country—in the south, in the slums of big cities, and in poor rural areas—
would be identified, and their most talented and motivated students would be
given supplemental instruction to prepare them for university admission
and "to contribute to their upbringing." The second phase of the program,
for those who qualified, would be a university education at a first-rate school
some distance from where they had grown up; the third phase, graduate
study at a university of their choice.

He talked and wrote to a number of people about his plan. Among them
was a Berkeley colleague who was much in sympathy and who he felt would
add influence to the proposal. This was the Nobel physicist Owen
Chamberlain.

A few months after Neyman's return from the MAA tour, he and
Chamberlain drafted a letter to Francis Keppel, the United States Commissioner of Education, describing Neyman's plan.

"...the benefits we expect from [this] program are not limited to the gain
of some 200 high grade intellectuals per year, at the cost of perhaps $50,000
per individual," they told Keppel. "We expect these 200 persons to be welded
into the general intellectual group of the country and, at the same time, to be
reasonably close to their original background. The benefit to be anticipated
from the program outlined lies in the influence that the yearly influx of 200
top level intellectuals from slums, combined with that of some 300 other
somewhat less successful scholars, will exercise both on the social group of
their origin and on the recipient group which they will join."

The Chamberlain-Neyman letter, as ultimately sent to Keppel, was also
signed by W.G. Cochran (Harvard), J.L. Doob (Illinois), and Charles M.
Stein (Stanford). The reply they received a month later was disappointing to
Neyman. Although he had not expected the government to support the plan
directly, he had hoped that Keppel would provide access to one of the big
foundations—he had been thinking of the Ford Foundation—but the
commissioner, although expressing his basic sympathy with the idea, had
merely suggested the National Scholarship Service and the Fund for Negro
Students as more experienced in the area than the Office of Education.

Discouraged, Neyman, "rightly or wrongly," let a month go by. The
assassination of President Kennedy on November 22, 1963, brought him into
action again. The day after the long weekend of national mourning, he and
some colleagues, lunching together at the Faculty Club, discussed the
possibility of the Academic Senate's establishing several fellowships for
students from neglected areas as a memorial to the dead president. To
Neyman the idea was "a reasonable one," but too limited in scope. He
proposed instead "the establishment by our University of a memorial on an
impressive scale and having a more than symbolic significance for a solution
of a big national problem."

He was almost superstitiously convinced that if he spoke in favor of
something in the Academic Senate it would be voted down, and in recent
years he had developed what he called "the trick" of having someone else put
forth an idea he supported. (Further insurance was to bet a martini that the

proposal would fail.) Thus it happened that at the meeting of the Academic Senate on January 13, 1964, it was Leon Henkin of the mathematics department who proposed the resolution (henceforth known as the Henkin Resolution) "that the Berkeley division of the Academic Senate establish special University of California scholarships designed to aid talented young people, who have been denied the stimulation and education required, to prepare themselves for and to obtain a university education." The resolution passed, and President Kerr appointed Neyman chairman of the newly created Special Scholarships Committee.

A disagreement soon developed in the committee. Although Neyman recognized that there were too few disadvantaged young people getting a university education and that a substantial effort should be made to increase their number, it was not this task that attracted him. He dreamed of creating a channel through which truly talented people—the Blackwells among the Negroes, the Le Cams in poor rural areas—could be drawn into the general intellectual community of the country.

"Is the goal of the scholarships to seek out in this land future Ph.D.'s or equivalent with a mediocre Bachelor's being an acceptable by-product or, vice versa, should the Bachelor's be our main goal, with Ph.D.'s considered as a welcome by-product?"

Other members of the committee, less optimistic about the number of Blackwells and Le Cams in any group, envisaged a more modest beginning and an ultimately broader program. Within a couple of months the assistant to the president who had been assigned to follow the deliberations of the Special Scholarships Committee was complaining to Kerr, "This is a more complicated assignment than might have appeared at first blush." A few days later Neyman submitted his resignation as chairman. He wanted to drop out completely but was persuaded to remain as a member of the group.

"As you know," he explained to Chamberlain, who succeeded him as chairman, "I am emotionally committed to the plan outlined in our letter to Dr. Keppel and it is difficult for me to change my allegiance."

In January 1964, when the Special Scholarships Committee was organized, Neyman was approaching his seventieth birthday. Although he had created at Berkeley "what everybody would agree," as D.G. Kendall was later to say, "is the most important and the largest statistical center in the world," he continued to be concerned with the development of statistics on an international scale. He had long endeavored to make the International Statistical Institute—the only international organization in the field—more representative; and in 1963 he had played an important role in the founding of the International Association for Statistics in the Physical Sciences, a group affiliated with the ISI but with open membership.*

In spite of travel and emotional involvements he had not neglected either his students or his scientific interests. On occasion he was overseeing more

---

*Twelve years later, under Kendall, the IASPS became the Bernoulli Society, which encompasses all mathematical statistics and probability.

theses than any other member of the department. The list of cosmological papers with Scott continued to lengthen. There was a growing interest in medical questions, especially cancer. (The paper "A stochastic model of epidemics," which his students were updating in 1978–79, was also produced during this period.) He and Le Cam were editing the papers presented during the three months of "B-B-L" seminars, and he was busy contacting prospective publishers. (None of the Russians had been able to attend, but Yaglom had contributed two papers.) From time to time letters arrived at the Lab inquiring as to the progress of his book on indeterminism. He also had in mind a collaborative effort with a biologist and a translation of some important Russian papers in probability and mathematical statistics. He continued to teach more classes than were required of him under the terms of his recall and to conduct a weekly seminar, which at that time met at night, preceded by dinner at the Faculty Club with the speaker. Afterwards everybody went to Edy's, a popular ice cream parlor, where Neyman bought a box of candy for each of the wives who had been left at home.

"The time when I was a theoretician is past," he confided to Egon. "Now it's either galaxies, or cell division, or carcinogenesis, etc." It was hard to believe that when he had come to Berkeley in 1938 he had been the most "modern" of the mathematicians on the faculty. Now when the members of the mathematics department began to "talk shop" he was soon lost. The same thing was true in his own department. "I keep promising myself to sit down for a couple of months and consume some modern book. However, nothing comes out of it. Presumably, one of the reasons is that my efficiency has declined (on occasion I have to have a nap in the afternoon). Otherwise, there is always plenty to do."

In spite of the occasional nap, he still retained his ability to get individuals—at times very much against their own inclination—to do what he wanted them to do. For some the pressure—"to come and have a little drink"—"to get this job done"—to give a talk to the seminar—to organize a symposium—seemed impossible to resist. "I know distinguished statisticians who will leave a room when Jerzy Neyman enters it," one man tells me, "—because they *know* he will get them to do something they don't want to do."

How is he able to exert such power?

The answer is most often a bewildered "I don't know"—"It's clear he will never give up"—"You're just swept away"—"He is convinced that, whatever it is, it is just so important"—"You want to please him."

"Jerzy likes to say, 'I never insist, I always agree,'" David Blackwell laughs. "He says he learned that in England. *But he never learned!*"

Surprisingly, however, after his "retirement" Neyman let his old department go pretty much its own way. He attended meetings, expressed his opinion when it was asked for, but never pressed his views. The department, on its side, tactfully recommended Le Cam to succeed Blackwell when the latter's term as chairman expired.

Neyman has always greatly enjoyed working with Le Cam. He has tremendous admiration for the Frenchman's "conceptual abilities" and is "appreciative" of his interest in applications, which is rare in the Berkeley department. On his side, Le Cam has a very high regard for Neyman.

"It is my personal belief that an objective look at the record will show that Fisher contributed a large number of statistical methods, but that Neyman contributed the basis of statistical thinking," he says.

Between Le Cam and Neyman there is not the easy relationship which exists betwen Neyman and Blackwell, who jokes with him and will on occasion shout, "Jerry, I'm just *not* going to do that!" Although Neyman isn't always easy to work with, Le Cam never shouts at him or calls him "Jerry," and Neyman murmurs at times that he still finds the French "difficult to approach." Since 1961, however, Le Cam has been the one senior member of the department, other than Scott, who has worked closely and continuously with Neyman.

In 1964 plans were already well advanced for the Fifth Berkeley Symposium, which was scheduled for the summer of 1965. In the past Neyman had served as chairman of the organizing committee and editor of the *Proceedings;* but the Fifth Symposium was planned as a joint project of the Department of Statistics and the Statistical Laboratory with Le Cam as chairman of the organizing committee and co-editor of the *Proceedings.*

The academic year 1964–65 which preceded the Fifth Berkeley Symposium was to be the most turbulent that the university had seen in its near century of existence. The seeds of "the Free Speech Movement" were sown that September when a dean of students decided to enforce a state law—long ignored—that university facilities could not be used in ways which would involve the university as an institution in controversial issues. Students would no longer be permitted to set up tables at the Bancroft and Telegraph entrance to the campus to advocate off-campus political action, solicit funds, or recruit students for such causes. Although conservative youth groups also protested, it was widely felt that the ruling was primarily directed at civil rights groups. On October 1, the dean of men summoned five students who had been running unapproved tables for SNCC and CORE to appear in his office for discipline. Some five hundred students followed the five offenders into Sproul Hall (the administration building of the university) and refused to leave unless they also were disciplined. At midnight Chancellor Edward Strong announced the suspension of the five along with the three leaders of the protesting students.

"The description of events before Sproul Hall as I read it in the *Chronicle* is unpleasant," Neyman wrote to Blackwell the next morning. "I wonder whether some faculty members should try to do something to stop the riots. We supported the students in a number of enterprises that seemed reasonable. Maybe we should try to stop them from doing that which appears unreasonable."

But the events of October 1 were to be only the beginning.

The same day that Neyman wrote to Blackwell, a former student was arrested by campus police for running a CORE table. When he went limp, a police car was brought into the plaza before Sproul Hall to remove him. Students threw themselves on the pavement, and for the next thirty-six hours the car, along with its prisoner (fed hamburgers and milkshakes through the window), was unable to move. By the evening of the following day seven thousand people had gathered, and five hundred police were being held in readiness on campus. Late that night an agreement was reached between the university administration and the leaders of the Free Speech Movement that charges against the arrested man would be dropped and a committee would be set up to deal with the question of advocacy on campus. The eight students previously suspended would remain so.

In spite of his initial feeling that the behavior of the students was "unreasonable," Neyman felt "a degree of responsibility" for those who had been suspended. Over "a substantial period of time," dating from his letter appealing for contributions to the Southern Christian Leadership Conference, he and a few of his colleagues had done their best "to ingrain in our students the idea that (i) many of the developments in the deep South are disgusting and (ii) that it is decent and proper to extend all possible help to the various Civil Rights groups working in the South and elsewhere." A number of students had worked in Mississippi during the past summer (when three civil rights workers had been murdered), and the repressions they had observed and even personally suffered had contributed to building "a fighting spirit" in them. "No one should be surprised to see our students decide that last summer was not just an interlude in their ordinary life. On the contrary, we visualize these young people deciding that the work they were doing during the summer should be continued, possibly throughout their lives, but certainly right now in Berkeley on the campus."

He and his colleagues had been raising money for what they believed in by buying advertisements in the *Daily Californian*. The students had also wanted to raise money for the same causes, and they had done this by setting up tables at Bancroft and Telegraph. When they had found that this procedure was illegal, they had quite naturally protested.

Their transgressions had not been dishonorable.

"...they were not caught stealing library books, or cheating at examinations, or striking children, or throwing eggs at colleagues whose opinions they did not like." Their actions had been motivated by a desire to improve society. "And they reacted to these forbidding regulations in the manner which was the only possible way to struggle with the inequities in the South: non-violent disobedience."

Neyman and four colleagues—the same who had paid for the advertisements in the *Daily Californian*—signed this letter, which was addressed to the chairman of the Committee on the Eight Suspended Students.

Although the eight students continued under suspension, the campus was relatively quiet during the rest of October and most of November. At the beginning of December it exploded again. The regents had approved a more liberal policy of political advocacy, but the leaders of the "FSM" now took the stand that it was not the place of the university but of the courts to regulate speech. On December 2, after a rally in the plaza, a thousand people poured into Sproul Hall and remained there until the early hours of the morning when Governor Edmund G. Brown ordered the building cleared. Although everyone was given an opportunity to leave before the police moved in, nearly eight hundred students were arrested.

A week later—the day after a convocation in the Greek Theater at which the FSM leader, Mario Savio, had been forcibly prevented from speaking—the Academic Senate of the Berkeley campus ratified, 824 to 115, a series of propositions recommending broad political freedoms on the campus as a solution to the controversy. Neyman, who had been a member of the drafting committee, personally sent letters to former students and colleagues all over the country urging them to write or wire Governor Brown in support of the faculty stand.

"The usual comments of the newspapers begin with the premises that there is no justifiable cause for discontent in Berkeley, that the student-demonstrators are a small fraction of the total led by a 22-year-old subversive individual anxious to cause trouble for trouble's sake....

"It is, of course, inconceivable that a young man could bamboozle 824 professors into an unprecedented action without a real issue. My hope is that, in these circumstances, you will question the premises and would admit the possibility of there being real deep causes for unrest at Berkeley which bother both the thinking section of our students (many of the demonstrators are our best graduate students) and of the faculty."

He continued to be "emotionally involved." He acted as chairman of the Faculty-Student Independent Legal Fund, which he had organized the day after the Sproul Hall sit-in to help pay the students' court costs—the legal services had been donated. He wrote and urged other faculty members to write to the trial judge testifying as to the scholastic standing and moral character of arrested students.

During the next few months the Free Speech Movement subsided. Six of the eight suspended students were reinstated, and the administration returned to what had been, essentially, the policy of the past in regard to advocacy on campus. The FSM got briefly back in the news at the beginning of March as the "Filthy Speech Movement" when a non-student was arrested for carrying an obscene sign on campus. The next day, four others were arrested for carrying similar signs to protest the first arrest.

"Lately we had some trouble between the students and the administration," Neyman wrote to Egon Pearson a few days after these events. "I am not sure now how I could have taken something of this sort say 15 years ago. Now, however, I was emotionally involved.... Some of the students are not quite angels too! F.N. [David] finds them 'awfully rude,' which they are, some of them."

During this same time he found himself also "emotionally involved" in an old scientific interest. In the fall of 1964 he had received a large package of mimeographed materials from Switzerland—the voluminous reports of three successive experiments in hail prevention conducted over a period of sixteen years by a special commission of the Swiss government and supported by certain businesses and industries. Although the experiments had been concerned only with the prevention of hail, the summaries contained a great deal of information relevant to any work in weather modification. They impressed Neyman by their thoroughness. "…contrary to what one sees in reports of many other experiments, all the relevant information was presented in full detail, without any signs of attempts at 'coverups' or 'cosmetic treatments.'" In addition, the third experiment—*Grossversuch III*—was an example of "carefully planned and meticulously executed randomization."

When, five years earlier, disenchanted with the Santa Barbara weather modification experiments, he had returned the Laboratory's grant for that work to the National Science Foundation, he had been regretful:

"Whoever has made airplane trips over Arizona and looked a little out of the window must have been impressed, as I was, by the vastness of desert lands interspersed with emerald-coloured, emerald-shaped and, apparently, emerald-sized spots of greenery. These are locations where there is some source of water, marked by magnificent vegetation. Similar impressions, on a bigger scale, are gained from air travel over many other lands… If only one could bring down more rain in these areas, what effect would this have on the lives of millions of people!"

Now—in the middle of his seventieth year—he found that he had become again "a believer in the possibility of learning how to modify the weather." He was eager to go to Switzerland immediately, that winter, for he was still convinced a statistician should always go to the scene of the experiment and see for himself what was being done; but such a trip would have to be postponed until he could obtain funding for a new weather modification study by the Lab. Also, he and his colleagues had been away from the subject for several years. Before they could do any new work, they would have to familiarize themselves with what had been done—and learned—in the interim, both in the United States and abroad. A conference—the kind of cross-view for which the Berkeley symposia had been conceived—was called for. He did not have time or money to set up such a conference for the coming summer session of the Fifth Symposium, but he began to plan a special session the following winter with a separate volume of the *Proceedings* devoted entirely to weather modification.

It is no wonder that, writing to colleagues about Neyman at seventy, Le Cam on occasion complained:

"He can still run the rest of us ragged!"

1964
—
1969
When Neyman passed the age of seventy, the procedures surrounding his post-retirement employment at the university changed. His yearly reappointment as "Professor Recalled to Active Duty and Director of the Statistical Laboratory"—which in the past had gone from the Department of Statistics to dean, chancellor, and president for approval—now had to go also to the Board of Regents. He thought, uneasily, that perhaps it was time for him to leave Berkeley and let the Laboratory have a new director. Several of the full professors, including Le Cam, were his former students. "They are talented and full of initiative and it is not unlikely that, with me out of the way, the Department and the Laboratory will gain in impetus and vigor," he wrote to a friend. But he didn't feel any more like retiring in 1964 than he had in 1961. When he was invited to set up a laboratory at the University of Arizona, he turned over the offer in his mind. Would he still be able to get the kind of grants he had been getting? Would there be a position for Betty Scott? In the end he decided, once again, to remain in Berkeley.

Attaining seventy was also the occasion for the inevitable honors, celebrations, and other tokens of appreciation which accompany that age. In April 1964—in fact, almost on the day of his birthday—he officially signed his name in "the great book" of the National Academy of Sciences. In May he went to Sweden to receive an honorary degree from the University of Stockholm. It amuses him to take from his cupboard the top hat he wore for the ceremony and the laurel wreath with which he was crowned. But he won't put them on. "Not very becoming," he explains.

The people at University College had wanted to do something for his seventieth birthday—"to express in some way the esteem in which we hold [him] as a statistician, the pleasure we have in him as a friend, and the pride which we have in recalling that it was while he was here that he published the theory of confidence intervals and, with Egon S. Pearson, the theory of testing statistical hypotheses." Opinions on what to do had differed until Evelyn Fix had arrived in London on a sabbatical. Then, with her assistance, F.N. David had organized a festschrift to which a number of mathematical statisticians and probabilists connected with the English period were invited to contribute. It was for the festschrift, at David's suggestion, that Egon Pearson wrote his affectionate account of the collaboration, "The Neyman-Pearson Story..." In June 1965, just before the Fifth Symposium, David presented Neyman with a bound volume of the festschrift manuscripts, bearing as an epigraph: "Hereux que comme Ulysses a fait un beau voyage."

Although David told Pearson that Jerzy was "touched" by what he had written, Neyman commented only, "I like your article and am very grateful for your writing it." He made no objections to Pearson's technical account of their work but asked for a few changes in personal details. He wanted it mentioned, for instance, that he had gone to London originally on a Polish government fellowship rather than on a Rockefeller fellowship. "This is a minor detail but, if it is not put right, it may hurt some people....

272

"I would [also] appreciate it if you would omit remarks suggesting my inimical feelings towards Russians in general and Bolsheviks in particular. It is, of course, true that some imperial Russians and some communist Russians persecuted Poles, including myself. On the other hand: (i) my aunt married a Russian and (ii) so did I. Also I have a few cordial friendships in Russia, some contracted in 1912, when I became a student at the university, and some contracted recently. I did not ask the people concerned whether they are communists and, if so, why.... For all I know some of them may be members of the party, but this does not prevent them from being nice people for whom I have warm feelings."

The people at Berkeley had also talked for a long time about doing something for Neyman's seventieth birthday. Originally they had wanted to publish his collected works, but there had been a number of obstacles. First the sheer volume of papers. Then the fact that many were in Polish and—a further complication—some of these had later been republished in English but never in quite the same form. It would be almost impossible to avoid repetition. Still another obstacle was the fact that Neyman did not look upon his work as completed. (The epigraph about Ulysses's voyage was omitted from the published festschrift.) Unforeseen, however, was Egon Pearson's obvious unwillingness to give permission for the joint papers to be published in Neyman's collected works.

All obstacles were overcome, after much negotiation, with the publication of *three* volumes as a collaborative effort of the University of California Press and the Cambridge Press:—*A Selection of Early Statistical Papers of J. Neyman, The Selected Papers of E.S. Pearson,* and *Joint Statistical Papers of J. Neyman and E.S. Pearson.*

"Our cooperation over a number of years did leave a mark, and its character should not be adulterated by occasional waywardness of either one of us," Neyman wrote to Pearson, expressing his approval of the final idea of "yours, mine, and ours."

The three volumes appeared at a time in the history of mathematical statistics when some of its practitioners were turning away from the Neyman-Pearson point of view and back to the ideas of Thomas Bayes ("which," one mathematical statistician ruefully commented, "a lot of us had thought were dead and buried"). To Harvard's A.P. Dempster, who reviewed the books for *Science,* they seemed to provide "a welcome platform, amid conflicting intellectual crosscurrents...."

The fundamental approaches and contributions of Neyman and Pearson— Dempster wrote—"should be viewed against the tension which has existed from the time of Laplace between the approaches of Bernoulli and Bayes." Both approaches came naturally to probabilistic thinkers, although Laplace had "rather confused the picture" by prescribing the Bayesian approach as a matter of principle and then often using sampling distributions, "apparently under the impression that they necessarily would yield the same results." Karl Pearson had also written in both modes. Then Fisher had come

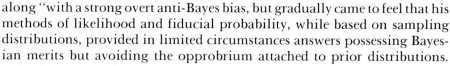

along "with a strong overt anti-Bayes bias, but gradually came to feel that his methods of likelihood and fiducial probability, while based on sampling distributions, provided in limited circumstances answers possessing Bayesian merits but avoiding the opprobrium attached to prior distributions.

"...Neyman and Pearson rode roughshod over the elaborate but shaky logical structure of Fisher and started a movement which pushed the Bernoullian approach to a high-water mark from which, I believe, it is now returning to a more normal equilibrium with the Bayesian view."

(Although during the next decade the new Bayesians sniped persistently at Neyman-Pearsonian ideas, particularly the concept of confidence intervals, Neyman "remained in his tent," as Egon put it, until 1977 when, in an article in *Synthese,* he defended "Frequentist Probability and Frequentist Statistics." He continued to say, as he had said forty years earlier, that the theory of probability in which a statistician works is a matter of personal taste. When asked by me in 1979 about his attitude toward the Bayesian point of view, he replies simply but firmly, "It does not interest me. I am interested in frequencies.")

The publication of the Neyman-Pearson papers in 1966 coincided with the selection of Neyman by the Royal Statistical Society for its Guy Medal in Gold. He was the first non-Englishman to be so honored. ("...for the purpose you are treated as an Englishman," Egon explained, "because all the early work centered around London, I suppose!") He received the medal in person the following year when he presented a paper to the society on "Experimentation with weather control."

By that time (1967) the subject had thoroughly taken hold of him again. It combined both the elements which had characterized his earlier "emotional involvements"—interested collaborators and challenging critics. In many cases he was suspicious of the motives of the latter—his conversation on the subject is sprinkled with references to "cheaters," "cover-ups," and "cosmetic treatments of data." "Uneasy," as he describes it, about the scientific scrupulosity of commercial cloud seeding operators, he suspected them—on occasion—even of covering rain gauges on non-seeded days. He was also "regretful" about the political pressure exerted on such scientific work as the report on weather and climate modification by the National Academy of Sciences and the National Research Council.

"...after some prodding from the cloud seeding industry," he noted in *Science* in 1968, "the report of the NAS-NRC Panel, and also Dr. [Gordon J.F.] MacDonald's testimony before a U.S. Senate Committee (delivered in his capacity of Chairman of the Panel), misplaced the emphasis [of the report] so much that the resulting picture of the present knowledge on cloud seeding became distorted."

The special weather control session of the Fifth Symposium had brought to Berkeley—in addition to many Americans—visitors from Australia, France, Germany, Israel, and Switzerland, all of whom were or had been involved in randomized cloud-seeding experiments in their own countries. Another guest had been Frank Yates who, as Fisher's successor at

Rothamsted, was now the authority on the randomization of experiments. For the *Proceedings*, data on as many randomized weather experiments as could be assembled had been compiled. This was an especially important contribution to future research, since in any substantial weather experiment many observations were made in addition to those on precipitation. Most, however, were not used in the evaluation of the experiment. They remained to be analyzed by the interested statistician—the only real possibility for uncovering the essence of what goes on in the clouds and the role "if any" of seeding.

Neyman's analysis of just such data had resulted in several surprising discoveries. As James Hughes of the Atmospheric Sciences Program of the Office of Naval Research describes it to me, "Neyman found things in the data that the people doing the experiment didn't even know were there."

"Our interest [in the problems of weather modification] was revived in 1964...," Neyman explained to the Royal Statistical Society in 1967. "Our originally casual inspection of the [Swiss] rain data suggested that here at last there may be evidence of the effectiveness of cloud seeding. This impression was later confirmed by a careful analysis and, as of now, it is my belief that cloud seeding with silver iodide smoke *can* affect precipitation. Furthermore, it appears that this effect is not uniform: in some as yet unidentified, but identifiable, conditions...cloud seeding appears to increase the rainfall, occasionally by a substantial factor, perhaps by 80 per cent. On the other hand, in some other conditions...the effect of cloud seeding is the opposite: a decrease in rainfall, possibly by a factor of 50 per cent. Presumably there are also intermediate conditions."

Neyman's further study of the Swiss data indicated that cloud seeding over a target twenty to fifty kilometers across could apparently affect precipitation in areas a couple of hundred kilometers from the boundary of the target.

"It came so surprisingly," he says to me, still with pleasure a dozen years later.

He is eager to explain the details of "faraway effects," as he calls them, and plods time and again to his study for another chart or document. When I object to his making so much physical effort—he is only recently up and about, and his leg is still giving him trouble—he insists, "Just a second. It is *so* interesting!"

It was, of course, Egon Pearson who proposed the vote of thanks after Neyman's presentation of his paper on weather control to the Royal Statistical Society. He recalled a time when, more than thirty years earlier, a much smaller society had met "in that very pleasant hall" of the Royal Society of Arts in the Adelphi and "new ideas on inference and experimentation, new applications of statistical methods in agriculture and industry, as well as in biology and medicine, were being introduced for the first time into the Society's discussions." Comments had been at times "a little destructive," they had also been at times "illuminating."

When he received a printed copy of Egon's remarks, Neyman was reminded of a little poem by Pushkin which he had read many years before.

He tried to translate it into English. "Unfortunately, after spending one full hour on an effort...,I feel compelled to give it up...." In "olden days" Egon had often helped him with his English. "What about helping me now?"

Egon had recently married his long-ago love, Margaret Turner, a widow since 1963; and it was she who helped Neyman revise his translation of the Pushkin poem which, a dozen years later, he is still quoting to me and others on what he feels are appropriate occasions:

> The heart lives with tomorrow
>   And grinds through a dreary today...
> Yet, when today drifts into the past,
>   It will shine with its own magic glow.

During the late nineteen sixties at Berkeley, it must indeed have been a comfort to think that those turbulent days, too, would shine eventually with their own magic glow. In Sproul Plaza representatives of the Black Panthers and the Students for a Democratic Society harangued noontime mobs from the steps of Sproul Hall. Ronald Reagan ran for governor on the promise that he would "clean up the mess at Berkeley." Within a month of his inauguration, the Board of Regents fired Clark Kerr.

Neyman had always been staunchly loyal to the "gentle and conciliatory" Kerr "because of my appreciation of what Kerr did over the period of his Chancellorship and subsequent Presidency to change the spirit of the Berkeley campus and of the University as a whole [after the crisis of the Loyalty Oath]." As soon as the regents announced their decision, he and a group of colleagues from various departments met in his home and drafted a telegram, dispatched to each regent, condemning the action of the board.

Like so many members of the beleaguered university, Neyman felt increasingly alienated from the majority of the citizens of the state, who had voted for and supported the new governor.

"...in spite of the many diversified currents and interests within the multiversity, there seems to be within it something of a common spirit which, briefly, might be summarized as follows: the universe in which [we] live is split in our minds into broad categories. One category is 'us,' including certainly the faculty and certainly the students and frequently the administration. All the rest is 'them.' The individuals belonging to 'them' may, and occasionally are, friendly to 'us.' Also the individuals belonging to 'us' may be personally obnoxious to other individuals of the same category. Nevertheless, I believe that in the minds of most members of the university community, the subdivisions into 'us' and 'them,' conscious or subconscious, exist."

Neither getting rid of Kerr nor cutting the budget "cleaned up the mess at Berkeley," since the University of California was only one of the more conspicuous centers of unrest in a young, restless world. As students and non-students chanted, "Fuck Ronald Reagan!", it was difficult to believe that only a couple of years earlier a non-student had been arrested for carrying onto the campus a sign bearing that same short word.

Since 1963 the civil rights movement had become increasingly militant— and more limited to blacks. Neyman, however, continued to be deeply committed to the nonviolent campaign of Martin Luther King, Jr. When King was murdered in 1968 and his "Poor People's March" floundered in financial difficulties, he organized another fund-raising letter which produced between four and five thousand dollars for the King group. If the money did not mean too much financially, at least "it [was] likely to have a beneficial effect on the development of inter-strata feelings of our society."

He had withdrawn completely from the Special Scholarships Committee in 1965, but he continued "to make an effort" whenever he was called upon to help the type of student for which the scholarships were intended. In 1965 Shyrl Miller, the young woman who had heard his lectures at Tougaloo, applied for graduate study at Berkeley. He arranged a clerical job for her in the Lab, assuring her (although she couldn't believe it) that her pay would cover her out-of-state fees and living expenses. When she got off the train, he was at the station to meet her and whisk her off to a dinner party at the Faculty Club arranged to welcome her. Later he volunteered for a group called the Faculty Fellows, which had been organized to provide faculty contact for students coming to the Berkeley campus under the new "opportunity" programs. For several years he went regularly to dinners at the residence hall assigned to him. But although his personal attitude was "most favorable," he found that he was "singularly unsuccessful" in establishing contact with these young people and reluctantly resigned as a Fellow in 1969. It has been his great regret, he tells me, that he has never had an American Negro Ph.D. student—"or a native American, an Indian."

He was also early and actively committed to the anti-war movement. In 1964 he was one of the signers and underwriters—twenty-nine of them Berkeley professors—who purchased a half-page advertisement in local newspapers showing a napalmed Vietnamese child and quoting Senator Wayne Morse: THE AMERICAN PEOPLE WILL BLUNTLY AND PLAINLY CALL IT MURDER....In 1965 he was one of eighty UC professors who issued a plea to President Johnson to reevaluate United States policy in Southeast Asia and "call for an end of the undeclared war in South VietNam."

When he was informed by the American Statistical Association that he had been selected as the recipient of its Samuel S. Wilks Medal, he declined to attend the presentation ceremonies, which were to be held at the Army Chemical Center in Maryland during the Fourteenth Conference on the Design of Experiments in Army Research Development and Testing.

"[Contrary to the situation during the Second World War] I do not work on war problems and, in fact, would like not to have any contact with them. This includes the attendance, either in person or through a representative, of any conference that may have something to do with the current war effort in Vietnam."

His name began to appear with that of Le Cam and others in a series of half-page advertisements in the *Notices* of the American Mathematical Society. These urged mathematicians to regard themselves as responsible for the uses to which their talents were being put, and concluded, "We believe this responsibility forbids putting mathematics in the service of this cruel war."

The advertisement ran several times, the list of signatures lengthening with each appearance until there were well over three hundred. In the summer of 1968 a constituent brought the ad to the attention of Senator Richard Russell of Georgia, who asked the Department of Defense to ascertain whether any of the signers were employed on government research contracts.

Neyman was in Yugoslavia at an FAO workshop when a letter from the Army Research Office arrived at the Statistical Laboratory addressed to himself and Le Cam:

"For the past sixteen years, your research has been supported by the U.S. Army. In turn, the results of your efforts have been utilized by the Army in various activities related to the current conflict in Vietnam...

"While you as individuals have every right to your own opinions and convictions, your present position vis-a-vis that of the Department of Defense must place you in a most uncomfortable, and perhaps untenable position....In view of this unfortunate circumstance, a mutually acceptable decision to terminate our present association when your present support expires appears to be consistent with both of our positions."

A much shorter letter from the Office of Naval Research, received at the same time, simply inquired whether Neyman and Le Cam still wished to have the contract for their work on weather modification renewed.

Two weeks later the story—"Mathematicians Are Queried on Anti-War Views"—broke in the *Washington Evening Star,* which quoted a representative of the Pentagon in rather strong terms on the government's position. His statement was toned down by a superior the next morning in the story in the *Washington Post;* and *Science* noted that the Pentagon, with a tradition of supporting basic research and of dealing with the "best people," was apparently reluctant to push for a confrontation.

Letters similar to the ones received by Neyman and Le Cam had gone to several people at other universities, but only the Berkeley people (who included Peter Bickel) were mentioned by name in the newspaper articles. None of the work being done by the three at Berkeley was "classified," and they all felt that whatever value it had for the military in Vietnam was outweighed by its general value.

Neyman responded to the ONR letter by writing that he would like to have his contract renewed. He was greatly distressed at the prospect of having the support for the Lab's weather modification project discontinued; and after the contract terminated on September 1, he fretted for some six weeks, trying to keep up his spirits by recalling Hughes's words to him over the telephone:

"All hope is not lost." Finally, in the second week of October, the Navy contract, retroactive to September 1, arrived. The Army contract was also ultimately renewed.

In spite of this much publicized opposition to the government's policy in Vietnam, Neyman was notified that same fall that he had been chosen, one of twelve scientists, to receive the Medal of Science, the nation's highest scientific award, from President Johnson, who had been forced to withdraw from the 1968 presidential campaign as a result of his continued support of the unpopular war. The presentation took place in a televised ceremony at the White House in January 1969, just three days before the inauguration of Richard M. Nixon. Neyman remembers that he was shown the exact spot where he was to stand—it was marked on the floor—and instructed not to upstage the President.

His citation for the Medal of Science did not have to be long:

"To Jerzy Neyman," it read, "for laying the foundations of modern statistics and devising tests and procedures that have become essential parts of the knowledge of every statistician."

He continued to oppose the war publicly, but there were no more advertisements in the *Notices*.

That fall the New Left, which was almost entirely white, and the Black Activists, on and off campus, joined forces for the first time over the issue of a course scheduled to be taught by Eldridge Cleaver, the Minister of Information of the Black Panther Party, an ex-convict and the author of the highly influential book, *Soul on Ice*. When the regents ruled that Cleaver could deliver only one lecture and that not for credit, protest escalated violently. Moses Hall was "occupied." There were efforts to "take" the computers in Campbell Hall (which lodged the mathematics and the statistics departments). A mysterious conflagration wiped out the auditorium of Wheeler Hall (the building in which Neyman had had his first office at Berkeley). The Telegraph Avenue approach to the campus—with windows boarded up and revolutionary slogans scrawled large on walls—looked like nothing so much as a street in Kharkov after the Revolution of 1917. Once more it was being said—ironically, during the centennial year of the university—that Berkeley was "through" as a first-rate institution of higher learning.

Mathematics and statistics, however, continued to prosper. The National Science Foundation had made a generous grant for the construction of a mathematics-statistics building, and work was already in progress on "Griffith C. Evans Hall." Some people thought it should be called "Evans-Neyman Hall," but they knew there was no point in sending their proposal to the naming committee, of which Neyman was a member. He heartily approved naming the new building after Evans. During this period he was asked to submit nominations for a centennial group to be known as the Berkeley Fellows who—as friends or faculty no longer active—had left a lasting imprint on the campus. He placed the name of Evans at the top of his list. Evans—he wrote—"using virtually dictatorial powers granted him by

President Sproul [had] changed the status of his Department from that of a secondary unit in a second rate university to the level of a first class center of research and instruction." Second on Neyman's list was Sproul, who "even though no scholar himself...had a remarkable ability of understanding scholars and judging correctly their merits...[and] had a vision of what a first class university should be."

For Berkeley, the climactic event of the nineteen sixties occurred on May 15, 1969, over the issue of "People's Park," a piece of university land which students and "street people" had cultivated and which the university now proposed to turn into a playing field. The violent events of that day—during which the campus and surrounding area were sprayed with gas from helicopters, one person was killed and scores injured—were described a few weeks later in a full-page advertisement in the *Los Angeles Times* (the most widely read newspaper in the state), which was signed by a number of faculty members of the Berkeley campus, including Neyman. The purpose of the advertisement, the signers explained, was to give the facts about the events of May 15:

"The threat to democratic institutions and to higher education in the State of California has never been greater. We call upon you, as the ultimate arbiters of authority in this state, to protect your sons and daughters, your University, your fellow citizens, and the heritage of American values."

"Well," Neyman shrugs helplessly, handing the yellowed sheet back to me, "you can read about it."

In the five years that had spanned the Free Speech Movement and the People's Park, academic life at Berkeley had not ceased. For Neyman there had been significant discoveries in weather modification and, with Scott, the development of a useful technique for dealing with "outliers"—the extreme elements of a sample which appear "too large" or "too small" as compared with the others. He had organized a number of scientific meetings and conducted his own brand of interdisciplinary classes and seminars. He had also guided half a dozen young people from several different countries through the doctorate. He has continued to be especially close to four of these: Prem Puri (Purdue University), Wolfgang Bühler (University of Mainz), Peter Clifford (Oxford University), and Robert Davies (Department of Scientific and Industrial Research of New Zealand), all of whom, working on problems in which he is deeply interested, are frequent visitors to the Statistical Laboratory. In 1969 his current Ph.D. student was John Oyelese of Nigeria.

That violent spring of 1969 Neyman was seventy-five years old, a little deaf and increasingly bothered by cataracts. As always each spring, he went through an unsettling period of suspense while he waited for his appointment as "Professor Recalled to Active Duty and Director of the Statistical Laboratory." Although the statistics department, sensitive to his feelings, petitioned for the appointment as early as it could, any new person in the chain of approval resulted in a delay. What was this? A seventy-five year old man to direct a university research facility? Strictly against regulations! On one occasion, Mike Neyman tells me, his father became literally sick with waiting and had to be hospitalized, recovering suddenly the moment he was notified he had been reappointed.

"Well, life is complicated," Neyman says with a smile, "but not uninteresting."

He chuckles and agrees when I quote Arthur Fiedler:

"He who rests rots."

1969
—
1979

"The campus has been a little disturbed lately," Neyman, Le Cam, and Scott wrote with a degree of understatement to mathematical statisticians and probabilists in the summer of 1969, "but we have decided to go ahead with plans for the Sixth Berkeley Symposium."

The Sixth would mark the "silver jubilee" of the symposia—twenty-five years since the little meeting which Neyman had organized to mark the end of the Second World War and the return to peacetime research. That first symposium had lasted a week and the $4,000 to support it had come entirely from the university. There had been twenty-five papers, including one by Hsu, who had arrived from war-torn China in time to contribute. (What had become of Hsu? No one had had word of him since he had returned to his native land in 1947.) The most recent symposium, the fifth, had been funded to over $125,000 by a variety of organizations, foundations, and government agencies in addition to the university, and had lasted for a month in the summer of 1965. The papers presented by the invited participants, almost half of whom had come from abroad, had resulted in five volumes of *Proceedings* and three thousand pages. Neyman had observed with satisfaction that every statistical center of significance "on the planet" had been represented—with the exception of the People's Republic of China. (Afterwards he made an effort to contact Hsu and learned that the Chinese statistician had died in 1970.)

The Sixth Symposium was social as well as scientific with plenty of opportunities for "shoulder rubbing." Accommodations in rented sorority houses and other facilities permitted the participants to get together in the evenings. There were weekly excursions to the natural beauties of the area, parties of all kinds, welcoming and closing banquets. Because of its significance as a special anniversary, the Sixth Symposium was also the occasion for calling up many memories. How, when the war had ended in the middle of the First Symposium, all the restaurants had closed and Evelyn Fix and Betty Scott had concocted emergency casseroles to feed the out-of-town guests. How, at a time when the serving of alcoholic beverages was prohibited on campus, even at the Faculty Club, a truck from a Sonoma winery had rolled up to Wheeler Hall with a dozen cases of wine contributed for the Second Symposium banquet at Mr. Neyman's request. How, during the Third Symposium, Paul Lévy had given everyone a bad fright when he had collapsed at a session in Dwinelle Hall. How at the Fourth Symposium, with the Russians participating for the first time, Linnik had composed a poem to "the indefatigable Miss Scott." How before the Fifth Symposium, with Neyman and Scott still at the ISI meeting in Tokyo, the Navy had called Le Cam to find out just how many destroyers Mr. Neyman needed to take his visitors on an excursion of the bay. (Le Cam had gulped and said that three would be enough.) It was also remembered with sadness how at the winter session of the Fifth Symposium, a few hours after the banquet which she had helped to arrange, Professor Evelyn Fix, a mainstay of statistics at Berkeley almost since the founding of the Lab, had died of a heart attack at the age of sixty-one.

Of course, as always, things did not go strictly according to plan at the Sixth Symposium. Funding from the National Institutes of Health arrived so late that the traditional session on biology and health, which was Neyman's special interest at the time, had to be postponed. Undaunted, he saw the postponement as an opportunity to organize, not just one, but three supplementary symposium sessions in the spring and summer of 1971, each to be represented by a separate volume in the *Proceedings*. On the eve of his seventy-seventh birthday, he presided over a four-day meeting which brought together statisticians and biologists to discuss the revolutionary new macromolecular studies in "non-Darwinian evolution." This subject had appeared on his horizon, as he puts it, only the previous year and fascinated him both because of its intrinsic interest and because he saw it as a potential source of novel problems in statistical theory. The postponed symposium session on biology and health followed two months later, and a conference on pollution and public health took place the following month.

Neyman's interest in pollution had been sparked by statistical studies being made on the effects of smoking. (I ask, as he stuffs the pockets of his overcoat with Marlboros, how many packs he smokes a day. "Never more than three," he says and then tells me how his grandmother, "who was reputed to be over a hundred," died in a rocking chair with a cigarette in her hand.) According to Le Cam, who is also a heavy smoker, "Neyman couldn't accept that it was smoking that caused all those deaths so he decided it must be *pollution*."

By 1971 he felt that the Berkeley symposia—conceived to present cross-views of contemporary statistical work in various fields—should devote a session specifically to the question of pollution and health. He saw in the multiplicity of pollutants and the difficulties of disentangling the complicated knot of their individual effects on the health of human beings a number of "delicate statistical questions"—a "big problem" which was "unprecedentedly interdisciplinary in character."

The purpose of the health and pollution conference was "to formulate a plan or, at least, to assemble the material for a plan of a comprehensive statistical study which could begin without delay, aiming at the estimation of health effects of several suspected pollutants as they occur in the selected localities for which it is possible to secure reliable data." Invitations went to all federal and state agencies, scholarly institutions, and individuals in the field and included a call for the submission of skeletal plans for the proposed comprehensive study.

Neyman presented such a plan, as did three other participants; but the meeting itself was a disappointment to him, as he indicated in an "Epilogue." There appeared to be available a wealth of factual information in the laboratory studies of biologists which had been reported, but "the performance of biometricians-statisticians—(who mostly did not honor the conference by their participation)—[had been] on a regrettable level."

Since then he has persistently advocated—abroad as well as at home—"a multipollutant/multilocality epidemiological study to evaluate real life public health hazards from environmental pollution." In the developing energy crisis—with new needs to produce power—it should be recognized that it is not only nuclear generators which will produce radioactive pollution, but also coal- and gas-burning plants. The effects on public health will depend upon the combination of pre-existing pollution with that produced by the new plant, whatever its type. Estimating these effects will require the development of a reliable statistical methodology to deal with delicate and interesting problems: those of "competing risks" ("If I am run over by an automobile," he explains, lighting another cigarette, "then I will not die of lung cancer.")—"synergisms" (such as the lethal interaction of the common household cleaners, ammonia and bleach)—and "dose-rate" effect (his favorite example here being sleeping pills, harmless when taken one at a time and infrequently but capable of killing when downed in large quantities).

In the spring of 1971, as he was organizing these three conferences—each of which was to be attended by one hundred to three hundred persons—he was also planning another, still more extensive project. On occasion, that spring, he asked to be forgiven by a correspondent for the "telegraphic" style of his communication—"... but there is so much that I have to do 'yesterday' that limitations on 'today' are unavoidable."

The year before, the National Academy of Sciences had received an invitation from the Polish government to assume some leadership in celebrating the 500th anniversary of the birth of Copernicus, which would fall on February 19, 1973. Zygmund and Neyman had been appointed chairman and vice chairman, respectively, of the committee to decide on the appropriate observance. Neyman was delighted to have an opportunity to work in collaboration with his old friend, for whom his affection and admiration had, if anything, grown over the years. Even before the other members of the committee had been chosen, he was writing "Dear Antos":

"I personally am preoccupied with the idea of a volume that our Academy might publish.... The book I visualize would be somewhat as described in the preface of the book by Thomas Kuhn.... Kuhn speaks of the mutiplicity of revolutions (plural) that Copernican work has produced in all domains of thought. In this, of course, the essential point was the detachment from long established routines of thought...."

The phrase "routine of thought," which does not occur in Kuhn's preface, echoed *The Grammar of Science* and the theme of the talk Neyman himself had given in the opera house in Bydgoszcz fifty years earlier to celebrate the 450th anniversary of the astronomer's birth. Copernicus had detached himself from a routine of thought which had dominated fourteen centuries. This one breach in one domain had then paved the way for breaches in others, and modern science had emerged.

In September 1971 the National Academy's *Letter to Members* carried the announcement of the project which Neyman had proposed—to be provisionally called "the Copernican volume" and to be addressed to the general educated public. The contents would consist of a number of essays on Copernican-type revolutions in recent centuries—such revolutions being characterized by the abandonment of widely held concepts and the replacement by dramatic new conceptualizations resulting in altered public attitudes and considerable development in science and technology.

Neyman was appointed editor of the volume and warned that the essays were not to give advice to the government nor to the general public. He threw himself into the project with a passion. It was a tremendous task; and, from time to time, as he submitted the minutes of his editorial committee to the National Academy, he heard the word "Impossible!" For a while he considered using professional science writers; but when Dennis Flannery, the editor of *Scientific American,* told him that there weren't more than ten good science writers in the country and five of them were working for that magazine, he turned to the scientists themselves for his authors. The next three years—for there were to be "most regrettable but unavoidable delays," including two operations for cataracts—were "the Copernican years" in the Statistical Laboratory.

Kay Kewley, who was Neyman's secretary during most of this period, found it "a very inspiring thing" to watch him at work on the volume.

"I think he was still deciding who the authors were going to be when I came. One of the main things I remember from that time was how much he wanted to learn. I remember he was reading Copernicus's papers in Latin and was just all inspired, his eyes just crinkling; and I said, 'Oh my god, this man just loves intellectual-type things.' I think he just loved learning about all that biology and chemistry and whatever other subject was being written about. He was just reading all the time. His eye was still patched up, and he was reading every piece of paper just two inches from his face. So the strain must have been phenomenal. When someone would come into his office, a student or some visitor, he would shove one of the manuscripts in front of him and say, 'Read this!' If that person could understand it, that was very good. Because the book was meant for educated laymen, you know."

Neyman wrote the introductory chapter about Copernicus's life and work, seeking out a colleague in the classics department to help him with the church Latin, which was different from the classical Latin in which he was fluent as he was in so many other languages. Copernicus—he pointed out—had generated a revolution in scholarly thinking in all domains by introducing "a completely novel yardstick for appraising a new theory: conformity with observation and intellectual elegance." As a "subjective note" he added that he personally had found the astronomer a very attractive person:

"...a tremendous intellectual horizon, including public affairs and economics, hard work in several domains, including some military endeavors, sense of humor and straight dealing with no falsity that I can see....[His] motivation for writing *De Revolutionibus* had nothing to do with the 'publish or perish' motive of much present research. He could not and did not expect personal material advantage from his work, only condemnation and ridicule. What drove him was pure scholarly interest in the phenomena themselves and the hope of explaining them in a manner 'pleasing to the mind.'"

He repeats when he presents me with a copy of the book that he found Copernicus a very attractive person and tells me again how he asked the illustrator to show him on the jacket "a little bit smiling."

All during this time when Neyman was working on the Copernican volume, he was also traveling extensively, often at the invitation of organizations wanting him to talk on Copernicus at their own celebrations of the anniversary year. But he was always careful never to miss more than two sessions of his class and one seminar. He was now approaching eighty. He felt increasingly uneasy about the tenuous character of his appointment at Berkeley and regularly listed himself in a register of retired university professors as "available for employment." At a seminar discussion at Michigan State University on "What is done in the Statistical Laboratory at U.C., Berkeley," he indicated that he would be open to the right kind of offer from another institution. The statisticians in East Lansing were excited at the prospect of adding Jerzy Neyman to their faculty, and by the end of the summer it appeared that Neyman would be receiving an offer with some assurance of tenure.

Lehmann, who was the new chairman of the statistics department, wrote worriedly to Dean Calvin C. Moore:

"Among the consequences for the University of losing Neyman, I feel that the following three are the most serious, in this order:

"1. Loss of morale in the Department of Statistics....

"2. Loss of prestige to the University....

"3. Nearly certain loss of close to $200,000 in grants...."

Perhaps some "statement of intent" by the administration would be in order?

Moore replied promptly to Neyman himself.

"I agree at this time, contingent upon your continued good health and contingent upon your vigor as a scholar and teacher, and contingent upon your own desire and that of your department to be recalled to active duty in the future, to support future recommendations for your recall."

(Ultimately, it turned out that the people at Michigan State were unable to persuade their provost to approve an offer to a man who was almost eighty. In January 1974 James Stapleton, the chairman of the East Lansing department, conceded to Neyman, "We must finally admit defeat." Neyman urged him not to feel bad. "Such things do happen from time to time, particularly when the individual concerned is afflicted by an incurable illness: old age.")

Having received the assurance from the dean, Neyman went happily off to the Soviet Union with Betty Scott to attend a conference on weather modification in Tashkent, where his maternal grandfather is buried. They had arranged to arrive early so that they would have a few days for rest and sightseeing. Neyman's "vacations" are always brief. "Four days in Dubrovnik" was one of the longest in the last twenty years. More usual is a trip down the Pacific coast to a place called Nepenthe, where there is a restaurant looking out over the ocean which he likes. Such "vacations" are measured in hours rather than days. In Tashkent, true to form, but to Professor Scott's disappointment, he shut himself up in his hotel room and began to translate his paper on rain stimulation experiments into Russian. He, nevertheless, tremendously enjoyed his visit, and there is in existence a recording of him and Soviet mathematicians delightedly reciting Russian poetry to one another. One day he hired a guide and went to the local cemetery to find the grave of his Grandfather Lutosławski. He thinks that he found it but, because of the worn lettering on the gravestone, he cannot be sure.

In 1974—the year that the Copernican volume appeared under the title *The Heritage of Copernicus: Theories "More Pleasing to the Mind"*— Neyman was eighty. There was another round of birthday celebrations. During the first part of April the mathematicians of the Polish Academy of Science organized a symposium "to honor Jerzy Neyman." When he got back to Berkeley, the birthday itself was observed with a special Berkeley-Stanford colloquium featuring talks—in the newly restored auditorium of Wheeler Hall—by the two younger statisticians whom he so much admires: Charles Stein and Herbert Robbins. Lehmann, as department chairman, announced the establishment of the "Jerzy Neyman Lectureships in Mathematical Statistics." Afterwards, on the tenth floor of the still new Evans Hall, with its magnificent 360 degree view of the Berkeley hills, the campus, and the Golden Gate, there was a champagne reception—and a tipsy toast by a guest "to all the ladies absent and some of those present." That evening Chancellor Bowker and his wife, Rosedith Sitgreaves, both statisticians, were hosts at a dinner in his honor. The April issue of the *International Statistical Review* and the May issue of the *Annals of Mathematical Statistics* were dedicated to him. In June he was made an honorary member of the London Mathematical Society and in December awarded an honorary degree from the Indian Statistical Institute in Calcutta.

At eighty, however, he did not deceive himself that he still contributed to mathematical statistics. *"The days of our youth are the days of our glory,"* he quotes to me. He frequently refers to his paper on *C-alpha* tests in the Cramér festschrift (1958) as "my last performance"; and on occasion, discussing some current controversy as he fingers the pages of the *New York Times,* he murmurs, "I don't know if I would be so interested in these things if I could do other."

If he could not do "other," he was confident that he knew what should be done. With all the force and persistence of which he is capable, he pressed on every possible occasion for two studies on a broad national, even international, scale—the one having to do with health and pollution and the other with weather modification. He became increasingly convinced that in both these fields the public and its lawmakers were being misled by scientists who give what he calls "a cosmetic treatment" to commercially unwelcome research results about the detrimental effects of radiation and the far from unambiguous effectiveness of cloud seeding. Such behavior is "understandable" in people who make their living in the affected fields, *but scientists*—"and that has happened!" When it was revealed that a cancer researcher had deliberately falsified his research results, Neyman proposed to the Council of the National Academy of Sciences that its bylaws be amended to include a provision for expulsion of members for cause:

"In cases when the fact of misrepresentation and its willfulness are established beyond reasonable doubt, the guilty person should be expelled, not just quietly phased out, but expelled, so to speak, with all the drums beating and the trumpets blaring."

The public and punitive aspects of this proposal did not sit well with the members of the Council of the Academy. They felt that "should such circumstances arise..." the member in question should be quietly taken aside and invited to resign. However, the following year the bylaws were amended to provide for the removal of officers for malfeasance.

Neyman did not consider this change satisfactory. It was limited to members of the council, and the term *malfeasance* seemed "so unspecific as to leave room for doubts...." In subsequent years he made similar proposals regarding expulsion for cause to the IMS and to other scientific organizations to which he belonged which did not already have such provisions. When I first meet him in 1978, he is still campaigning—writing to the chairman of the Committee on Science and Technology of the House of Representatives "to suggest certain legislation [to] effectively discourage public institutions from publishing evaluations of particular technologies formulated with [such] a substantial degree of enthusiasm [that] they tend to be unrealistic and misleading." He also writes, this time to Sacramento, suggesting that all scientific testimony be given *under oath.* A definition appears chalked in large letters on the long blackboard in his office: *"Intentional misrepresentation of a material existing fact that induces another to act to his damage in reliance thereon is FRAUD."*

He shakes his head regretfully over a paper on weather modification which he has just finished commenting on. He says that Betty Scott is angry at the implication of what he has written: "Betty stamped her foot and screamed, 'You can't write that—you are saying that the man is ignoring his own results for money!'" But he will not change what he has written and remains unperturbed by the angry letter he shortly receives from the author in question, who objects that for one thing he is being quoted out of context—a not always unjustified complaint as, reading line by line, he marks in thick black ink any word or phrase which catches his eye. The letter which comes back is strongly worded, in fact almost insulting, but Neyman quotes to me a line from a story he first heard in his Latin class in Kharkov— "You are angry, Jupiter. *You must be wrong.*"

During these past years he has devoted much time and effort to obtaining honors for others and to thinking of ways "to show appreciation." The Evelyn Fix Prize, which is awarded each year at graduation to a student who has shown exceptional promise in research, was his suggestion. In the spring of 1978 he is delighted when Lehmann, whom he has backed, is elected to membership in the National Academy of Sciences; and when Doob, who was one of his personal nominees, is selected for the President's Medal of Science, he gathers signatures in the statistics and mathematics departments for a message of congratulation.

Although he has on occasion reminded his combative friend Joe Berkson that he who fights alone is not a fighter to be reckoned with, he still gets himself into positions where he stands pretty much by himself. In 1978 a long, unsigned letter by a Russian mathematician describing the difficulties under which Jewish mathematicians work in the Soviet Union appears in the *Notices* of the American Mathematical Society under the sponsorship of a group that includes such admired colleagues as Doob and Zygmund. Ultimately a number of mathematicians and statisticians publicly pledge themselves to reduce cooperation with the scientific community of the USSR until such policies are changed. Neyman flatly refuses to join them. He remains unmoved by suggestions that he is motivated by anti-Semitism, communist leanings, or simply a selfish desire to continue to be invited to scientific meetings in the Soviet Union.

"There are elements in this country that I don't approve of, and I don't associate myself with them," he says. "I am a believer in the existence of an *international intellectual community,* and I am a member of that community. After we have our seminars—my seminars—we go to the Faculty Club for a little drink and we have toasts. The first is Polish toast—to all the ladies present, some of those absent. Second is to the speaker. And the third is what?"

To the international intellectual community.

"All right. And so I try to rub it into the minds of young people who are working with me that there is such a thing. Something like what they are talking about can affect, can hurt the position of members of the international intellectual community *there,* and so I am against it."

All during 1978–79 he continues to propagandize assiduously for the broad multidisciplinary pollution study and the new, properly designed large-scale experiment in weather modification which he sees as urgently necessary. At every opportunity he does what he can to bring these studies into being; but if plans do not go as he wants them to, he is inclined to withdraw completely. At the same time, whenever he has the funds, he invites interested younger people to Berkeley to work on the theoretical problems which will be involved in such studies. These visits are paid for out of the two grants for which he is principal investigator: one from the Office of Naval Research and the other from the National Institutes of Health.

In June 1978, as he is getting ready to leave for Brazil, he is concerned about the latter grant, which will terminate at the end of September. He apologizes for not having time to talk to me. "But I am very busy getting money for the Lab!" A "site visit," scheduled by the NIH for the end of June, has been postponed until the end of August. What does this mean for the Lab? He has to leave for Brazil with the question unanswered.

Upon his return, at the end of July, he begins to pull his people together for the presentation to the "site visit" committee. Later, he hears that the meeting "has gone well." No decision is made, however, and the Lab's grant expires. His secretary gets a job in another department—better for her, he explains, because she will be paid with "hard" money which is not dependent upon grants—but now he must begin to train a new secretary—"a borrowed young lady." He makes several attempts to get some sort of commitment from the NIH. When these fail, he asks the university for an "emergency" grant "to enable us to carry on certain unpopular studies we have been conducting"; but no funds are forthcoming. The problem of money is much on his mind while he is laid up in bed in December 1978 with his bad leg. Finally, in January, when he is well enough to return to school but still having difficulty getting around, he receives the long awaited letter informing him that the NIH grant, retroactive to September 30, has been renewed for three years. He begins to invite people to come to work in Berkeley during the summer.

One of these is Robert Bohrer of the University of Illinois who, with his assistant Mark Schervish, has made a significant advance toward a solution of the problem of synergisms. Neyman has great admiration for the way in which Bohrer has risen above a debilitating diabetes which has finally blinded him. On his side, Bohrer is tremendously impressed by Neyman and his activities as "a statesman of science."

"I have been impressed, even astonished that several of Mr. Neyman's colleagues have commented how much he has slowed down since the early seventies, before I knew him. If so, then surely he stands as foreshadow and proof of that ultimate goal we seek—the permanence and victory of mind, will, emotion, and spirit over body."

A number of "pleasant" things happen to Neyman after the NIH grant is renewed. He receives word that a volume honoring the astronomer Ambartsumian, for which he wrote a piece entitled "A Revolutionary Period in Cosmology," has finally been published. This paper will bring the number of items in his bibliography to two hundred. His *Synthese* paper on "Frequentist Statistics and Frequentist Probability" has been translated into Portuguese. He finds it very satisfying to see his work in yet another language. The chairman of the statistics search committee on the Davis campus of the university writes to ask for recommendations for a position there and suggests that he himself may be interested. He enjoys composing his reply:—"I am not sure that you are familiar with my present status and a detail of my vita: I was born on April 16, 1894. (Yes, eighteen hundred and ninety-four!)"

As his birthday approaches, he begins to correct me whenever I refer to him as eighty-four.

"Eighty-five in a few more days."

There is discussion in the department whether the passing of another half decade should be observed in some special way. Another big celebration perhaps—like the one on the eightieth birthday.

Neyman determines the answer. He will not be in Berkeley on April 16 but in Chapel Hill, where he will open a two-day celebration of the sixty-fifth birthday of Wassily Hoeffding. He will not be in Berkeley on his "other" birthday either—on April 29 (the birthdate according to the Gregorian calendar) he will be leaving for Storrs, where he will give another talk (this one on the concept of clustering and its many applications) as a feature of the University of Connecticut's long-planned colloquium on the history of statistics. Egon has also been invited, but he has declined to go. He marvels at his old co-worker. "To think of his having the energy!"

In Chapel Hill, Neyman is delighted to see Herbert Robbins, who has come down from New York for the meeting. They talk for several hours in his hotel room.

Friends and acquaintances from all over the country, hearing that it is Neyman's birthday, come up to wish him many happy returns of the day.

"Not too many," he says, smiling. To Robbins he adds, "I have lived long enough."

1979   In Berkeley the statisticians, discussing the observance of the eighty-fifth birthday with colleagues in Palo Alto, decide to dedicate to Neyman the joint symposium they have been planning. Nothing could please him more than resumption of the symposium tradition. He has spoken to me several times, regretfully, of its apparent abandonment, pointing out that the last one was in 1970 and that there was to be one every five years.

He accepts an invitation to be a member of the symposium organizing committee; but then, after his return from Chapel Hill, an "unpleasantness" occurs. A number of Russian mathematicians will, of course, be invited to participate, including Kolmogorov, whom he has tried on so many occasions to bring to Berkeley. Perhaps, this time, Kolmogorov will actually come! However, several members of the organizing committee, including Peter

Bickel, who is currently chairing the Berkeley department, have signed the pledge that they will not cooperate with the Russians in view of their policies toward their scientific community. They feel that the symposium should—in its inviting—take a stand against the treatment by the Soviet government of Russian mathematicians who are Jewish and of the substitution at international meetings of scientific mediocrities who are politically acceptable for outstanding scientists who are not.

"We would like to say that there can be no political selection of participants," Bickel explains to me. "We would like to make it rather clear. If any mathematician invited is prevented by the government from coming, then the others are no longer welcome."

Neyman objects strenuously.

He cites, as he has before, the case of Linus Pauling, Nobel Prize winner, who was not permitted by a president of the United States to go to England to accept an honor. How would Americans have responded if the English had staged a meeting and announced that unless Pauling could come no one could come?

Bickel and Jack Kiefer, a recent addition to the Berkeley department, who has signed the same pledge, feel that unless some sort of condition is laid down when inviting the Russians they will have to resign from the organizing committee.

Neyman says that if there is such a condition he will resign.

Bickel has been planning to make the announcement of the 1980 symposium at the statistics department's annual spring picnic the day before Neyman leaves for Storrs. Now he isn't sure whether there will be a symposium in 1980.

Happily, the day before the picnic, the unpleasantness is temporarily forgotten in the general excitement over the arrival of a telegram announcing that Neyman has been elected a Foreign Associate of the Royal Society. It is a particularly prestigious honor and one that means a great deal to Neyman personally.

Bickel decides—literally at the last moment—to go ahead with the announcement of the planned symposium. He thinks that somehow things will work out.

"Jerry will grumble, but he will go along."

Older members of the department are not so sure.

"It is one of the things he feels most strongly about."

The department's picnic takes place in a little glen behind the merry-go-round in Tilden Park, and the music churns constantly in the background as people talk and joke. Neyman sits quietly, since he cannot hear over the music, and accepts congratulations.

I sit with my fried chicken and potato salad beside a French girl. Next to her is an Iranian couple. We discuss whether there is anyone in the department who is so alone that he has no one with whom he can converse in his own language. Everybody agrees that there is not.

While we are eating, Betty Scott arrives from a meeting on skin cancer, still wearing her name tag and carrying a gaily decorated cake that says "F.A.R.S."—Foreign Associate of the Royal Society.

After the cutting and eating of the birthday cake, Neyman would like to go back to his house. I offer to drive him, since I also must leave.

On the way down the hill, he asks if I will be coming as usual the following Saturday to talk with him. I say no—the time has come that I must stop talking and begin to write.

"Too bad!"

He is regretful. He will miss our Saturdays. They have become a part of the rhythm of his life, and when that rhythm is broken he is uncomfortable for a while.

We stop at the bookstore to pick up the copy of the *New York Times* which has been saved for him. He looks happy coming back to the car with the paper under his arm. A glance at the headlines reveals that the news of the day is bizarre, as usual. As we start back up the hill to his house, he reminds me of the last chapter of *Penguin Island*, which describes the end of the civilization of the Penguins: "The scene on the hill where they are waiting for the city to blow up…" He talks about stopping in Chicago on Tuesday, on his way home from Storrs, to see "my cordial friend" Zygmund, who will come to the airport to have a visit with him. He tells me that he has started work on the monograph on weather modification. "Rather long." Parts of it will be controversial—he will call a spade a spade and pull no punches—some people will not care for it. He would like to think of a way to encourage Herbert Robbins to write a book about "empirical Bayes." Then there is the question of inviting Joshua Lederberg, a geneticist at Stanford, also a new Foreign Associate, to give a talk to his Wednesday seminar. How to do it in a nice way—he is not sure how old Lederberg is. "But not many people are as old as I am." The election to the Royal Society brings new responsibilities. He does not want to be simply a figurehead member. Perhaps he can interest the society in sponsoring an international study on cancer.

"Life is complicated," he reflects as he lifts himself out of the car in front of his house, "but not uninteresting."

He asks me to wait a minute while he breaks off a camelia which has just come into bloom. He hands it to me with a little bow. And so I leave him, at the gate to his garden, a little bit smiling.

JERZY NEYMAN

1894–1981